Piperonyl Butoxide

The Insecticide Synergist

Piperonyl Butoxide
The Insecticide Synergist

Edited by

Denys Glynne Jones

San Diego   London   Boston
New York   Sydney   Tokyo   Toronto

This book is printed on acid-free paper.

Academic Press
24–28 Oval Road, London NW1 7DX, UK
http://www.hbuk.co.uk/ap/

Academic Press
525 B Street, Suite 1900, San Diego, California 92101-4495, USA
http://www.apnet.com

ISBN: 0–12–286975–3

A catalogue record for this book is available from the British Library

Library of Congress Catalog Card Number: 98–86854

Typeset by J&L Composition Ltd, Filey, North Yorkshire
Printed in Great Britain by Redwood Books, Trowbridge, Wiltshire
98 99 00 01 02 03 RB 9 8 7 6 5 4 3 2 1

Contents

Plate section appears between pages 52 and 53.

Contributors

Mervyn Adams Consultant Entomologist. Previously responsible for development of insecticides for protection of food at Wellcome Environmental Health, Berkhamsted, UK

David J. Arnold Head of Environmental Sciences, AgrEvo UK Ltd. Main interests are fate of pesticides in soil and water. Member of SETAC Europe. AgrEvo UK Ltd, Chesterford Park, Saffron Waldon, UK

Rory Breathnach Lecturer in small animal clinical studies, Faculty of Veterinary Medicine, University College Dublin, Dublin, Eire

John E. Casida Director, Environmental Chemistry and Toxicology Laboratory, and Professor of Entomology, College of Natural Resources, University of California, Berkeley, California, USA

Andrew Cockburn Worldwide Head of Toxicology for AgrEvo based at AgrEvo UK Ltd, Chesterford Park, Saffron Walden, UK

Ian Denholm Principal Research Scientist B.E.C. Department, IACR Rothamsted, Harpenden, UK

Greg Devine Entomologist at Department of Zoology, University of Aberdeen, Scotland, UK

Giovanna Di Blasi Senior Chemist, Technology Department, Endura Spa, Bologna, Italy

Alan L. Devonshire Head of the Insecticide Resistance Group, IACR, Rothamsted, Harpenden, UK, and Professor of Genetics at Nottingham University, UK

Andrew W. Farnham Previously worked at the Department of Insecticides and Fungicides, Rothamsted Experimental Station, UK, on insecticide resistance and the relationship between the chemical structure and biological activity of pyrethroids.

Antonia Glynne Jones Technical Director, Gooddeed Chemical Company, Aylesbury, UK

Denys Glynne Jones Consultant to Endura Spa. Formerly Technical Director Pyrethrum Board of Kenya. Editor *Pyrethrum Post*

Robin V. Gunning Senior Research Scientist, Tamworth Centre for Crop Improvement, NSW, Australia. Specialist studies on factors affecting insecticide resistance in cotton pests

Christopher A. J. Harbach Senior Mass Spectroscopist at M-Scan Ltd. Main interest is problem solving by GC-MS. M-Scan Ltd, Silwood Park, Sunninghill, Ascot, UK

James F. Hobson Arcadis Geraghty, Miller Ink, Millersville, Maryland, USA. Previously served for six years as a corporate Environmental Toxicologist for the agrichemical and chemical sectors of two major international chemical companies, Rhône-Poulenc and FMC Corporation

Ernest Hodgson William Neal Reynolds Professor and Head, Department of Toxicology, College of Agriculture and Life Sciences, North Carolina State University, Raleigh, North Carolina, USA

Paul Keane Chairman Pyrethrum Task Force USA, Consultant Biologist, Paul Keane Associates, Chicago, Illinois, USA. Formerly on staff of Pyrethrum Board of Kenya

Robert Large Director, M-Scan Ltd. Main interests are environmental, petroleum and petrochemical analysis. M-Scan Ltd, Silwood Park, Sunninghill, Ascot, UK

Patricia E. Levi Associated Professor of Toxicology (Retired), North Carolina State University, Raleigh, North Carolina, USA. Biochemical toxicologist, particularly in the area of agricultural chemicals. Well known teacher of toxicology and co-author of two toxicology textbooks

Donald Maciver Previously Research Scientist for the Pyrethrum Board of Kenya and Senior Chemist, AgrEvo Environmental Health, Montvale, New Jersey, USA

Richard A. Miller President of Prentiss Incorporated, Floral Park, New York, USA. Prentiss has had a long-term association with the marketing of pyrethrum and PBO in addition to many other insecticides used in the home, horticultural and the food industry

Graham D. Moores Biochemist at Rothamsted, UK. Studies the biochemical basis of insecticide resistance. IACR, Rothamsted, Harpenden, UK

David Needham Head of Toxicokinetics, AgrEvo UK Ltd. Worked for 25 years in this field in the pharmaceutical and agrochemical industries. AgrEvo UK Ltd, Chesterford Park, Saffron Walden, UK

Thomas G. Osimitz Toxicologist. Manager of Regulatory Affairs. S.C. Johnson and Son Inc., Racine, Wisconsin, USA

K. Ropkins Environmental Chemist. M-Scan Ltd, Silwood Park, Sunninghill, Ascot, UK

Sami Selim Formerly Director of Toxicology, Biological Test Centre. Specialized in metabolism and residue studies in animals and plants. Biological Teast Center, Irvine, California, USA

Lindsay R. Showyin Previously Technical Director, Reckitts Household Products, Sydney, Australia

Duncan Stewart Previously Principal Research Scientist at Wellcome Environmental Health, Berkhamsted, UK, and later Roussel. Specialized in studies on methods used for the laboratory and field bioassay of insecticides

Robert Testman Analytical Chemist at Biological Test Center, Irvine, California, USA

Antonio Tozzi Organic Chemist. Involved in the manufacture of speciality chemicals. President of Endura Spa, Bologna, Italy. Remains active in research and development

John C. Wickham Previously Head of Entomology Department, Wellcome Environmental Health, Berkhamsted, UK. Undertook biological evaluation and development of the Rothamsted pyrethroids and usage of natural pyrethrum

Foreword

Piperonyl butoxide (PBO) science is half a century old. With five decades of use, it must do something important and do it well. As the first major insecticide synergist, PBO established the principle and practice of making the best use of valuable pest control agents. The term insecticide synergist is synonymous with PBO. It always dominated the field and still leads the way.

Pyrethrins are synergized by PBO. Pyrethrum extract is much more effective with PBO and pyrethroids for protection of humans, households, livestock and some crops are more economical with PBO. Pyrethroids must combine bio-degradability with effectiveness and synergism with PBO allows this goal to be achieved. Other insecticides also benefit from PBO, but the history and use are dominated by the pyrethroid/PBO duo. This is a partnership where each member of the pair does a very different job and together they are far more effective than either one alone – a truly beneficial synergism.

Organic insecticides have progressed from botanicals to chlorinated hydro-carbons, organophosphates, methylcarbamates, pyrethroids, insect growth regu-lators and nicotinoids as major groups, plus a plethora of other types used in small amounts. These compounds generally work at different biochemical targets as required to keep pace with development of resistance and best balance scheduling in pest and pesticide management. The biochemical target changes but PBO remains a consistent supplement in many cases for improved activity and to overcome resistance. PBO is used primarily with the pyrethroids for economic reasons, but it acts on a fundamentally weak link to improve the effectiveness of many pest control chemicals.

Cytochrome P450 is the target for PBO. Many years of PBO use preceded the understanding of its mechanism of action. PBO was here before P450 was named or conceptualized or even imagined. In fact, PBO played an important part in early studies on insect P450s and their role in insecticide oxidative detoxification and cross-resistance phenomena. PBO metabolism by P450 is intimately linked to its action as an inhibitor. PBO is now the additive used to test if P450 limits insecticidal activity. I take particular pride in my role in establishing these relationships.

PBO uses and safety evaluations, as for many other major pesticides, should be available for regulatory consideration and public education and scrutiny. This openness was achieved in a 1996 symposium in Florence held during the XX International Congress of Entomology, bringing together experts in all facets with representation from many parts of the world. Dedicated attention was given to safety issues and test protocols. The presentations ranged from firmly estab-lished facts to controversial and speculative topics, as should be the case to advance PBO science.

This volume presents the panorama of PBO chemistry, toxicology and uses and the detailed evidence and approaches needed to evaluate the state-of-the-art and the state-of-the-science. It brings together the published and unpublished information on PBO for you, the reader, to relate to your personal interest in effective and safe pest management. Denys Glynne Jones brings to the task of editing this volume the extensive experience he has gained working with pyrethrum flowers at the Kenya Pyrethrum Board and with PBO at Endura Spa in Bologna. His knowledge is combined with diplomacy and patience so critical in bringing a volume of this scope to fruition so that PBO science can be understood by all interested parties.

John E. Casida
Environmental Chemistry and Toxicology Laboratory,
University of California, Berkeley, USA

John E. Casida

Preface

Fifty years ago the editor of *Pyrethrum Post* reviewed an article in the US trade journal *Pest Control and Sanitation* which described the powerful synergistic and stabilizing effects that the then novel compound piperonyl butoxide (PBO) had on pyrethrins. The author claimed that the past popularity of aerosols was now likely to be greatly increased as the new combinations of pyrethrins and PBO had produced equal 'knockdown' but greater mortality than any of the formulations previously used.

As we approach the millenium, we are aware of a rapid upsurge in the incidence of insect- and tick-borne diseases in humans and animals, with malaria, dengue and many others again causing an annual death toll of millions of people. PBO has always been used in Public Health formulations and today use is also made of its ability to overcome resistance to insecticides.

In 1983 a group of interested companies representing the manufacture and usage of PBO formed a Task Force in the USA to finance and organize a completely new appraisal of its toxicology, chemistry, environmental fate and effects on wildlife. This extensive programme of work, costing over 12 million dollars, is now almost complete and this volume aims to present key data from these studies together with new information on present and future use patterns. The contents of each chapter is the sole responsibility of the author(s), and each chapter has been peer reviewed.

I wish to thank all the busy scientists who willingly gave of their time to speak at the Symposium on PBO held in Florence in 1996 and then to write their chapters together with additional authors not attending the Symposium. I am particularly grateful to Professor John Casida for chairing the opening of the Symposium and for writing the Foreword to this book.

The Directors of Endura SpA, in Italy, have generously given the essential financial support to the whole project and what should have taken two years has actually taken five to complete. May I also thank various members of my family who have been very helpful during this prolonged gestation period. The support of various members of the PBO Task Forces I and II is also appreciated. Lastly, I must thank the staff of the Academic Press Ltd, who saw each deadline vanishing into a black hole, for staying with me until the manuscript was complete.

I met Herman Wachs, the inventor of PBO, during my visit to the USA in 1964, and later met Lyle Goodhue, the inventor of the aerosol, and John Fales, who tested the prototype on that Easter Sunday in 1941. All three were very modest men. I wish to dedicate this book to these three innovative scientists.

Denys Glynne Jones
April 1998

1

A Brief History of the Development of Piperonyl Butoxide as an Insecticide Synergist

ANTONIO TOZZI

1. INTRODUCTION

Piperonyl butoxide (PBO) was invented and developed in the USA in response to fears for public health. In this instance one can truly say that 'Necessity was the Mother of the Invention'.

2. RENEWED THREAT TO PUBLIC HEALTH BY INSECT VECTORS

From 1937 onwards, commercial air travel between North and South America increased dramatically with many refuelling stops in tropical locations. As the USA itself had almost brought malaria (originally introduced from Britain and Africa) and other insect-transmitted diseases under control, great concern was expressed that these diseases could be reintroduced and again become endemic, following the rapid transport by air of live infected mosquitoes.

At that time pyrethrum was considered to be the strategic insecticide for the control of mosquitoes and other insect vectors. Japan was the major producer of pyrethrum, with East Africa under active development. The possibility of imports being disrupted in a future war was regarded as a serious risk, as existing supplies of the insecticide were barely adequate for public health uses.

PIPERONYL BUTOXIDE
ISBN 0-12-286975-3

3. NEW RESEARCH

Research and development was therefore directed towards the following goals.

- Developing more efficient methods of applying insecticides to control flying insects.
- Searching for chemical substitutes for pyrethrum with the same valuable properties of rapid 'knockdown', kill and complete safety to humans.
- Developing 'pyrethrum extenders' – chemicals which could be added to existing pyrethrum formulations to improve their efficiency. Later such compounds which had little or no intrinsic action on their own were called *synergists*.

From February 1938 to June 1941, 75 patents were granted in the USA listing 1400 synthetic or natural organic compounds which were proposed as pyrethrum substitutes, 'extenders', 'boosters', 'activators' or 'synergists'. One of these, N-isobutyl-undecylenamide (Weed, 1938), was used commercially.

In a US patent C. Eagleson (1940a) described the use of sesame oil as a 'synergist' or 'activator' when combined with insecticides containing pyrethrum or rotenone. Further details were published in the same year (Eagleson, 1940b). Eagleson later examined 42 other plant and animal oils (Eagleson, 1942), but none showed the effect produced by adding sesame oil.

Haller *et al.* (1942a), working for the Bureau of Entomology and Quarantine of the US Department of Agriculture, realized the importance of Eagleson's discovery and fractionated sesame oil in a molecular still. The various fractions were separately added to standard amounts of pyrethrum extract in kerosene and tested for 'knock down' and kill against house flies. Two fractions showed the most activity; from these a crystalline solid was produced that was identified as sesamin. When tested against house flies with pyrethrum and alone, both the sesamin and the noncrystalline residue were shown to behave as synergists.

Haller *et al.* (1942b) studied the structure and synergistic effect of sesamin and several related natural compounds. The data is summarized in Table 1.1. Isosesamin and asarinin showed similar synergistic activity to sesamin and the authors noted a similarity of chemical structure in the presence of methylene-dioxy groups. Later Gertler and Haller (1942) claimed that the amides of 3,4-methylenedioxy cinnamic acid were effective synergists of pyrethrins, but these products had low solubility in petroleum solvents and were not exploited commercially.

It is of interest to note that all these scientists recorded that the test compounds were not appreciably toxic to insects by themselves.

Parallel to the work on identifying pyrethrum synergists, Dr Lyle Goodhue had prepared the first liquefied gas aerosol on Easter Sunday morning, 13 April 1941, and arranged with his colleagues in the US Department of Agriculture to test it that afternoon. This aerosol contained dichlorodifluoromethane, pyrethrum extract and sesame oil as a synergist. J.H. Fales, the entomologist who performed the test with house flies, has recorded the spectacular performance of

Table 1.1. Synergistic effect of sesamin and related compounds (Haller *et al.*, 1942b)

No.	Material	Concentration (%)	Average mortality after 24 h (%)
	Sesamin and its isomers		
1	Sesamin	0.2	4
2	Sesamin + pyrethrins	0.2 + 0.05	84
3	Isosesamin	0.2	5
4	Isosesamin + pyrethrins	0.2 + 0.05	87
5	Asarinin	0.2	14
6	Asarinin + pyrethrins	0.2 + 0.05	88
7	Pyrethrins (control)	0.05	25
	Pinoresinol and derivatives		
8	Pinoresinol	0.18	1
9	Pinoresinol + pyrethrins	0.18 + 0.05	12
10	Dimethyl pinoresinol	0.2	1
11	Dimethyl pinoresinol + pyrethrins	0.2 + 0.05	17
12	Diacetyl pinoresinol	0.3	2
13	Diacetyl pinoresinol + pyrethrins	0.3 + 0.05	11
14	Pyrethrins (control)	0.05	19

the first aerosol (Fales, 1991) and also mentions the ladies present which included Lyle Goodhue's wife who at all times supported her husband, particularly with the limitations set by his restricted vision. The development of the aerosol had a tremendous impact on methods of applying domestic and public health insecticides.

O.F. Hedenburg of the Mellon Institute, Pittsburg, USA, worked independently of Haller and his colleagues, investigating the use of compounds containing the methylenedioxyphenyl group as insecticides. The results obtained with these chemicals alone were not encouraging because they had poor 'knockdown' properties when compared with pyrethrins. However, when tested with pyrethrins very positive and encouraging results were obtained. One such product with outstanding synergistic properties was produced by the condensation of the alkyl-3,4-methylene dioxystyryl ketones with ethyl acetoacetate. It was called piperonyl cyclonene, and made some commercial sales in spite of its limited solubility in the solvents used with insecticides.

Herman Wachs, working for Dodge & Olcutt, Inc., co-operated with Hedenburg in the search for an improved compound. He synthesized PBO (Wachs, 1947), which had many obvious advantages over piperonyl cyclonene, including complete miscibility with freon and petroleum solvents. Details regarding the chemistry of PBO are given in other chapters. The key raw material is naturally occurring safrole, which supplies the methylenedioxyphenyl portion of the molecule.

Initial appraisals of the mammalian toxicity of PBO were made by Dove (1947) and by Sarles *et al.* (1949). Sarles and Vandergrift (1952) reported on chronic oral studies with rats, dogs, goats and monkeys and their general findings of a low order of toxicity to mammals still remain valid today. The studies were made with a technical product said to contain not less than 80% of the active ingredient and also with mixtures of PBO and pyrethrins.

In spite of the recognition that PBO and piperonyl cyclonene were powerful synergists with pyrethrins, they were originally described as 'new safe insecticides' (Wachs *et al.*, 1950) and the authors claimed 'they had definite insecticidal properties in themselves'.

4. COMMERCIAL DEVELOPMENT

From 1952 onwards, significant tonnages of PBO were manufactured and sold in the USA. The relevant patents, which were originally assigned to the US Industrial Chemicals, were later owned by the Fairfield Division of the Food Machinery Corporation (FMC). Licences were granted to Cooper McDougall and Robertson in the UK, Samuel Taylor in Australia, Rexolin in Sweden and Tagasako in Japan. In 1960, A. Tozzi started manufacture in Italy and now under the name of Endura produces over 60% of the world's demand for PBO.

Piperonyl cyclonene was not developed beyond 1954 but sulfoxide prepared from isosafrole and n-octyl mercaptan was almost as active a synergist as PBO and was used in many formulations until 1973. Several other synergists such as Bucarpolate Tropital, Safroxan and MGK 264, all less active than PBO, were developed, but only MGK 264 is still available.

The demand for PBO continues to expand and it is used with pyrethroids and carbamates as well as with pyrethrum. The predominant position of PBO has been attributed to a variety of factors.

- It has high efficacy as a synergist with a wide range of insecticides.
- It has excellent stability during storage and in insecticide formulations.
- The PBO Task Force (members are Agrevo, M.G.K., Prentiss, Tagasako, S.C. Johnson Wax and Endura) has developed a most extensive technical information package confirming its safety to humans, animals and birds; full data on residue levels in plants, animals, soil and water are given.
- There was a planned programme for continued improvements in the purity of the technical product from 80% in 1952 to over 90% in 1997.
- It has a very wide range of uses on humans, animals, birds and plants as well as in food storage, in addition to its classical use in household and industrial pest control.

REFERENCES

Dove, W.E. (1947). Piperonyl butoxide: a new and safe insecticide for the household and field. *Am. J. Trop. Med.* **27**, 339–345.
Eagleson, C. (1940a). Synergists for Insecticides. U.S. Patent 2202145 (28 May, 1940).

Eagleson, C. (1940b). Insect powder labels. *Soap* **16**, 97–101, 115, 129.

Eagleson, C. (1942). Sesame in insecticides. *Soap* **18**(12): 125–127.

Fales, J.H. (1991). A chronology of early aerosol development. *Spray Technology & Marketing* July 1991, 27–31.

Gertler, S.I. and Haler, H. (1942). U.S. Patent 2202145 (May 28, 1940).

Haller, H., McGovran, E., Goodhue, L. and Sullivan, W. (1942a). The synergistic action of sesamin with pyrethrum insecticides. *J. Org. Chem.* **7**, 183–184.

Haller, H., La Forge, F. and Sullivan, W. (1942b). Effect of Sesamin and related compounds on the insectidal action of pyrethrum on houseflies *J. Econ. Entomol.* **35**, 247–248.

Sarles, M.P. and Vandergrift, W.B. (1952). Chronic oral toxicity and related studies on animals with the insecticide and pyrethrum synergist, piperonyl butoxide. *Am. J. Trop. Med. Hyg.* **1**, 862–863.

Sarles, M.P., Dove, W.E. and Moore, D.H. (1949). Acute toxicity and irritation tests on animals with the new insecticide, piperonyl butoxide. *Am. J. Trop. Med.* **29**, 151–166.

Wachs, H. (1947). Synergist insecticides. *Science* **105**, 530–531.

Wachs, H., Jones, H. and Bass, L. (1950). New safe insecticides. *Advances in Chemistry Series* 1, 43–47.

Weed, A. (1938). New insecticide compound. *Soap* **14**, 133–135.

2

The Safety of Piperonyl Butoxide

RORY BREATHNACH

1. INTRODUCTION

During the initial development of piperonyl butoxide (PBO) in the late 1940s, a large amount of toxicology work was performed. Over the intervening years, the relative lack of adverse reactions and general high-standing of this substituted methylenedioxyphenyl compound led to a very high level of acceptance of its use on or near humans, animals and food in storage. The low order of toxicity associated with PBO meant that this compound did not attract much attention from regulatory authorities. More recently, further data have been required principally by the Environmental Protection Agency (EPA) in the USA, and as such, a large degree of further toxicity work has been performed, principally by the PBO Task Force (PBTF). A significant degree of knowledge of both the pharmacodynamics and toxicity profile of PBO is now available. This chapter aims to combine the knowledge gained from a diverse group of studies, performed over several decades in many different laboratories. Particular emphasis is also placed on current knowledge relating to the effects of PBO in humans and any potential risks posed to consumers from the various commercial formulations in worldwide use. This review concentrates on data generated since 1970. However, some data generated earlier, particularly that by Sarles et al. (1949) and Sarles and Vandegrift (1952), is still relevant today and is quoted where appropriate.

2. TOXICOLOGY

2.1. Acute Toxicity Studies

A variety of tests have been performed with PBO, by different routes and in different species. An outline of the results of the various studies is included in Table 2.1. From these data it can be concluded that the acute toxicity of PBO is of a low order. Whilst the recommended therapeutic dose used in most animal

PIPERONYL BUTOXIDE
ISBN 0-12-286975-3

Table 2.1.   Acute toxicity of PBO in animals by various routes

Species	Route	LD$_{50}$ or LC$_{50}$ value (g per kg body weight or g per L of air)	Reference
Mouse	Oral	4.0	Negherbon (1959)
Mouse	Oral	8.3	Draize *et al.* (1944)
Rat	Oral	4.57 (male) 7.22 (female)	Gabriel (1991a)
Rat	Oral	8.0–10.6	Sarles *et al.* (1949)
Rat	Oral	12.8	Lehman (1948)
Rat	Inhalation	> 5.9	Hoffman (1991)
Rat	Subcutaneous	> 15.9	Sarles *et al.* (1949)
Rabbit	Oral	2.7–5.3	Sarles *et al.* (1949)
Rabbit	Dermal	> 2.0	Gabriel (1991b)
Cat	Oral	> 10.6	Sarles *et al.* (1949)
Dog	Oral	> 8.0	Sarles *et al.* (1949)

species at a clinical level is of the order of 1–2 mg per kg body weight, the available data shows that several thousand times such levels would be required to reach the LD$_{50}$ values. The clinical signs of acute toxicity with PBO varied slightly between species. In rats, deaths occurred on the second or third day post-dosing and were preceded by overt clinical signs including loss of appetite, ataxia, dark faeces, bloody ocular discharges and prostration. Post-mortem examination revealed haemorrhage into the alimentary tract as well as pale and enlarged livers and kidneys (Sarles *et al.*, 1949). However, clinical signs of inappetence, vomiting and weight loss were more evident in the acute studies performed in the rabbit, cat and dog.

Following exposure by inhalation in the rat, transient secretory and respiratory signs were recorded during the treatment period. Effects on body weight gain were minimal. However the parenteral administration of PBO to rats (subcutaneous) resulted in inappetence, weight loss, dyspnoea and discoloration of the urine.

Based on studies performed in rabbits and guinea pigs, PBO was found to be nonirritating to eye and skin and is not a dermal sensitizing agent (Sarles *et al.*, 1949; Romanelli, 1991a,b,c).

2.2. Subchronic Toxicity Studies

2.2.1. Oral

(a) Mice   Fujitani *et al.* (1993) reported on a study where CD-1 mice were fed PBO at dose levels of 0, 1000, 3000 and 9000 ppm for 20 days. Ten animals of each sex were included at each dosage level. No mortalities occurred during this trial. Body weights were reduced in the high-dose groups (about 15% after 20 days), as well as in mid-dose females (8% reduction at termination). There was a dose-dependent increase in liver weights, with a 79% increase being evident in high-dose males as compared with control mice. Kidney and spleen weights were reduced at the highest dose level. On biochemical analysis, high-dose animals exhibited increased levels of cholesterol, phospholipids and total proteins in both sexes and increased levels of γ-glutamyl transpeptidase in females. At post-mortem, histopathological examination of the liver revealed hypertrophy of hepatocytes, single cell necrosis and inflammatory cell infiltration. These changes were most marked in the centrilobular area of the lobules of the high-dose group. Kidney examinations were unremarkable. The no observable adverse effect level (NOAEL) for liver toxicity was set at 1000 ppm (approximately equivalent to 150 mg kg^{-1} day^{-1}). The high-dose level incorporated in this study was well above the maximum tolerated dose (MTD).

A longer-term 90-day oral toxicity study was performed with CD-1 mice. Fifteen male and 15 female mice were utilized per dosage group. Dose levels were set at 0, 10, 30, 100, 300 and 1000 mg kg^{-1} day^{-1}. No treatment related clinical signs, alterations in food intake or mortalities occurred in any test group. Decreases in mean body weight and body weight gain (especially during week 1) were evident in males at the highest dose level throughout the study: the mean weight gain for high-dose males was 34% lower when compared with controls. There were no statistically significant corresponding decreases in the high-dose females. Whilst male mice had decreased kidney weights at 1000 mg kg^{-1} day^{-1} (this was considered to be related to decreases in overall body weight), the liver was again the only organ to show definite treatment-related changes. Liver tissue appeared to be the primary target site for PBO, with findings generally being dose-dependent. Prominent effects in this organ were seen at dosage levels ≥ 300 mg kg^{-1} day^{-1}. Besides the expected increases in liver weights, histopathological findings included necrosis, centrilobular hypertrophy and polymorphonuclear cell infiltrates. The incidence of such hepatocyte hypertrophy was statistically significantly increased for male animals treated at a dose level of 100 mg kg^{-1} day^{-1} when compared to control animals. Based on the findings of this study, the dose of 300 mg kg^{-1} day^{-1}, which resulted in moderate hepatic changes, was selected as the high-dose level for a mouse oncogenicity study (Chun and Wagner, 1993).

Tanaka (1993) reported a subacute oral toxicity study in male CD-1 mice where dosages of 0, 1500, 3000 and 6000 ppm were fed daily for 7 weeks. Besides a treatment-related decrease in food intake during week 1 in the mid- and high-dose groups, the main aim of this study was to investigate any possible treatment-related changes in behavioural patterns. No consistent or significant effect was observed in the T-maze or exploratory behaviour tests at the 4-week time point, but by week 7 some parameters in the exploratory behaviour test were altered in treated animals.

(b) Rats   Bond *et al.* (1973) fed PBO to 20 rats for 90 days at a dose level of 1857 mg kg^{-1} day^{-1}. Forty per cent of the rats died. At post-mortem, the most significant finding was an increase in liver weights, which in rats dosed with PBO was an average of 2.4 times heavier than in untreated controls. In this same article, reference is also made to a 17-week study where rats were fed 500 mg kg^{-1} day^{-1}. Liver and kidney damage was present at post-mortem in this study.

A 4-week dose range-finding subacute oral toxicity study was performed in albino rats by Modeweg-Hausen *et al.* (1984). Dose levels corresponded to 62.5, 125, 250, 500, 1000 and 2000 mg kg^{-1} day^{-1}, and 10 male and 10 female Sprague-Dawley rats were incorporated at each dosage level. Histologically detectable adaptive liver changes were seen at all dose levels, consisting primarily of eosinophilic infiltration and loss of vacuolation of hepatocytes. Such changes were more significant in mid- and high-dose groups. Signs of possible degenerative liver changes were observed at the two highest dose levels. Changes in liver weight were detected in male rats at dose levels of 250 mg kg^{-1} day^{-1} and in female rats at dose levels of 500 mg kg^{-1}day^{-1}. Kidney weights were increased at the 2000 mg kg^{-1} day^{-1} dose level. There was a marked and dose-related decrease in body weight gain in females at dose levels of 500 mg kg^{-1} day^{-1}. Haematological and biochemical examinations were unremarkable, except for an increase in alkaline phosphatase levels at the highest dose level. The NOAEL for this study was set at 125 mg kg^{-1} day^{-1}, based on the absence of any significant effects on the liver.

A dose range-finding study for a teratogenicity trial in rats demonstrated marked clinical and post-mortem signs of subacute PBO toxicity in the dams. In this trial, female CD rats were dosed with PBO by gavage on days 6–15 of pregnancy at dose levels of 250, 500, 1000, 2000 and 4000 mg kg^{-1} day^{-1}. Overt clinical signs were seen at dosage levels of 500 mg kg^{-1} day^{-1} and above, and consisted of urogenital area wetness and perinasal encrustation in some but not all dams. At higher dose levels of 2000 mg kg^{-1}day^{-1} and above, such clinical signs consisted of ataxia, twitching, prostration, dyspnoea, gasping and lacrimation, as well as periocular and perinasal encrustation. These signs were most evident within the first 3 days of dosing. At post-mortem, ulceration of the lining of the glandular portion of the stomach as well as haemorrhage and sloughing of the lining of the nonglandular portion of the stomach was seen (Chun and Neeper-Bradley, 1992).

A 13-week subacute oral toxicity study was performed in Fischer F344 rats (10 of each sex/dosage group) at levels of 0, 6000, 12 000 and 24 000 ppm in the diet (Fujitani *et al.*, 1992). No mortalities occurred. Nasal bleeding and dose-related abdominal distension were reported. A significant decrease in body weight was evident in the high-dose group at study termination (36% decrease in males as compared with controls, 24% decrease in females). Blood haemoglobin levels were reduced in both sexes in the high-dose groups and in mid-dose females. Biochemical changes in the high-dose groups consisted of increases in albumin, cholesterol, γ-glutamyl transpeptidase and urea. Liver and kidney weights were increased in a dose-dependent manner. Histopathological examination revealed hypertrophic hepatocytes (containing a basophilic granular substance) and vacuolation of hepatocytes in periportal areas. Coagulative necrosis and oval cell proliferation were occasionally seen. Atrophy of the epithelial lining of the proximal convoluted tubules in the renal cortex was present in some male rats. A NOAEL for PBO could not be established in this study owing to the presence of liver and kidney effects induced by the clearly excessive dose levels.

Further information on the subchronic toxicity of PBO in the rat was reported by JMPR (1995).

(c) Other Species   Sarles *et al.* (1949) performed a subacute oral toxicity experiment in rabbits. A 5% PBO emulsion was fed once weekly in the diet to three rabbits over a 3-week period. The dosage used varied between 1.0 and 4.0 mL kg^{-1} week^{-1}. There were no mortalities. No clinical signs of toxicity occurred. The rabbit that received the highest dosage was sacrificed 1 week after the last treatment, but no lesions were detected at post-mortem examination.

A 4-week oral toxicity study was performed in monkeys in 1952 by Sarles and Vandegrift. Two African Green Monkeys were fed PBO by capsule, 6 days a week (for 4 weeks), at a dosage level of 0.03 or 0.1 mL kg^{-1} day^{-1} (one monkey at each dosage level). The total volumes of PBO fed to these two animals during the entire experiment were 0.94 mL and 3.36 mL, respectively. No gross pathological lesions were evident in the treated monkeys' livers. On histopathological examination of the liver, the monkey on the higher dose level showed evidence of minimal dystrophy and dysplasia, occasional acidophilic and hyalin-necrosis cells, as well as hydropic swelling.

PBO was administered in an 8-week oral toxicity study to dogs. The dosage levels were 0, 500, 1000, 2000 and 3000 ppm of PBO in the diet, and each dose group consisted of two male and two female dogs. All dogs survived to study termination. All animals in the 3000 ppm dose group had decreased appetites and reduced defaecation during the first week. No other abnormal clinical signs were present. Treatment did have an effect on body weight, with three out of four dogs losing weight in the highest dose group. Even at 1000 ppm PBO in the diet, weight gains were lower than the negative control group. Food intake was similar between treated and control groups, except for a slight reduction in some animals at the 3000 ppm dosage level. There were no treatment-related effects on haematological parameters at any dose levels, but slight increases in alkaline phosphatase values and slight decreases in cholesterol at doses ≥ 2000 ppm. No treatment-related microscopic changes were noted at post-mortem in any group, but a compound-related increase in absolute and relative liver and gall bladder weights was recorded in males. There was a decrease in the absolute and relative weights of the testes and epididymis in the groups treated with 2000 and 3000 ppm. On histopathological examination, hypertrophy of hepatocytes was noted in males of all dose levels and in females at dosages above 2000 ppm. This finding was consistent with the increases in liver weights and serum alkaline phosphatase levels described above. No other treatment-related microscopic changes were evident. The dose level of 500 ppm can be set as a NOAEL for this study as the changes recorded in the liver were adaptive in nature rather than adverse, and were not accompanied by any systemic signs of toxicity (Goldenthal, 1993a).

The results from these subchronic oral toxicity studies are summarized in Table 2.2.

Table 2.2 Summary of results of sub-chronic oral toxicity studies with PBO

Species	Dose	Duration	Comments	Reference
Rat	2.5–5 mL kg^{-1}	31 days	Anorexia, loss of weight, coma and deaths reported	Sarles and Vandegrift (1952)
Rat	1857 mg kg^{-1}	90 days	40% of animals died Increased liver weights	Bond et al. (1973)
Rat	62.5–2000 mg kg^{-1}	28 days	Increased liver and kidney weights NOAEL = 125 mg kg^{-1} day^{-1}	Modeweg-Hausen et al. (1984)
Rat	250–4000 mg kg^{-1}	10 days	Ataxia, twitching, dyspnoea Gastric ulceration at post-mortem	Chun and Neeper-Bradley (1992)
Rat	6000–24 000 ppm	13 weeks	Nasal bleeding and abdominal distension Increased liver and kidney weights No mortalities	Fujitani et al. (1992)
Mouse	1000–9000 ppm	20 days	No mortalities Hepatocyte hypertrophy, necrosis and inflammatory cell infiltration in liver	Fujitani et al. (1993)
Mouse	10–1000 mg kg^{-1}	90 days	No treatment-related mortalities At post-mortem, liver changes included necrosis, centrilobular hypertrophy and polymorphonuclear cell infiltrates	Chun and Wagner (1993)
Mouse	1500–6000 ppm	7 weeks	Decreased food intake Some minor alterations in behavioural tests	Tanaka (1993)
Rabbit	1–4 mL kg^{-1}	Once a week for 3 weeks	No signs of toxicity	Sarles et al. (1949)
Monkey	0.03–0.1 mL kg^{-1}	4 weeks	No clinical signs of toxicity Minor changes in liver at post mortem.	Sarles and Vandegrift (1952)
Dog	500–3000 ppm	8 weeks	No mortalities Decreased body weights Increased liver weights and hepatocyte hypertrophy	Goldenthal (1993a)

2.2.2. Dermal

A 21-day subchronic dermal toxicity study was performed in rabbits by Goldenthal (1992). Dosages of 100, 300 and 1000 mg kg^{-1} were applied topically once a day, 5 days a week, for 3 consecutive weeks. There were five male and five female rabbits in each group, and a control group to which mineral oil only was applied. No mortalities occurred. Treatment-related effects were limited to minor skin changes at the application site. Dermal irritation was present in all three treatment groups, although to a lesser extent and incidence at the 100 mg kg^{-1} dose level. Dermal lesions consisted of very slight erythema and oedema. This irritation usually appeared by day 5 and persisted for the remainder of the study. Desquamation and fissuring of the skin appeared in the 300 mg kg^{-1} and 1000 mg kg^{-1} groups. At post-mortem, redness and discoloration of the skin, as well as thickening and scab formation, were recorded. On histopathological examination, moderate acanthosis, hyperkeratosis and chronic inflammation of the epidermis were present. The severity of these lesions increased with increasing dosage. Body weights were comparable with those in the control group, and food intake was only slightly lower in treated animals. No treatment-related changes were seen on haematology and serum biochemistry, and no signs of systemic toxicity were present at any dosage level. The dose level of 100 mg kg^{-1} can be set as a NOAEL for this study.

2.2.3. Inhalation

A subchronic (3-month) inhalation toxicity study was performed in the rat using whole body exposure. This study was performed by Newton (1992). The PBO was applied as an aerosol to Charles River CD rats. Concentrations corresponding to 15, 74, 155 and 512 mg PBO per m^3 of air were used in this study. Thirty rats per dosage group were utilized and a similar negative control group. Ninety-seven per cent of particles were 10 μm or less in diameter. The rats were exposed to the aerosol for 6 hours per day, 5 days a week, for 13 weeks. No rats died during this study. Body weight gain and food intake were not affected by treatment. No ocular changes due to treatment were noted. Nasal discharges and anogenital staining were reported at higher dosage levels (above 155 mg m^3). At the highest dosage level, serum levels of alanine transaminase, aspartate transaminase and glucose were decreased, whilst blood urea nitrogen, total protein and albumin values were increased. There was a statistically significant increase in absolute and relative liver and kidney weights at the highest dosage level. Equivocal changes were seen on histopathology of the liver and included vesiculation and vacuolation of hepatocytes. These changes are of unknown significance and were not dose-related. Minimal to slight laryngeal irritation was also noticed on post-mortem examination. Such laryngeal irritation was associated with inflammation, congestion, oedema and debris in the lumen. Squamous metaplasia of the epithelial lining of the larynx was also recorded and was most marked in high-dose groups. The severity of mucosal inflammation was slightly greater in both males and females exposed to high dose levels. These changes are considered to represent a localized response indicative of laryngeal irritation and not systemic toxicity. Based on this study, a NOEL for systemic toxicity was set at 155 mg PBO m^{-3} of air. Further details of this study were reported by JMPR (1995).

Lorber (1972) exposed normal and splenectomized dogs to PBO fog for four periods of 5 minutes each with an 8-minute rest between each exposure, for 4 days. In splenectomized dogs, the treatment was associated with a reduction in the platelet count and occasional reticulocyte elevation.

2.3. Chronic Toxicity Studies

2.3.1. Oral

(a) Mice   Bond *et al.* (1973) reported on chronic oral toxicity tests in mice, fed either 45 or 133 mg kg^{-1} of PBO daily in the diet for 18 months. No abnormal effects were detected in the test animals in this study.

A combined chronic toxicity/carcinogenicity study was performed in CD-1 mice by Hermanski and Wagner (1993), over a 78-week period using PBO with a purity > 90%. The dosages were 0, 30, 100 and 300 mg kg^{-1}day^{-1} and 60 male and 60 female mice were dosed at each level. There were no treatment-related signs of toxicity nor any such changes in food intake. There was no increase in palpable masses nor were any changes in haematological parameters recorded. In the high-dose group, the mean body weight and body weight gains were slightly reduced for both sexes, and especially so in the case of males. The mean absolute and relative liver weights of the low-dose group of male mice were slightly increased, primarily due to an increase in the size of the nodules that were found in the livers of this group. Overall, there was a dose-related increase in both the mean absolute and relative liver weights in the mid- and high-dose groups (both sexes).

(b) Rats   A combination chronic toxicity/oncogenicity study was performed with PBO over a 2-year period in albino rats (Graham, 1987). Whilst the major details of this study will be presented later in the section on carcinogenicity, some of the chronic toxicity data is relevant to this section. PBO was fed to Sprague-Dawley rats at dose levels of 0, 30, 100 and 500 mg kg^{-1}day^{-1}. Sixty male and female rats were used at each dosage level. Two additional low-dose groups (70 male and 70 female rats) were added to the early part of this study and were treated at dose levels of 15 and 30 mg kg^{-1}day^{-1}. These animals were sacrificed early in the study, so that a no-effect level could be identified with regard to minor alterations in liver cell morphology. As no abnormal findings were discovered in the initial phase of the trial at either of these two lower dosage levels, the 15 mg kg^{-1}day^{-1} group was dispensed with for the 2-year study.

No treatment-related mortalities occurred in this trial, and no abnormal treatment-related clinical signs were recorded. There were slight reductions in food intake and body weight gain in the 500 mg kg^{-1} group. No changes in haematological or urinary parameters occurred in any treatment group over the 2 years. On serum biochemical analysis, increases in both cholesterol and total protein were recorded in the highest dosage group, and this group also recorded decreased levels of alanine transaminase, aspartate transaminase and CPK. Occasional female rats at the highest dosage level also had small increases in blood urea nitrogen values. Examination of the liver in the 100 and 500 mg kg^{-1} day^{-1} dosage groups, revealed organ enlargement, as well as the presence of pale and raised areas on the surface of the liver in some animals. The most prevalent histological lesion was regional or generalized hypertrophy of hepatocytes in rats from high-dose groups. This was characterized by a centrilobular location of enlarged eosinophilic cells, which occasionally contained brownish cytoplasmic pigment. Changes in the thyroid gland included increased pigment in colloid, as well as follicular hyperplasia, especially in the highest dosage group. The incidence of bilateral atrophy of the testes was statistically significantly increased at all dose levels, but the total incidence of atrophy (i.e. unilateral and bilateral) was normal. Furthermore, testes weight did not differ when compared with other organs or control animals. There was a slight increase in the incidence of ovarian masses and general ovarian enlargement seen grossly in females at the 500 mg kg^{-1} day^{-1} dosage level. There was no evidence of any dose-related microscopic lesions on histopathological examination. Overall, based on the above findings, where no significant toxicity was evident at the 30 mg kg^{-1} day^{-1} dosage level, this figure was taken to indicate the NOAEL for this study (Graham, 1987). Further information on this study is reported by JMPR (1995).

Maekawa *et al.* (1985) also performed a 2-year chronic oral toxicity trial in F344/Du Crj rats. Dosage levels in this study were set at 0, 5000 or 10 000 ppm of technical grade PBO (89% purity). Fifty male and 50 female rats were included at each dosage level. Treatment duration was 104 weeks, with all surviving animals being sacrificed 6 weeks later, after being fed on a basal diet for those final 6 weeks. There was a dose-dependent increase in mortality values (both males and females) for the different dosage groups. Anaemia and the presence of blood in the faeces were evident prior to death in many animals. There was a dose-dependent decrease in body weight in treated animals, without a concomitant decline in food intake. A dose-related increase in the incidence of ulcers, regenerative hyperplasia and ossification in the ileocaecal mucosa was evident, as well as frank haemorrhage in both the caecum and colon. Histopathological evaluation revealed ulceration, inflammatory cell infiltration and excess granulation tissue, sometimes with ossification of the surrounding mucosa.

Takahashi *et al.* (1994b) performed a combined chronic oral toxicity/carcinogenicity study in F344/Du Crj rats. Diets containing 0, 6000, 12 000 or 24 000 ppm PBO (approximately 94.5% purity) were fed to male rats for 95 weeks and to female rats for 96 weeks. In total, 30–33 animals of both sexes were incorporated at each dosage level. Mortality figures were significantly elevated in the 12 000 ppm dosage male group. These deaths occurred as early as week 4 and were statistically significant from week 45. Such fatalities were most commonly associated with caecal haemorrhage. Body weights were also reduced in a dose-dependent manner, but such changes were only statistically significant in mid- and high-dose groups. Male and female rats from the high-dose group had body weights equal to approximately 50% of control values at study termination. Absolute liver weights were increased in the 12 000 ppm dosage group (males +22%; females +51%) and in females of the 6000 ppm dosage group (+29%). During the first month of treatment, male and female rats at the 24 000 ppm dosage level displayed lethargy, decreased food consumption, rough hair coats and epistaxis. Haematological examination revealed a dose-dependent hypochromic, microcytic anaemia. This finding was significant at all dose levels in both sexes. Biochemical evaluation revealed decreased cholinesterase activity and a decreased triglyceride content at all dose levels, with an increase in BUN values, especially in high-dose animals and all dose level females. Thrombocythaemia was evident in treated male rats, but was not dose-dependent. There was an increased incidence of gastric haemorrhage in both sexes treated at the high-dose level. An increased incidence of haemor-rhage and/or oedema of the caecum was seen in all treated males and mid-dose females. There were also increased incidences of black-coloured kidneys and white spotting of the lungs in mid- and high-dose males. Histopathological examination revealed distension of Bowman's space, tubular dilatation and interstitial fibrosis in kidneys from the high-dose treated animals.

(c) **Dog**   PBO was administered to dogs in capsule form for a 1-year chronic dietary toxicity study (Sarles and Vandegrift, 1952). Groups of four dogs each were treated at dose levels corresponding to 0, 3, 32, 106 and 320 mg kg^{-1} day^{-1}. The dosage changed in accordance with any alteration in body weight, except for one individual animal per dosage group, which received a constant dose level throughout the trial. All dogs belonging to the two highest dosage groups lost weight (meaningful comparisons between the lower-dose groups and control animals were not possible owing to large variations in body weight gains and the small numbers of animals involved). All dogs at the high-est dosage level died. However, no toxic reaction was seen at 3 mg kg^{-1} day^{-1}. Red blood cell (RBC) and white blood cell (WBC) counts were unremarkable at all dose levels. There was a dose-dependent increase in liver, kidney and adrenal weights. Microscopic changes were quite similar to those in other long-term toxicity studies performed in rats, with the liver again being the major target organ for toxicity. Hydropic swelling was evident in the mid-dose group, with hepatic dystrophy and dysplasia becoming more obvious at the two high-est dosage levels. Acidophil and hyalin-necrosis cells increased in number as the dystrophy increased. The conclusion of these authors was that dosages up to 32 mg kg^{-1} day^{-1} for 1 year were well tolerated by dogs.

A 1-year chronic dietary toxicity study was conducted with PBO in the dog (Goldenthal, 1993b). Beagle dogs were fed PBO for 1 year at dosages of 100, 600 and 2000 ppm of active ingredient in the diet. Four male and four female dogs were incorporated in each group, and a similar number were included in a negative control group. All animals survived to study termination. A reduction in body weight gain and food intake was evident in the 2000 ppm group. Physical examinations were otherwise normal throughout the test period. Whilst no treatment-related changes were noted on haematological examination, biochemical analysis showed increases in serum alkaline phosphatase levels at 6 months and 12 months in the highest dosage group. Female beagles showed a decrease in serum cholesterol at the 2000 ppm dosage level. Increased liver and gall bladder weights, with mild hypertrophy of hepatocytes, were also recorded at this highest dosage level. There was also a small increase in thyroid gland and parathyroid gland weights in this highest-dose group. However no microscopic abnormalities were detected in the thyroid gland. The dose level of 600 ppm can be set as a NOEL for this study.

(d) Goat   Sarles and Vandegrift (1952) also reported on a chronic oral toxicity experiment where a mature goat was fed a daily dose of 2.0 mL PBO by capsule, 6 days a week, for 1 year. This dose equated approximately to 0.1% PBO in the diet. The dose started 4 days after the goat gave birth to a female kid, with both dam and offspring being observed for 1 year to ascertain any signs of direct or indirect (i.e. PBO in dam's milk) toxic effects. The general health of both dam and kid was unaffected by treatment. The kid was nursed by the treated dam for approximately 6 months and continued to grow and thrive as expected. RBC and WBC counts were unremarkable. At post-mortem, the dam's liver revealed slight dystrophy and dysplasia, with central hydropic swelling and slight fatty accumulation. No abnormalities were detected in the organs of the kid goat.

Table 2.3 summarizes the results obtained from chronic oral toxicity studies with PBO.

2.4. Reproductive Toxicity and Teratogenicity

2.4.1. Reproductive Toxicity

Specific data on PBO is available in the areas of teratogenicity and fertility from studies performed in rats, mice and rabbits. Some background information is also available on the effects of long-term administration of PBO on reproductive tissues from chronic dietary toxicity studies performed in the rat and the dog.

(a) Mice   A three-generation, one litter per generation, reproductive toxicity study was performed in Crj:CD-1 mice by Tanaka *et al.* (1992). Ten animals of either sex were incorporated at each dosage level and dosage rates were set at 0, 1000, 2000, 4000 and 8000 ppm of PBO in the diet (purity not specified). The details of the study protocol were reported by JMPR (1995). Food intake was reduced in the F_0 generation at the 8000 ppm dosage level, except during the mating period, and was also reduced during the lactation period in the F_1 generation, at the same dosage level. The 4000 ppm treatment groups of both the F_0

Table 2.3. Summary of results of chronic oral toxicity studies with PBO

Species	Dose	Duration	Comments	Reference
Mouse	45–133 mg kg^{-1}	18 months	No abnormal clinical findings	Innes et al. (1969)
Mouse	30–300 mg kg^{-1}	78 weeks	No treatment-related signs of toxicity Liver weights increased in male mice	Hermanski & Wagner (1993)
Rat	100–25 000 ppm	2 years	Deaths at high dosages from liver failure Liver and kidney weights increased at high dose levels	Sarles and Vandegrift (1952)
Rat	30–500 mg kg^{-1}	2 years	Hypertrophy of hepatocytes and follicular hyperplasia in thyroid gland	Graham (1987)
Rat	5000–10 000 ppm	2 years	Dose-dependent increase in mortality figures Anaemia and blood in faeces	Maekawa et al. (1985)
Rat	6000–24 000 ppm	95–96 weeks	Mortalities often associated with caecal haemorrhages Increase in absolute liver weights; gastric haemorrhages	Takahashi et al. (1994b)
Rat	Combination study: PBO = 2000 ppm; pyrethrins = 400 ppm	2 years	No clinical signs of toxicity No significant abnormalities at post-mortem	Hunter et al. (1977)
Dog	3–320 mg kg^{-1}	1 year	No mortalities All dogs at highest dosage level died of liver damage No toxicity at lowest dosage level	Sarles and Vandegrift (1952)
Dog	100–2000 ppm	1 year	No mortalities Decreased body weight gain at mid- and high-dosages Increased liver weights, with hypertrophy of hepatocytes Increased thyroid gland weight at highest dose level	Goldenthal (1993b)
Goat	2 mL PBO (~0.1% of diet)	1 year	No clinical signs of toxicity Slight liver pathology at post-mortem	Sarles and Vandegrift (1952)

and F_1 generations also had a reduction in food intake during the lactation period. The mean F_1 litter size was not significantly decreased in the 8000 ppm treatment group. Mean F_1 litter weight was decreased at the 8000 ppm dosage level by a statistically significant amount (38%). The mean F_1 litter weight was reduced by 18% at the 4000 ppm treatment level. Pups born in the 8000 ppm treated group of the F_1 generation had a lower survival index at post-partum day 21 (63% versus 91% for males of control group; 79% versus 89% for females of control group). Pup weights in the F_1 generation were decreased for all dosage groups, but there was no dose-related response at lower or mid-dose levels. Neurobehavioural tests in treated F_1 generation animals gave different results in certain instances from control mice, but such an effect was not dose-related. The mean F_2 litter size was significantly decreased at the 4000 and 8000 ppm treatment levels, whilst mean F_2 litter weights were decreased in all treated groups. Pups in the 8000 ppm treated group of the F_2 generation had a lower survival index than controls at day 21 post-partum (59% in males and 79% in females). Pup weights in the F_2 generation were decreased at dosage levels of 2000 ppm and above. At 1000 ppm, pup weights were reduced on days 4 and 7 post-partum, when compared with controls. Again, some neuro-behavioural tests were altered in treated animals, but as a clear dose–response relationship was not evident, such findings do not indicate developmental neurotoxicity. Owing to a recorded effect on pup weights at all dosage levels, a NOAEL could not be set for this study.

A further reproductive toxicity study using PBO in groups of Crj:CD-1 mice was reported by Tanaka (1992). Ten mice of either sex were utilized at each dosage level and such dosage levels were set at 0, 1500, 3000 or 6000 ppm in the diet for a 4-week period prior to mating (F_0), during gestation and up until the F_1 generation was 8 weeks old. In order to measure general activity, the open field test was performed on the F_0 generation, and results demonstrated a dose-dependent decrease in ambulation and rearing in male mice. Because of the excessive dosage levels incorporated in this study, pup body weights were reduced at birth in all treated animals. By day 21 post-partum, the mean pup body weight in the mid-dose group was 7% lower than controls, with a 41% reduction being reported for pups derived from the high-dose group. The survival index for pups at day 21 post-partum was 79.2% (controls), 92.9% (low-dose group), 80.0% (mid-dose group) and 51.7% (high-dose group). There were no significant differences in the behavioural tests during the lactation period, except for a reduction in olfactory orientation in mid- and high-dose group animals. The open field test and multiple water T-maze tests were not significantly altered by treatment, except for a few non-dose-related differences in some parameters.

(b) Rats A two-generation reproduction study was performed in rats with PBO by Robinson *et al.* (1986). Groups of 26 male and 26 female Sprague-Dawley rats were utilized and adults of the F_0 and F_1 generations were treated at dose levels of 0, 300, 1000 or 5000 ppm PBO in the diet. Such animals were treated for 83 to 85 days prior to placement for mating, and treatment continued throughout the mating, pregnancy and lactation periods where appropriate. There were no treatment-related deaths in any of the adult or pup generations studied. The only consistent finding throughout the study period was a lower body weight gain at the highest dosage level. This tendency was partially reversed during the lactation period, when females at this dose level showed higher weight gains when compared with control rats. Parental performance in terms of the mean day of mating, mating and fertility indices, and conception rates was unaffected by treatment. Oestrous cycles, gestation index, length of gestation, duration of parturition, numbers of live and dead pups at birth and sex ratios, all parameters of maternal performance, were not significantly different from control values. In the case of the F_1 and F_2 generation pups, the viability, survival and lactation indices were unaffected by treatment. There were no treatment-related abnormal findings for the pups, and weanlings did not reveal any treatment-related adverse effects. Body weights of pups born to dams treated at the highest dose level were reduced in the early post-partum period. The NOAEL for parental toxicity and pup development was thus set at 1000 ppm PBO in the diet. The NOEL for reproductive toxicity was set at 5000 ppm PBO in the diet in this study.

In a reproductive toxicity study reported by Sarles and Vandegrift (1952), groups of rats were fed diets containing 0, 100, 1000, 10 000 and 25 000 ppm of PBO (technical grade, 80% purity) for three generations. None of the female rats at the highest dose level were fertile and there were marked reductions in the incidence of pregnancies, numbers of litters per dam, general health of the progeny and in the average weanling weights of pups born to dams treated at the 10 000 ppm dosage level. Such findings are clearly explained by the excessive dosage levels utilized in this study. No adverse effect on reproductive performance was observed in three generations of progeny fed diets containing up to 1000 ppm PBO on a daily basis.

(c) Other Studies   In a 2-year chronic oral toxicity study in Sprague-Dawley rats (Graham, 1987), test animals were dosed daily at levels of 0, 30, 100 and 500 mg kg^{-1} day^{-1}. At the end of the 2-year study, detailed post-mortem studies were performed in which increases in ovarian weights and decreases in testicular and seminal vesicle weights were detected in some treated animals. These findings were not consistent for all treated animals, and in both the 30 mg kg^{-1} and 100 mg kg^{-1} treated groups, such macroscopic or organ weight changes were not associated with any histopathological findings that could be related to treatment.

In an 8-week dietary toxicity study in dogs (Goldenthal, 1993a), PBO was fed to beagles at dose rates of 500, 1000, 2000 and 3000 ppm of active ingredient in the diet each day. Again, there was an increase in the absolute and relative weights of the testes and epididymis of male beagles at the two highest dosage levels. However, there were no microscopic abnormalities recorded in the testes in this study. Spermatozoa were being produced in the testes and no evidence of any toxic reaction was noted.

In both the above studies, it is postulated that the changes noted in reproductive organs were similar to those seen in other hormone-dependent tissues in the same studies. As PBO is a known inducer of hepatic enzymes, the changes recorded in organs such as the thyroid gland and testes were considered to reflect changes in circulating hormone levels, with consequent effects on hormone-dependent tissues.

2.4.2. Teratogenicity

(a) **Mice**   Teratogenicity data was reported from a study utilizing Crj:CD-1 mice performed by Tanaka *et al.* (1994). Twenty animals per dosage group were included in this trial and PBO was administered by gavage on day 9 of gestation. The doses utilized were set at 0, 1065, 1385 and 1800 mg kg^{-1} body weight (>95% purity; PBO dissolved in olive oil). No abnormal behaviour or mortality patterns were observed in dams. Three abortions occurred between the mid- and high-dose groups. Four litters were resorbed between the two higher dosage groups, but maternal body weights were comparable between all groups. Total resorption rates were significantly increased in the mid-dose (26%) and the high-dose groups (32%) when compared with control values (6%). The number of viable fetuses per dam was comparable between all dosage groups. There was a significant decrease in body weights for male and female fetuses derived from treated dams, which did appear to be dose-dependent. Certain external malformations such as exencephaly, craniochisis, open eyelids, omphalocele, kinky tail and *talipes varus* were observed in all groups (including controls) and oligodactyly was recorded in the forelimbs of some fetuses derived from treated dams. The incidence of this latter defect was 6% in those fetuses derived from the highest dosage group. The authors concluded that a single oral very high dose of PBO, i.e. 1065 mg kg^{-1} body weight or above, when given to pregnant mice on day 9 of gestation could cause embryo–fetal toxicity with associated oligodactyly of the forelimbs. The significance of this finding is difficult to interpret as the dosage levels investigated were excessively high.

(b) **Rats**   A developmental toxicity study with PBO was performed in Sprague-Dawley rats by Chun and Neeper-Bradley (1991). Timed pregnant rats were administered PBO by gavage on gestation days 6 to 15. The dosage groups incorporated were 0, 200, 500 and 1000 mg kg^{-1}day^{-1}and 25 animals were included in each group. In this study, the pregnancy rate was equivalent among groups and ranged from 88% to 96%. No females aborted, delivered early or were removed from the study. Gestational body weight and body weight gains were reduced in the 500 and 1000 mg kg^{-1} groups, as was food intake for the first 7 days. Other clinical signs of toxicity were evident in dams at the highest dosage level, including urogenital area wetness and discharges, urine staining and perinasal encrustation. Two dams in the mid-dose group, i.e. 500 mg kg^{-1} day^{-1}, also displayed similar signs of toxicity. There were no effects of treatment on gestational parameters including resorption, pre- and post-implantation losses, percentage of live fetuses and sex ratios. Treatment did not affect the fetal body weights. There were no effects of treatment on the incidence of fetal malformations. However, two common skeletal variations (non-ossification of centrum of vertebrae 5 or 6) had a higher incidence in the two highest dosage groups. These findings were not considered treatment related, as adjacent vertebrae did not have delayed ossification. In this study, therefore, the NOEL for maternal toxicity in the rat was set at 200 mg kg^{-1} day^{-1} and the NOEL for developmental toxicity was at least 1000 mg kg^{-1} day^{-1}.

Kennedy *et al.* (1977) performed a teratogenicity study with PBO in pregnant rats. The animals were dosed by gavage, at levels corresponding to 0, 300 or 1000 mg kg^{-1} day^{-1} (purity not specified). Group size consisted of 20 animals at each dosage level and PBO dissolved in corn oil was administered from day 6 to day 15 of gestation. All animals were sacrificed on day 20 of gestation. Besides a decline in body weight gain in both treated groups (especially in the later stages of gestation), there were no other treatment-related signs of toxicity. The reproductive parameters of the dams were not significantly affected by treatment. One female from each treatment group resorbed most or all of their litters. The fetuses derived from both treatment groups exhibited no internal, external or skeletal malformations that could be related to treatment. A NOAEL for maternal toxicity could not be set from this study, owing to the decline in maternal body weight recorded at both dosage levels. As PBO was not found to be teratogenic in this study, the NOAEL for embryo–fetal toxicity for this compound in rats was set at 1000 mg kg^{-1} day^{-1} by the authors.

Pregnant female Wistar rats (17–20 per dosage group) were dosed with PBO at levels of 0, 62.5, 125, 250 or 500 mg kg^{-1} day^{-1} from day 6 to day 15 of gestation in a study performed by Khera *et al.* (1979). The types and incidences of anomalies in fetuses derived from treated dams were comparable with those of the control group and it was concluded that dosage levels of PBO as high as 500 mg kg^{-1} day^{-1} produced no signs of either maternal or embryo–fetal toxicity. Further information on this study was reported by JMPR (1995).

(c) **Rabbits** New Zealand White female rabbits were dosed with PBO at levels of 0, 50, 100 or 200 mg kg^{-1} body weight by gavage between day 7 and day 19 of pregnancy. Sixteen animals were included at each dosage level, and the PBO was administered in corn oil (0.5 mL per kg body weight). Caesarian sections were performed on day 29 of gestation. Maternal toxicity was evident at 100 and 200 mg kg^{-1} day^{-1} and was manifested by decreased defaecation and a dose-dependent weight loss during the treatment period (these weight losses were recovered post-treatment). Common developmental defects were recorded in all dose groups, including an increase in the number of full ribs and the presence of more than 27 presacral vertebrae. The number of litters in the treated groups with these observations was not increased when compared with control values. The NOEL for maternal toxicity was set at 50 mg kg^{-1} day^{-1}, whilst the NOEL for embryo–fetal toxicity was set at 200 mg kg^{-1} day^{-1} from this study. Further information on this and a previous dose range-finding study in the rabbit was reported by JMPR (1995).

2.5. Mutagenicity

A variety of studies have been performed with PBO to assess any mutagenic potential. A summary of the salient features and results of such studies is presented in Table 2.4. More detailed accounts of some of these studies are provided by Butler *et al.* (1996).

From the above results, although positive or equivocal results were recorded in a minority of the test systems, the weight of evidence indicates that PBO does not demonstrate any significant potential for mutagenicity, and should be classified as nongenotoxic.

2.6. Carcinogenicity

2.6.1. Mice

Innes *et al.* (1969) reported on two oncogenicity studies in (C57BL/6 × C3H/Anf) F_1 and (C57BL/6 × AKR) F_1 mice. From 4 weeks of age, the test mice in the first study were fed 300 ppm of PBO in the diet, whilst those in the second study were fed a level of 1112 ppm in their ration. Both studies terminated at 70 weeks of age. There was no statistically significant increase in tumour incidence between treated and control mice in either study.

The US National Cancer Institute (1979) reported on a study involving groups of 50 male and 50 female (B6C3) F_1 mice, 6 weeks old, who were fed PBO (purity value 88.4%) in the diet. A low-dose group received 2500 ppm daily for 30 weeks and thereafter 500 ppm daily for an additional 82 weeks, whilst a high-dose group received 5000 ppm daily for 30 weeks followed by 2000 ppm daily for 82 weeks. A matched control group (20 males and 20 females) received no PBO. The time-weighted average doses were 0 (controls), 1036 ppm (low-dose) and 2804 ppm (high-dose). There was a dose-dependent decrease in mean body weight for treated animals when compared with controls. At necropsy, there was no dose-related increase in the incidence of hepatic tumours, nor in other microscopic findings in this organ other than nodular hyperplasia and focal necrosis. Hepatocarcinomas were reported, especially in male mice, but the incidence was highest in the control group. Adenomas of the lacrimal gland were observed in males at the high-dose level, but such a finding was not statistically significant. Although variations in other tumour incidences were observed, no significant treatment-related differences were found for any tumour type. The authors concluded that PBO was not carcinogenic for this strain of mouse, under the conditions of this study.

A more recent oncogenicity study was performed in CD-1 mice by Hermanski and Wagner (1993). Target doses fed in the diet were 0, 30, 100 and 300 mg kg^{-1} day^{-1} for at least 78 weeks. The PBO utilized was 90.8% pure and 60 male and 60 female mice were incorporated at each dosage level. Two control groups were included. There were no treatment-related clinical signs of toxicity, nor any such related changes in food intake or haematological parameters. There was an increased incidence of masses or nodules recorded in the livers of

Table 2.4. Summary of results of mutagenicity studies with PBO

End point	Test system	Concentrations tested	Result
Reverse mutation	*Salmonella typhimurium* strains TA98, 100, 1535, 1537 and 1538	Six dose levels tested between 100 and 5000 μg per plate, with and without metabolic activation	No significant increase in the number of histidine revertants, either in the presence or absence of S$_9$ [a]
Gene mutation	Mouse lymphoma L5178Y cells	6.3–100 μg mL^{-1}	Positive[b]
HGPRT gene mutation assay	Chinese hamster ovary (CHO)	10–100 μg mL^{-1} in the absence of metabolic activation 25–500 μg mL^{-1} in the presence of metabolic activation	No significant increase in mutation frequency[c]
Chromosomal abberations	Chinese hamster ovary (CHO)	9.99–49.9 μg mL^{-1} (10-h assay) and 49.9–99.9 μg mL^{-1} (20-h assay) in the absence of metabolic activation 25.1–251 μg mL^{-1} in 10- and 20-h assays in the presence of metabolic activation	No significant increase in the number of cells with chromosomal abberations was detected at any concentration tested[d]
Unscheduled DNA synthesis (UDS)	Rat primary hepatocytes	2.5–49.9 μg mL^{-1}	No significant increase in UDS detected at any of the test concentrations[e]
Dominant lethal mutation (*in vivo*)	ICR/Ha Swiss mice	0, 200 or 1000 mg i.p. or 1000 mg per kg bodyweight orally × 5	Equivocal[f]

[a] Lawlor (1991); [b] McGregor *et al.* (1988); [c] Tu *et al.* (1985); [d] Murli (1991); [e] McKeon and Phil (1991); [f] Epstein *et al.* (1972)

some treated animals. There was a corresponding mean absolute and relative increase in liver weights, which was dose-related in the two highest dosage groups. Male mice of the low-dose group also had a slight increase in liver weight. The higher incidence of combined hepatic nodules and masses was recorded in males at 100 mg kg^{-1} day^{-1} and in both sexes at 300 mg kg^{-1} day^{-1}. The author's opinion is that such lesions should be classified as hyperplasia or adenomas of the liver, which in fact are commonly seen in CD-1 mice on such protocols. Spontaneous hepatic adenomas are seen primarily in male CD-1 mice and are relatively uncommon in females. The typical appearance of these spontaneous adenomas includes the presence of small to medium sized cells, which are well differentiated and are predominantly basophilic when stained. The lesions observed in this study did, however, differ from this normal pattern in certain ways. They were, in general, larger than expected and often encompassed entire lobes of the liver. Multiple growths were commonly observed and a few male mice exhibited malignant as well as benign liver tumours. Eosinophilic staining of the cytoplasm has been observed in liver cells following the use of compounds known to be mixed function oxidase (MFO) inducers. This effect has been attributed to an increase in the amount of smooth endoplasmic reticulum present in the cytoplasm of these cells (Burger and Herdson, 1966). Such eosinophilic lesions are not precursor lesions for hepatocarcinomas and do not represent promotion of previously initiated clones. The lesions in PBO-treated mice are very similar to those seen with other compounds which are known to be hepatic enzyme inducers. Hepatocellular hyperplasia (particularly in males) and hypertrophy (males and females) were both recorded at the highest dosage level. The neoplastic findings were most likely secondary to the significant degree of toxic damage (haemorrhage, hyperplasia and hypertrophy) caused by the high-dose PBO treatment. Although hepatocellular necrosis was slightly increased in the high-dose group, most such lesions were graded as minimal or mild. Overall, it was postulated that the observed hyperplastic foci of eosinophilic cell type probably represented hepatocytes exhibiting a treatment-related response to MFO induction. It is possible that such foci would regress when treatment was removed, as has been shown for other compounds such as phenobarbital. They are considered an end-stage lesion and not part of a continuing neoplastic process. Certain characteristics of such nodules, i.e. failure to grow in semi-solid agar, failure to transplant efficiently into nude mice, as well as their lack of activated oncogenes, distinguish them from the spontaneous adenomas or carcinomas otherwise seen in murine livers. The review by Butler (1996) of the hepatic tumours related to MFO induction in the mouse supports these views.

A 1-year hepatic carcinogenicity study was performed in male CD-1 mice by Takahashi *et al.* (1994a). Animals were fed PBO (94.3% purity) in the diet at dosage levels corresponding to 0 (52 mice), 6000 ppm (52 mice) and 12 000 ppm (100 mice) of diet. Mortality was 6%, 2% and 19% for the control, 6000 ppm and 12 000 ppm treatment groups respectively. Terminal body weights were reduced by 17% (6000 ppm group) and 29% (12 000 ppm group) when compared with control values. Treated animals exhibited hepatic adenomas and hepatocarcinomas. Hepatocellular hyperplasia and haemangio-endothelial sarcomas were also observed. The author's opinion is that the results of this study should be interpreted with caution as raw data were unavailable, dosages used were excessively high, and the resultant hepatic and systemic toxicity was severe at both treatment levels.

The results of oncogenicity studies in mice with PBO are summarized in Table 2.5.

2.6.2. Rats

A study was performed in the USA by Cardy *et al.* (1979), in which F344 rats were fed PBO at dosage levels of 0, 5000 or 10 000 ppm over a 2-year period. Fifty male and 50 female Fischer rats were included in each treated group (20 animals of both sexes in the control group). The PBO was technical grade (purity 88.4%). Whilst there was a dose-related increase in the incidence of hepatocytomegaly found on histopathology of the liver, there was no such dose-related increase in the incidence of tumours. There was minimal distortion of lobular architecture in the foci of hepatocytomegaly detected in the liver, and such lesions corresponded morphologically to those described as 'eosinophilic foci' by Squire and Levitt (1975). Female rats exhibited a dose-related increase in the incidence of lymphomas (control groups 1/20, low-dose group 7/50 and high-dose group 15/50). However, there was a background historical control incidence of lymphomas, leukaemias and reticuloses in female Fischer 344 rats at this same laboratory (incidence was 19/191). The authors of this study thereby concluded that PBO was not carcinogenic to F344 rats under the conditions of this bioassay.

Graham (1987) performed an oncogenicity study in Sprague-Dawley rats, in which PBO was fed in the diet at dosage levels of 0, 30, 100 and 500 mg kg^{-1} day^{-1} for 2 years. Following cessation of the study, post-mortem evaluation of the test rats revealed increases in the sizes of the liver, adrenals,

Table 2.5 Results of oncogenicity studies in mice with PBO

Dose	Duration	Findings	Reference
300 ppm	69 weeks	No significant increase in tumour incidence	Innes *et al.* (1969)
1112 ppm	69 weeks	No significant increase in tumour incidence	Innes *et al.* (1969)
500–2000 ppm	112 weeks	Dose-related increase in nodular hyperplasia and focal necrosis of the liver	US National Cancer Institute (1979)
0–300 mg kg^{-1} day^{-1}	18 months	Significant increase in hepatic eosinophilic foci or adenomas in males at 100 and 300 mg kg^{-1} day^{-1} Females had similar findings at 300 mg kg^{-1} day^{-1}	Hermanski and Wagner (1993)
0–12 000 ppm	12 months	Hepatic adenomas and hepatocarcinomas	Takahashi *et al.* (1994a)

thyroid and ovaries of some treated animals. The testes and epididymis of some treated rats were reduced in size. Increases in liver and kidney weights were also observed in some test animals at the end of the treatment period. In the $30 \text{ mg kg}^{-1} \text{day}^{-1}$ treatment group, none of the above findings were associated with any histopathological changes that could be treatment related. In the higher dosage groups, histopathology of the liver revealed hypertrophy of hepatocytes, as well as focal cell infiltrates and eosinophilic infiltrates in some treated rats. Eosinophilic cell infiltrates tended to be found in the male rats only. Hepatocyte hyperplasia was also evident. These liver changes were considered to indicate a metabolic response to treatment, compatible with hepatic enzyme induction. Thyroid gland changes included increased levels of pigment in colloid material and follicular hyperplasia (especially at $500 \text{ mg kg}^{-1} \text{day}^{-1}$). Thyroid gland enlargement was seen in both males and females at the highest dose level, with the follicular hyperplasia being both focal and generalized in nature. There was a generalized increase in the numbers of follicles present and a large variation in follicle size. This focal response of the thyroid is considered to reflect the known physiological heterogeneity of follicles and such hyperplastic changes have been shown to be reversible. Although a slightly higher incidence of ovarian masses was observed grossly in females at the highest dosage level, there was no evidence of any dose-related microscopic abnormal findings. Histopathological analysis of other tissues did reveal secondary and tertiary changes in endocrine and endocrine-dependent tissues. The changes in the incidence and severity of background lesions were considered to reflect changes in circulating hormone levels. Overall, there was no increase in tumour incidence or tumour type beyond the normal historical findings for this species and age range. Based on this study, there was no evidence of carcinogenicity for the PBO treatment.

Maekawa *et al.* (1985) performed a 2-year oncogenicity study in F344/Du Crj rats, feeding diets containing 0, 5000 or 10 000 ppm of technical grade PBO (89% purity) on a daily basis. Following treatment for 104 weeks, animals were fed a basal diet for a further 6 weeks, prior to being sacrificed. There was a dose-dependent increase in mortality (both males and females) for the different dosage groups. However, PBO was not found to be carcinogenic in this study.

A chronic toxicity/oncogenicity study was performed by Takahashi *et al.* (1994b) with F344/Du Crj rats. Diets containing levels of 0, 6000, 12 000 or 24 000 ppm of PBO were fed to 30–33 animals of each sex per treatment group. Male rats were fed for 95 weeks and female rats for 96 weeks. Absolute liver weights were increased in the 12 000 ppm group (males +22%; females +51%) and were also increased in females at the 6000 ppm dosage level. Mortality figures were significantly elevated in males of the 12 000 ppm group. Hepatic tumours were seen in rats that died from week 74 onwards. Nodular lesions of the liver were only seen in treated animals. Their incidence, number per rat and size were found to be dose-related. Hepatic adenomas and hepatocarcinomas were found in rats at mid- and high-dosage levels only. Focal hepatocellular hyperplasia was recorded in the low-dose group. There was no difference in the incidence of other tumour types between the various groups, with the exception of a lower incidence of interstitial cell tumours of the testes in treated animals as compared with controls. The authors concluded that PBO did cause hepatic tumours. Again, the scientific interpretation of this study is open to doubt in the author's opinion as the dosage levels employed were clearly excessive. The relevance of the liver findings are undermined by the severe systemic and hepatic toxicity recorded in this study at such high dosage levels.

The results of oncogenicity studies in rats with PBO are summarized in Table 2.6.

2.6.3. Dogs

A 1-year chronic dietary toxicity study performed in beagle dogs at dosages ranging from 100 to 2000 ppm PBO day^{-1} failed to show any increase in tumour incidence (Goldenthal, 1993b). Post-mortem examinations of the 24 treated beagles did reveal increased liver and gall bladder weights at the highest dosage level, as well as a small increase in thyroid and parathyroid gland weights in females at this same dosage level, i.e. 2000 ppm day^{-1}. There were no abnormalities detected in the thyroid gland on microscopic evaluation. Histopathological analysis of the liver revealed mild hypertrophy of hepatocytes, but no evidence of neoplasia.

Table 2.6 Results of oncogenicity studies in rats with PBO

Dose	Duration	Findings	Reference
0–25 000 ppm	2 years	No significant increase in tumour incidence	Sarles and Vandegrift (1952)
0–10 000 ppm	2 years	Dose-related increase in hepatocytomegaly or eosinophilic foci Dose-related increase in lymphomas in female rats, but incidence in controls was also high	Cardy et al. (1979)
0–10 000 ppm	2 years	No evidence of carcinogenicity Decreased incidence of thyroid C-cell adenomas in high-dose males	Maekawa et al. (1985)
0–500 mg kg^{-1} day^{-1}	2 years	Hypertrophy of hepatocytes and follicular hyperplasia in thyroid gland Overall, no increased tumour incidence	Graham (1987)
0–24 000 ppm	95–96 weeks	Hepatic adenomas and hepatocarcinomas at mid and high dosage levels Severe general and hepatic toxicity	Takahashi et al. (1994a)

2.7. Pharmacodynamics

The primary function of PBO in commercial preparations, e.g. pyrethrin/PBO combinations, is to act as a synergist and thus enhance the potency of the insecticide.

Yamamoto (1973) suggests that the primary function of synergists for pyrethrins is to provide an alternative substrate for the MFO enzyme system, which would normally metabolize such pyrethrins. This author states that inhibition of oxidation of the transmethyl groups and the alcohol moiety on the pyrethrin molecule is the most important function of PBO. Inhibition of ester hydrolysis is also an important step.

PBO is known to be an alternative substrate for the liver microsomal NADPH system. Thus, this synergist can be used to inhibit the metabolism of several drugs and pesticides. Brown (1970) reported that the detoxification of certain drugs such as pentobarbital, zoxazolamine, antipyrine and benzopyrene were inhibited by PBO, presumably due to the inhibition of oxidative processes which metabolize such drugs. Conney *et al.* (1972) investigated the inhibition of antipyrine metabolism in rats and mice. Both species were treated intraperitoneally (i.p) with a single dose of PBO, followed by a further i.p injection of antipyrine (200 mg kg^{-1}) 1 hour later. The resultant NOEL for antipyrine metabolism in the mouse was 0.5–1.0 mg PBO kg^{-1}, whilst the NOEL for the rat was 100 mg PBO kg^{-1}. This huge difference in sensitivities between rats and mice may be accounted for by differences between liver microsomes for both species.

The effects of PBO on the metabolism of benzopyrene in Sprague-Dawley rats were studied by Falk *et al.* (1965). PBO was administered by the oral, i.p or intravenous (i.v) routes at various times before the i.v injection of labelled benzopyrene. The level of radioactivity was then measured in bile at frequent intervals up to 4 hours. This author demonstrated marked inhibition of benzopyrene metabolism when PBO is administered i.v at 262 mg kg^{-1}, some 5 minutes to 16 hours before the administration of benzopyrene. However, this effect is much reduced at 121 mg kg^{-1} body weight and virtually no effect is seen at 25 hours post-dosing. This implies that single large doses of PBO are quickly metabolized by rats. Administration of PBO by the oral and i.p routes resulted in a greatly reduced effect when compared with the i.v route. A second similar study performed by Conney *et al.* (1972), where the effects of i.p administration of PBO on the metabolism of benzopyrene were investigated, showed less sensitive results when rats of lighter weight were used (approximately 180 g versus 400 g in Falk's study). It was postulated by Brown (1970) that the extra fat in the animals in Falk's study could possibly act as a reservoir of PBO and lead to a longer duration of action. It would appear from these studies in rats that 250 mg of PBO per kg body weight is the minimum oral dose required to give any significant effect on benzopyrene metabolism.

Brown (1970) reports on an experiment in rats where a single dose of PBO (333–1000 mg kg^{-1} body weight) increased the sleeping time of the animals following administration of pentobarbital. However, the i.p administration of eight injections of 50 mg kg^{-1}, each at 12-hour intervals, followed by the injection of pentobarbital some 18 hours later, causes a reduction in sleeping time in rats. This study served to underline that whilst a single dose of PBO could inhibit the metabolism of pentobarbital, repeated PBO doses could stimulate rats' livers to metabolize this drug. The administration of 50 mg PBO per kg body weight i.p to rats (Anders, 1968) and mice (Graham *et al.*, 1970) prior to treatment with hexobarbital, approximately doubled the sleeping time of both species.

CD-1 mice given a single i.p dose of 600 mg PBO per kg body weight were found to have suffered less hepatotoxicity when treated with acetaminophen (600 mg kg^{-1} body weight p.o.) at either 2 hours prior to or 1 hour following PBO administration. This reduced hepatotoxicity was measured via GSH and sorbitol dehydrogenase levels, as well as subsequent histopathology of the liver. Since the hepatic MFO system metabolizes acetaminophen to a toxic metabolite, the decreased toxicity seen in this experiment is likely due to inhibition of such oxidase enzymes by PBO (Brady *et al.*, 1988).

A study was performed by Phillips *et al.* (1997) into the hepatic effects of PBO in the rat. Four groups of 16 male F344 strain rats were fed PBO in the diet at nominal dosage levels of 100, 550, 1050 and 1850 mg kg^{-1} day^{-1}. Animals were treated for either 7 or 42 days, after which time point they were sacrificed and detailed studies of the livers performed. In summary, treatment of rats with 100 mg kg^{-1} day^{-1} of PBO led to increased relative liver weights and microsomal protein content (42 days treatment), increased cytochrome P-450 levels (7 days treatment), increased GGT levels (42 days) and increases in certain MFO enzyme activities. Treatment at higher dosage levels also results in similar findings. PBO caused a minor, but statistically significant, decrease in alkaline phosphatase, alanine transaminase and aspartate transaminase levels at both 7 and 42 days, particularly at the two intermediate dosage levels. PBO is a clear inducer of hepatic xenobiotic metabolism in rats.

Many other studies have been undertaken relevant to the pharmacodynamics of PBO and are reported elsewhere (Skrinijaric-Spoljar *et al.*, 1971; Conney *et al.*, 1972; Goldstein *et al.*, 1973).

2.7.1. Observations in Humans

PBO has not been used as an internal medicine in humans. This compound has, however, been extensively used both as a synergist with pyrethrins or pyrethroids for the control of household pests and as an agricultural insecticide. PBO has been incorporated in millions of aerosols utilized in homes, offices and factories. Furthermore, PBO has had very extensive dermal application to humans in combination with both pyrethrum and pyrethroids (e.g. permethrin) for the control of head and pubic lice, as well as scabies infestations. Such exposure has been on-going for over 40 years, with most applications being to children. The absence of untoward effects with the above uses serves to underline the wide safety margin of this compound.

A study performed in Austria by Wintersteiger and Juan (1991), investigated the absorption of pyrethrin and PBO combination sprays across the skin of six healthy subjects. The spray was applied over a wide area of the back with a total dose of approximately 3.3 mg pyrethrum extract and 13.2 mg PBO being applied. No untoward clinical signs occurred. Cutaneous absorption of PBO was shown to be extremely low, with plasma samples containing no more than 10 ng PBO per 60 μL. A second study was performed in the US with pyrethrins and PBO applied to the skin of six human volunteers. This trial, performed by Wester *et al.* (1994), investigated the percutaneous absorption of both compounds across the skin of the ventral forearm. In this report, based on the recovery of radioactivity in the urine, it was calculated that 2.1% ± 0.6% of the dose of PBO was absorbed through the skin. However, higher levels of absorption of PBO were achieved when the compound was applied to the skin of the scalp (8.3%). There was no evidence of any local or systemic toxicity to the use of PBO as a topical agent in humans.

Selim (1995) investigated the absorption, excretion and mass balance of ^{14}C PBO from two different formulations following dermal application to healthy volunteers. The first preparation applied was a 4% (w/w) solution of PBO in an aqueous formulation. This product was applied to four healthy human volunteers. The second preparation tested was a 3% (w/w) solution of PBO in isopropyl alcohol. In the former case, the mean amount of PBO applied was 3.8 mg per volunteer (approximately 39.9 μCi of radioactivity per volunteer), whilst the average exposure was 3.0 mg PBO per volunteer (approximately 40 μCi of radioactivity per volunteer) in the case of the isopropyl alcohol solution. Results from this study show that there was a similar dermal absorption pattern for both formulations. The principal route of excretion of absorbed radioactivity was via urine. Faecal samples from volunteers contained negligible levels of radioactivity. Dermally applied PBO was rapidly excreted from the volunteers. The majority of the applied radioactivity remained at the application site, with less than 3% of the applied dose being absorbed during the 8-hour test period. Radioactivity did not accumulate in the skin.

Three double-blind experiments were performed on human subjects, where eight men were either given a placebo or 50 mg of PBO orally, corresponding to an average dose of 0.71 mg kg^{-1} body weight. Two hours later, all subjects received a dose of 250 mg antipyrine. Samples of blood were taken at regular intervals up to 31 hours post-dosing, and were later analysed. The mean half-life of antipyrine in those volunteers receiving the placebo treatment was 14.57 hours, whilst the corresponding figure for the PBO-treated subjects was 14.16 hours. There was no evidence that an oral dose of 50 mg PBO in humans (or 0.71 mg kg^{-1} body weight) has any effect on the metabolism of antipyrine (Conney *et al.*, 1972).

Although PBO had proved negative in the majority of short-term mutagenicity tests, the finding of a recent mouse oncogenicity study warranted investigation into the ability of PBO to produce DNA repair in cultured human liver slices, using the unscheduled DNA synthesis (UDS) assay. Previous studies had shown that human liver slices were suitable for use in such studies of UDS (Beamand *et al.*, 1994, 1995). The effects of PBO on UDS in cultured human liver slices was thus investigated in a study performed by BIBRA International (Beamand *et al.*, 1996). Precision-cut liver slices were prepared from liver samples derived from four male human donors (ranging in age between 16 and 51 years). Such liver slices were cultured for 24 hours with 0.05–2.5 mmol PBO. Liver slices were also cultured with three known genotoxins, namely:

- 0.02 and 0.05 mmol L^{-1} 2-acetylaminoflourine (2-AAF);
- 0.002 and 0.02 mmol L^{-1} aflatoxin B_1 (AFB_1);
- 0.005 and 0.05 mmol L^{-1} 2-amino-1-methyl-6-phenyl-imidazo (4,5-b) pyridine (PhIP).

UDS in cultured human liver slices was quantified as the net grain count in centrilobular hepatocytes and as the percentage of centrilobular hepatocyte nuclei with >5 and >10 net grains. Control human liver slices were also cultured with dimethyl sulfoxide (DMSO) alone, in order to act as a negative control.

Treatment with 0.05–2.5 mmol L^{-1} PBO had no significant effects on the net grain count of centrilobular hepatocytes. Furthermore, PBO had no significant effect on the percentage of centrilobular hepatocyte nuclei with > 5 net grains. No specimen from any of the four donors had centrilobular hepatocyte nuclei with >10 net grains, irrespective of the concentration of PBO assayed. At the concentrations tested, neither PBO, 2-AAF nor PhIP had any significant effect on replicative DNA synthesis in 24-hour cultured human liver slices. In cultured liver slices treated with 0.02 mmol L^{-1}, but not 0.002 mmol L^{-1} AFB_1, a significant decrease in the rate of replicative DNA synthesis was observed. This reduction (down to 6.7% of control values) was postulated by the test authors to be possibly due to a cytotoxic effect.

These results demonstrate that, at the concentrations used, PBO did not induce UDS in cultured human liver slices. The fact that the three known genotoxins did produce a significant increase in UDS is testament to the functional viability of the human tissue used in this study, as all three compounds have been shown to undergo metabolic activation by human liver cytochrome P-450 isoenzymes (Gonzalez *et al.*, 1991).

A meeting of a joint FAO/WHO Codex Alimentarius in 1987 discussed the exposure to humans via dietary input. They calculated that the average total diet (1.408 kg) might contain 1 ppm of PBO, i.e. 1.4 mg. This calculation is an over-estimation as cooking destroys up to 90% of PBO present. In the case of aerosols, assuming a very large exposure time to PBO, the average amount ingested or inhaled would be approximately 0.63 mg day^{-1}. The total intake from food and aerosols, therefore, would be approximately 2.03 mg day^{-1}. As 50 mg PBO had no effect on antipyrine metabolism in humans, a wide safety margin was evident even in worst-case scenarios. JMPR (1995) set an ADI for humans of 0–0.2 mg per kg body weight for PBO. Crampton (1994) calculated that the daily exposure of individuals to PBO residues in food was 0.0037 mg kg^{-1} day^{-1}. As this figure is over 50 times lower than the new ADI proposed by JMPR (1995), it would seem that such low levels in the diet pose little risk to MFO enzyme activity and microsomal function in man.

3. CONCLUSIONS

PBO is an extremely widely used compound on humans, animals and on certain foodstuffs. PBO has been utilized for many decades in humans and domestic animals, with an extremely low order of toxicity. Many reports detail its efficacy as a synergist when combined with pyrethrins (Yamamoto, 1973).

A large battery of toxicity data is available both from the literature and from specific testing performed by sponsoring companies. Acute toxicity tests show that the LD_{50} values are many times greater than field exposure rates, by a variety of routes. Such wide safety margins in both laboratory and non-laboratory species allow for a high degree of confidence in its safety. Repeated dose toxicity studies also show very large safety factors, even though this compound would normally only be used on an intermittent basis. The dermal LD_{50} values are of particular interest, as this is the intended clinical route of exposure for ectoparasite control in animals. From the data it can be seen that the average exposure faced by a human or animal on an intermittent basis is a fraction of the acute dermal LD_{50} value in the rabbit.

PBO was tested in both primary eye irritation and skin irritation studies in rabbits, dogs and cats. In all such studies, the compound was found to have minimal or no irritating effects. Sensitization studies performed on guinea pigs and rabbits (dermal exposure) showed that PBO was not a sensitizing agent.

The target tissue in most chronic toxicity experiments is the liver. Whilst low dosage levels are associated with minimal changes in this organ, higher dosage levels for any prolonged period are associated with more severe changes. As well as alterations in enzyme and total protein levels, hepatocyte changes in the form of hypertrophy and hyperplasia were also evident. In certain experiments where excessive dosages were utilized, mortalities resulted from severe hepatic and systemic toxicity.

Reproductive toxicity and teratogenicity studies have been performed in rats, mice and rabbits. Despite the presence of some common developmental defects in a small proportion of newborns, e.g. skeletal defects and non-ossification of cervical vertebrae, no treatment-related teratogenic effects were attributed to this chemical by the test authors. In one study performed by Tanaka *et al.* (1994), a single very large dose of PBO (1065 mg per kg body weight) administered by gavage on day 9 of pregnancy in Crj:CD-1 mice caused embryo–fetal toxicity with associated oligodactyly of the forelimbs. In general, PBO had no effect on the fertility of rabbits and rats in double generation studies. Although ovarian and testicular weights were affected in chronic toxicity tests, no abnormalities were detected in such organs on histopathological examination, and spermatogenesis was unaffected. PBO administration had no effect on the percentage of animals becoming pregnant, gestation length, or litter sizes. Furthermore, the percentage of test animals carrying their young to full term and the percentage of live young was unaffected by the test compound. The study performed in rats by Sarles and Vandegrift (1952), where excessive dosage levels were utilized (up to 25 000 ppm), did show seriously affected reproductive performance and reduced pup weights at weaning. However, fertility and teratogenicity studies performed at lower dosage levels and in keeping with currently acceptable scientific and regulatory protocols indicate that this compound has no adverse effects on reproductive function nor on fetal development. It is considered safe during pregnancy and lactation.

Several different tests for mutagenicity were performed with PBO. These combined both bacterial and mammalian cell culture assays, as well as a single *in vivo* study. In three tests (*Salmonella* Ames test, chromosomal aberrations in CHO cells and UDS in rat primary hepatocytes), there was no evidence of any mutagenic/genotoxic effects of PBO. In an assay for gene mutation in mammalian cells (CHO/HGPRT), there was an equivocal response. A more recent development in the area of genotoxic studies has been the investigation of the effect of PBO on human tissues. Recent studies performed on cultured human liver slices derived from male donors showed that PBO did not induce UDS in this tissue. The direct use of human tissue in these studies allows extra confidence in the extrapolation of data derived from such sources in assessing any significant risks following human exposure. Overall, it is strongly felt that the weight of the evidence indicates that PBO is not mutagenic.

Much of the recent work on PBO has focused on the area of carcinogenicity. Five separate studies were performed in rats over periods of 2 years, each at varying dosage levels. An American study performed by Cardy *et al.* (1979), using dosages as high as 10 000 ppm in Fischer 344 rats, showed no evidence of an increased incidence in liver tumours. A Japanese study in F344 rats (Takahashi *et al.,* 1994b) utilized very high dosage levels of PBO in its 2-year oncogenicity study (up to 24 000 ppm). Despite all liver slides not being available for later examination, there did appear to be an increased incidence of both hepatic adenomas and hepatocarcinomas in both sexes, particularly at the high dose level. However, histopathological evaluation of the available liver sections did also reveal serious background pathology indicative of pronounced hepatic toxicity, e.g. necrosis, hyperplasia and oval cell proliferation. Such hepatic injury is likely to be linked to the long-term usage of excessively high doses of PBO in rats, and would seem to have served as an initial platform for future tumour formation. The latest rat study performed by Graham (1987) (dosages of $30–500$ mg kg^{-1}day^{-1} over 2 years) showed no evidence of any increase in tumour incidence or type beyond the historical incidence for this species and age group in the performing laboratory. A consistent finding in this study, however, was hypertrophy of hepatocytes, with occasional hyperplasia also being recorded. This effect was considered to be due to the enzyme induction properties of PBO and has been commonly seen with many other compounds with the same propensity for enzyme induction, e.g. phenobarbital. Secondary and tertiary effects on endocrine and endocrine-dependent tissues were also evident and again would appear to be due to alterations in circulating hormone levels.

In mice, three initial oncogenicity studies performed by Innes *et al.* (1969) and the US National Cancer Institute (1979), where diets containing up to 300, 1112 and 2804 ppm, respectively, of PBO were fed for a minimum of 69 weeks, revealed no statistically significant difference in tumour incidence between treated and control animals. However, a fourth, more recent, study performed by

Hermanski and Wagner (1993), found a higher than expected incidence of hepatic adenomas. The nature of the pathology of these hyperplastic foci or adenomas was such that they were easily differentiated from the spontaneous hepatic adenomas normally seen in CD-1 mice. These lesions were not precursors for carcinomatous changes and appear to be due to the enzyme induction properties of PBO. There was no statistically significant increase in the incidence of hepatocarcinomas in treated mice in this study. At sufficiently high dosage levels, PBO is known to be hepatotoxic. Present data indicate that the production of liver tumours in rats and mice at dosage levels greater than the maximum tolerated dose (i.e. the dose that produces $\geq 10\%$ loss in body weight) is most likely to be associated with regenerative hyperplastic changes in this organ resulting from PBO-induced toxicity.

In the case of the earlier mouse studies, the degree of liver damage reported was much lower than in Hermanski and Wagner (1993), and no significant increase in tumour incidence was recorded for the liver. The tumours observed in mice were seen at high dose levels only (100 and 300 mg kg^{-1} day^{-1}) and were associated with reduced weight gain and a dramatic increase in liver weights. The eosinophilic nature of these hyperplastic foci or adenomas is commonly seen with compounds known to be MFO inducers and often regresses when such treatment is removed. For example, hepatic adenomas seen with phenobarbital treatment did not progress to carcinomas upon cessation of treatment, whilst the number of eosinophilic nodules decreased in animals returned to a control diet for 20 weeks after 60 weeks of treatment (Evans *et al.*, 1992). The activation of proto-oncogenes is often seen in spontaneous and genotoxic carcinogen-induced mouse liver tumours. However, a decreased incidence of H-*ras* oncogene activation has been observed in mouse liver tumours induced by phenobarbital (Pedrick *et al.*, 1994) and oxazapam (Cunningham *et al.*, 1994). Furthermore, Pedrick *et al.* (1994) showed that phenobarbital-induced eosinophilic mouse liver tumours were less transplantable into nude mice than those induced by genotoxic mouse liver carcinogens. These latter type studies indicate that the liver tumours seen with phenobarbital treatment (and most likely other MFO inducers such as PBO) are less aggressive in nature than those induced by genotoxic agents. It must be emphasized again that there was no statistically significant increase in malignant hepatic tumours recorded in Hermanski and Wagner's (1993) mouse study. Furthermore, the available evidence indicates that PBO is not a genotoxic compound.

A recent mouse oncogenicity study performed by Takahashi *et al.* (1994a) resulted in increased incidences of hepatic adenomas and hepatocarcinomas when high dosage levels of 6000 and 12 000 ppm PBO in the diet were fed to CD-1 mice. Hepatocyte hyperplasia and haemangio-endothelial sarcomas were also observed in this study. The severe systemic and hepatic toxicity evident at the high dosage levels incorporated in this study were again considered to be the initial causal factors of such findings.

PBO does not constitute a significant hazard or risk to humans. Besides avoiding contact with the eyes and mucous membranes, this compound is otherwise very safe. The acute oral LD_{50} values for PBO in a number of species are extremely high, so even ingestion of a large volume of this compound is most unlikely to pose a serious threat to humans. PBO is present in many commercial formulations for the treatment of lice infestations on humans.

In conclusion, the safety of PBO appears to be extremely high. The extensive toxicity data generated and its widespread use and long-term good standing lend considerable weight to this argument. The huge safety margins in a battery of toxicity tests, by a variety of routes, allow for considerable confidence in its use. The lack of toxicity of this compound in humans suggests that it poses little or no risk.

REFERENCES

Anders, M.W. (1968). Inhibition of microsomal drug metabolism by methylenedioxy-benzenes. *Biochem. Pharmocol.* **17**, 2367–2370.

Beamand, J.A., Price, R.J., Blowers, S.D., Wield, P.T., Cunningham, M.E. and Lake, B.G. (1994). Use of precision-cut liver slices for studies of unscheduled DNA synthesis. *Food Chem. Toxicol.* **32**, 819–829.

Beamand, J.A., Wield, P.T., Price, R.J. and Lake, B.G. (1995). Effect of cooked food mutagens on unscheduled DNA synthesis (UDS) in precision-cut rat, Cynomolgus monkey and human liver slices. *Toxicologist* **15**, 285.

Beamand, J.A., Price, R.J., Phillips, J.C., Butler, W.H., Glynne Jones, G.D., Osimitz, T.G., Gabriel, K.L., Preiss, F.J. and Lake, B.G. (1996). Lack of effect of piperonyl butoxide on unscheduled DNA synthesis in precision-cut human liver slices. *Mutation Res.* **371**, 273–282.

Bond, H., Mauger, K. and DeFeo, J.J. (1973). The oral toxicity of pyrethrum, alone and combined with synergists. *Pyreth. Post* **12**, 59.

Brady, J.T., Montelius, D.A., Beierschmitt, W.P., Wyand, D.S., Khairalla, E.A. and Cohen, S.D. (1988). Effect of piperonyl butoxide post-treatment on acetaminophen hepatoxicity. *Biochem. Pharmacol.* **37**, 2097–2099.

Brown, N.C. (1970). Report A28/52. Research and Development, The Wellcome Foundation, UK.

Burger, P.C. and Herdson, P.B. (1966). Phenobarbitol-induced fine structural changes in rat liver. *Am. J. Pathol.* **48**, 793–809.

Butler, W.H. (1996). A review of the hepatic tumours related to mixed-function oxidase induction in the mouse. *Toxicol. Pathol.* **24**, 484–492

Butler W.H., Gabriel, K.L., Preiss, F.J. and Osimitz, T.G. (1996). Lack of genotoxicity of PBO. *Mutation Res.* **371**, 249–258.

Cardy, R.H., Renne, R.A, Warner, J.W. and Cypher, R.L. (1979). Carcinogenesis bioassay of technical-grade piperonyl butoxide in F344 rats. *J. Nat. Cancer Inst.* **62**, 569–578.

Chun, J.S. and Neeper-Bradley, T.L. (1991). Developmental Toxicity Evaluation of piperonyl butoxide administered by Gavage to CD® (Sprague-Dawley) Rats. Unpublished report no. 54-586 from Bushy Run Research Center. Undertaken for the PBO Task Force, Washington DC, USA.

Chun, J.S. and Neeper-Bradley, T.L. (1992). Developmental Toxicity Dose Range-Finding Study of piperonyl butoxide Administered by Gavage to CD® (Sprague-Dawley) Rats. Unpublished report no. 54-578 from Bushy Run Research Center. Undertaken for the PBO Task Force, Washington DC, USA.

Chun, J.S. and Wagner, C.L. (1993). 90 Day Dose Range-finding Study with PBO in Mice. Study no. 91 N0052, Bushy Run Research Centre, Union Carbide Chemicals and Plastics Company Inc., Export, FA. Undertaken for the PBO Task Force, Washington DC, USA.

Conney, A.H., Chang, R., Levin, W.M., Garbut, A., Munro-Faure, A.D., Peck, A.W. and Bye, A. (1972). Effects of piperonyl butoxide on drug metabolism in rodents and man. *Arch. Environ. Health* **24**, 97–106.

Crampton, P.L. (1994). Piperonyl Butoxide Human Intake Estimates. (Europe) G.R. 94–0009.

Cunningham, M.L., Maronpot, R.R., Thompson, M. and Bucher, J.R. (1994). Early responses of the liver of B6C3 F1 mice to the hepatocarcinogen oxazapam. *Toxicol. Appl. Pharmacol* **124**, 31–38.

Draize, J.H., Woodard, G. and Calvery, H.O. (1944). *J. Pharmacol Exp. Ther.* **82**, 377.

Epstein, S.S., Arnold, E., Andrea, J., Bass, W. and Bishop, Y. (1972). Detection of chemical mutagens by the dominant lethal assay in the mouse. *Toxicol. Appl. Pharmacol.* **233**, 288–325.

Evans, J.G., Collins, M.A., Lake, B.G. and Butler, W.H. (1992). The histology and development of hepatic nodules and carcinomas in C3H/He and C57BL/6 mice following chronic phenobarbitone administration. *Toxicol. Pathol.* **20**, 585–594.

Falk, H.L., Thompson, S.J. and Kotin, P. (1965). Carcinogenic potential of pesticides. *Arch. Environ. Health* **10**, 847–858.

Fujitani, T., Ando, H., Fujitani, K., Ikeda, T., Kojima, A., Kubo, Y., Ogato, A., Oishi, S., Takahashi, O. and Yoneyama, M. (1992). Sub-acute toxicity of piperonyl butoxide in F344 rats. *Toxicology* **72**, 291–298.

Fujitani, T., Tanaka, T., Hashimoto, Y. and Yoneyama, M. (1993). Sub-acute toxicity of PBO in ICR mice. *Toxicology* **83**, 93–100.

Gabriel, D. (1991a). Acute Oral Toxicity, LD50 – Rats. Unpublished report no. 91-7317A from Biosearch Inc., Philadelphia, Pennsylvania, USA. Undertaken for the PBO Task Force, Washington DC, USA.

Gabriel, D. (1991b). Acute Dermal Toxicity, Single Level – Rabbits. Unpublished report no. 91-7317A from Biosearch Inc., Philadelphia, Pennsylvania, USA. Undertaken for the PBO Task Force, Washington DC, USA.

Goldenthal, E.I. (1992). 21-Day Repeated Dose Dermal Toxicity Study with Piperonyl Butoxide in Rabbits. Unpublished report no. 542-007 from International Research and Development Corp., Mattawan, Michigan, USA. Undertaken for the PBO Task Force, Washington DC, USA.

Goldenthal, E.I. (1993a). Evaluation of Piperonyl Butoxide in an Eight Week Toxicity Study in Dogs. Unpublished report no. 542-004 from International Research and Development Corp., Mattawan, Michigan, USA. Undertaken for the PBO Task Force, Washington DC, USA.

Goldenthal, E.I. (1993b). Evaluation of Piperonyl Butoxide in a One Year Chronic Dietary Toxicity Study in Dogs. Unpublished report no. 542-005 from Internal Research and Development Corp., Mattawan, Michigan, USA. Undertaken for the PBO Task Force, Washington DC, USA.

Goldstein, J.A., Hickman, P. and Kimbrough, R.D. (1973). Effects of purified and technical piperonyl butoxide on drug metabolizing enzymes and ultrastructure of rat liver. *Toxicol. Appl. Pharmacol.* **26**, 444–458.

Graham, P.S., Hellyer, R.O. and Ryan, A.J. (1970). The kinetics of inhibition of drug metabolism *in vitro* by some naturally occurring compounds. *Biochem. Pharmacol.* **19**, 769–775.

Graham, C. (1987). 24-month Dietary Toxicity and Carcinogenicity Study of piperonyl butoxide in the Albino Rat. Unpublished report no. 81690 from Bio-Research Ltd. Laboratory, Seneville, Quebec, Canada. Undertaken for the PBO Task Force, Washington DC, USA.

Gonzalez, F.J., Crespi, C.I. and Gelboin, H.V. (1991). cDNA-expressed human cytochrome P450s: a new age of molecular toxicology and risk assessment. *Mutation Res.* **247**, 113–127.

Hermanski, S.J. and Wagner, C.L. (1993). Chronic Dietary Oncogenicity Study with Piperonyl Butoxide in CD-1 Mice. Unpublished report no. 91N0134 from Bushy Run Research Center, Pennsylvania, USA. Undertaken for the PBO Task Force, Washington DC, USA.

Hoffman, G.M. (1991). An Acute Inhalation Toxicity Study of Piperonyl Butoxide in the Rat. Unpublished report no. 91-8330 from Bio/dynamics, East Millstone, New Jersey, USA. Undertaken for the PBO Task Force, Washington DC, USA.

Hunter, B., Bridges, J.L. and Prentice, D.E. (1977). Long Term Feeding of Pyrethrins and Piperonyl Butoxide to Rats. Unpublished report from Huntington Research Centre, Huntington, Cambs, UK. Undertaken for the PBO Task Force, Washington DC, USA.

Innes, J.R.M., Ulland, B.M., Valerio, M.G., Petrucelli, L., Fishbein, L., Hart, E.R., Bates, R.R., Falk, H.L., Gart, J.J., Klein, M., Mitchell, I. and Peters, J. (1969). Bioassay of pesticides and industrial chemicals for tumorigenicity in mice: a preliminary note. *J. Nat. Cancer Inst.* **42**, 1101–1114.

JMPR (1995). *Piperonyl Butoxide – A Monograph Prepared by the Joint FAO/WHO Meeting on Pesticide Residues*, Geneva.

Kennedy, G.L., Smith, S.H., Kinoshita, F.K., Keplinger, M.L. and Calandra, J.C. (1977). Teratogenic evaluation of piperonyl butoxide in the rat. *Food Cosmet. Toxicol.* **15**, 337–339.

Khera, K.S., Whalen, C., Angers, G. and Trivett, G. (1979). Assessment of the teratogenic potential of piperonyl butoxide, biphenyl and phosalone in the rat. *Toxicol. Appl. Pharmacol.* **47**, 353–358.

Lawlor, T.E. (1991). Piperonyl Butoxide in the Salmonella/Mammalian-Microsome Reverse Mutation Assay (Ames test) with a Confirmatory Assay. Unpublished report no. 14413-0-401R from Hazleton, Washington, Kensington, Maryland, USA. Undertaken for the PBO Task Force, Washington DC, USA.

Lehman, A.J. (1948). The toxicity of the newer agricultural chemicals. *Q. Bull Assoc. Food Drug Off.* **12**, 82–89.

Lorber, M. (1972). Hematotoxicity of synergised pyrethrum insecticides and related chemicals in intact and totally and subtotally splenectomized dogs. *Acta Hepato-Gastroenterol* **19**, 66–78.

Maekawa, A., Onodera, H., Furuta, K., Tanigawa, H., Ogiu, T. and Hayashi, Y. (1985). Lack of evidence of carcinogenicity of technical-grade piperonyl butoxide in F334 rats: selective induction of ileocaecal ulcers. *Food Chem. Toxicol.* **7**, 675–682.

McGregor, D.B., Brown, A., Cattanach, P., Edwards, I., McBride, D., Riach, C. and Caspary, W.J. (1988). Responses of the L5178 tk+/tk− mouse lymphoma cell forward mutation assay. III. 72 coded chemicals. *Environ. Mol. Mutag.* **12**, 85–154.

McKeon, M.E. and Phil, M. (1991). Piperonyl Butoxide in the Assay for Unscheduled DNA Synthesis in Rat Liver Primary Cell Cultures with a Confirmatory Assay. Unpublished report no. 14413-0-447R from Hazleton Washington, Kensington, Maryland, USA. Undertaken for the PBO Task Force, Washington DC, USA.

Modeweg-Hausen, L., Lalande, M., Bier, C., Lossos, G. and Osborne, B.E. (1984). A Dietary Dose Range-finding Study of Piperonyl Butoxide in the Albino Rat. Unpublished report no. 81820B from Bio-Research Laboratories, Edgewater, Maryland, USA. Undertaken for the PBO Task Force, Washington DC, USA.

Murli, H. (1991). Piperonyl Butoxide: Measuring Chromosomal Aberrations in Chinese Hamster Ovary (CHO) Cells with Multiple Harvests under Conditions of Metabolic Activation. Unpublished report no. 14413-0-437C from Hazleton Washington, Kensington, Maryland, USA. Undertaken for the PBO Task Force, Washington DC, USA.

Negherbon, W.O. (1959). *Handbook of Toxicology,* vol. 3. W.B. Saunders, Philadelphia.

Newton, P.E. (1992). A Subchronic (3-month) Inhalation Toxicity Study of Piperonyl Butoxide in the Rat via Whole-body Exposures. Unpublished report no. 91-8333 from Bio/dynamics, East Millstone, New Jersey, USA. Undertaken for the PBO Task Force, Washington DC, USA.

Pedrick, M.S., Rumsby, P.C., Wright, V., Phillimore, H.E., Butler, W.H. and Evans, J.G. (1994). Growth characteristics and Ha-*ras* mutations of all cultures isolated from chemically induced mouse liver tumors. *Carcinogenesis* **15**, 1847–1852.

Phillips, J.C., Cunningham, M.E., Price, R.J., Osimitz, T., Gabriel, K.L., Preiss, F.J., Butler, W.H. and Lake, B.G. (1997). Effect of piperonyl butoxide on cell replication and xenobiotic metabolism in rat liver. *Fund. Appl. Toxico.* **38**, 64–74.

Robinson, K., Pinsonneault, L. & Procter, B.G. (1986). A Two-generation (two-litter) Reproduction Study of Piperonyl Butoxide Administered in the Diet to the Rat. Unpublished report no. 81689 from Bio-Research Laboratories Ltd., Montreal, Quebec, Canada. Undertaken for the PBO Task Force, Washington DC, USA.

Romanelli, P. (1991a). Primary Skin Irritation – Rabbits. Unpublished report no. 91-7317A from Biosearch Inc., Philadelphia, Pennsylvania, USA. Undertaken for the PBO Task Force, Washington DC, USA.

Romanelli, P. (1991b). Primary Eye Irritation – Rabbits. Unpublished report no. 91-7317A from Biosearch Inc., Philadelphia, Pennsylvania, USA. Undertaken for the PBO Task Force, Washington DC, USA.

Romanelli, P. (1991c). Guinea Pig Dermal Sensitization – Modified Buehler Method. Unpublished report no.91-7317A from Biosearch Inc., Philadelphia, Pennsylvania, USA. Undertaken for the PBO Task Force, Washington DC, USA.

Sarles, M.P. and Vandegrift, W.B. (1952). Chronic oral toxicity and related studies on animals with the insecticide and pyrethrum synergist, piperonyl butoxide. *Am. J. Trop. Med. Hyg.* **1**, 862–883.

Sarles, M.P., Dove, W.E. and Moore, D.H. (1949). Acute toxicity and irritation tests on animals with the new insecticide, piperonyl butoxide. *Am. J. Trop. Med,* **29**, 151–166.

Selim, S. (1995). Absorption, Excretion, and Mass Balance of ^{14}C Piperonyl Butoxide from two Different Formulations after Dermal Application to Healthy Volunteers. Unpublished report no. PO594006 from Biological Test Center, Irvine, California, USA. Undertaken for the PBO Task Force, Washington DC, USA.

Skrinijaric-Spoljar, M., Mathews, H.B., Engel, J.L. and Casida, J.E. (1971). Response of hepatic microsomal mixed-function oxidases to various types of insecticide chemical synergists administered to mice. *Biochem. Pharmacol.* **20**, 1607–1618.

Squire, R.A. and Levitt, M.H. (1975). Report of a workshop on classification of specific hepatocellular lesions in rats. *Cancer Res.* **35**, 3214–3223.

Takahashi, O., Oishi, T., Fujitani, T., Tanaka, T. and Yoneyama, M. (1994a). Piperonyl butoxide induces hepatocellular carcinoma in male CD-1 mice. *Arch. Toxicol.* **68**, 467–469.

Takahashi, O., Oishi, T., Fujitani, T., Tanaka, T. and Yoneyama, M. (1994b). Chronic toxicity studies of piperonyl butoxide in F344 rats: induction of hepatocellular carcinoma. *Fundam. Appl. Toxicol.* **22**, 293–303.

Tanaka, T. (1992). Effects of piperonyl butoxide on F1 generation mice. *Toxicol. Lett.* **60**, 83–90.

Tanaka, T. (1993). Behavioural effects of piperonyl butoxide in male mice. *Toxicol. Lett.* **69**, 155–161.

Tanaka, T., Takahashi, O. and Oishi, S. (1992). Reproductive and neurobehavioural effects in three-generation toxicity study of piperonyl butoxide administered to mice. *Food Chem. Toxicol.* **30**, 1015–1019.

Tanaka, T., Fujitani, T., Takahashi, O. and Oishi, S. (1994). Developmental toxicity evaluation of piperonyl butoxide in CD-1 mice. *Toxicol. Lett.* **71**, 123–129.

Tu, A.S., Coombes-Hallowell, W.E., Pallotta, S.L. and Catanzano, A. (1985). Evaluation of piperonyl butoxide in the CHO/HGPRT Mutation Assay With and Without Metabolic Activation. Unpublished report no. 53906 from Arthur D. Little Inc., Cambridge, Massachusetts, USA. Undertaken for the PBO Task Force, Washington DC, USA.

US National Cancer Institute (1979). Bioassay of Piperonyl Butoxide for Possible Carcinogenicity. DHEW Publ. no. 79-1375, Bethesda, Maryland, USA, US Department of Health, Education and Welfare.

Wester, R.C., Bucks, D.A.W. and Maibach, H.I. (1994). Human in vivo percutaneous absorption of pyrethrin and piperonyl butoxide. *Food Chem. Toxicol.* **32**, 51–53.

Wintersteiger, R. and Juan, H. (1991). Resorption Study of Tyrason after Dermal Application (study performed on healthy subjects). J.S.W. – Experimental Research, Studie Analytik. 01/91 p.l.

Yamamoto, I. (1973). Mode of action of synergists in enhancing the insecticidal activity of pyrethrum and pyrethroids. In: *Pyrethrum – Natural Insectide 195–210.* Academic Press Ltd, London, and New York.

3
Interactions of Piperonyl Butoxide with Cytochrome P450

ERNEST HODGSON and PATRICIA E. LEVI

1. INTRODUCTION

The initial suggestion that piperonyl butoxide (PBO) functioned by inhibition of oxidative metabolism of insecticides was made by Sun and Johnson (1960) on the basis of *in vivo* toxicity tests using house flies. Hodgson and Casida (1960, 1961) demonstrated that both PBO and sesamex inhibit the oxidation of N, N-dimethyl-p-nitrophenyl carbamate and its diethyl analogue by rat liver microsomes. Subsequently, it was shown that PBO and other methylenedioxyphenyl compounds (Casida, 1970; see Hodgson and Philpot, 1974, for additional references) inhibited the microsomal oxidation of many insecticides and other xenobiotics in a number of mammalian and insect species, regardless of whether the PBO was administered *in vivo* prior to the preparation of the microsomes or both substrate and inhibitor were administered *in vitro*.

The initial studies demonstrating that this inhibition of microsomal oxidations was due to a direct effect on P450 were carried out by Perry and Bucknor (1970) and Matthews and Casida (1970) using house flies and Matthews *et al.* (1970) using mice. These studies demonstrated an apparent reduction in P450 levels in microsomes prepared from animals treated with PBO. A similar reduction was shown in an *in vitro* system, using mouse liver microsomes, by Philpot and Hodgson (1971), the apparent loss of P450 being due to the formation of a metabolite-inhibitory complex which blocked CO binding to the cytochrome.

Not only has PBO continued to be used as an insecticide synergist but, in addition, it has become a diagnostic tool for two important aspects of insecticide toxicology: to determine if *in vivo* metabolism of an insecticide is oxidative and to determine if cases of insecticide resistance involve oxidative metabolism by P450.

PIPERONYL BUTOXIDE
ISBN 0-12-286975-3

2. CYTOCHROME P450 MONO-OXYGENASE ENZYMES

The P450 mono-oxygenases are a large and functionally diverse family of enzymes that carry out the initial oxidation of a wide variety of lipophilic compounds. These enzymes play a major role in the metabolism of xenobiotics such as drugs, pesticides, carcinogens and other environmental chemicals. In addition P450s are involved in the metabolism of many important endogenous compounds including steroids, fat-soluble vitamins, fatty acids and prostanoids. The P450 enzymes, discovered in the 1950s, are haem-containing proteins whose reduced carbon monoxide complex gives a UV-visible spectra with Soret maxima around 450 nm (Omura and Sato, 1964). To date over 200 individual P450 isozymes have been characterized in 27 gene families; individual isozymes within a family have at least 40% protein sequence homology (Nelson *et al.*, 1993).

Although metabolites of P450 oxidations are usually less toxic, there are a number of known examples of activation reactions (Miller and Miller, 1985; Monks and Lau, 1988; Guengerich, 1992, 1993, 1994; Gonzalez and Gelboin, 1994; Levi, 1994; Smith *et al.*, 1994). Examples of familiar compounds that are activated by P450 to toxic species include: aflatoxin, acetaminophen, vinyl chloride, parathion and carbon tetrachloride. Cytochrome P450s occur both as constitutive and inducible enzymes and a number of foreign compounds are known to induce specific P450 isozymes. In many cases the compounds that are P450 substrates serve as inducing agents, inducing the isozymes active in their metabolism. Because of the importance of the P450 enzymes in metabolism as well as the significant health-related consequences associated with inducing or repressing specific P450 isozymes, there is considerable interest in studying P450 regulation, the activities of specific isozymes, and the chemicals involved in the interactions with P450.

3. METHYLENEDIOXYPHENYL COMPOUNDS

PBO belongs to a group of chemicals known as methylenedioxyphenyl (MDP) compounds or benzodioxole (BD) compounds. These chemicals occur also as natural products in plant tissues and are widely used in commercial chemical preparations. Structures of some naturally occurring and some synthetic MDPs that have been studied are shown in Fig. 3.1. Parsnips, carrots, parsley, nutmeg, sesame seeds, pepper and sassafras are among plant sources of MDP compounds (Hodgson and Philpot, 1974). Safrole, a naturally occurring plant MDP, has been shown at high doses to be a hepatic carcinogen in rodents (Ioannides *et al.*, 1981). A comprehensive review of the literature on MDP compounds, including toxicity, inductive effects and synergism, was published by Hodgson and Philpot (1974), and another review dealing with PBO was published by Haley (1978). More recently there has been an increased interest in the role of MDP compounds in the regulation of P450 enzymes, and a review summarizing the current work in that area has been published (Adams *et al.*, 1995). In mammals, MDP compounds affect multiple enzyme pathways, including the P450 system, and MDPs have been studied extensively for their abilities to inhibit P450 metabolism as well as to induce P450 enzymes.

4. THE BIPHASIC RESPONSE

It has been known for some time that MDPs, as well as some other xenobiotics, can act as both inhibitors and inducers of P450 activity (Hodgson and Philpot, 1974; Adams *et al.*, 1995). Following a single dose, the time course for the two activities differs, with inhibition being relatively rapid and induction slower. Thus the effect on enzyme activity is biphasic: first a decrease below control levels, followed by an increase above control levels. Finally there is a return to control level. An example of this response is shown in Fig. 3.2 (Kinsler *et al.*, 1990).

While it might be speculated that a similar biphasic effect on metabolism could occur whenever a lipophilic xenobiotic capable of forming a stable inhibitory complex is administered, few cases have been well documented. Even in the case of MDP interactions, the isoform specificity of either inhibition or induction has not been well defined. Since both detoxification and activation of chemicals are, to a greater or lesser extent, isoform-specific, more recent studies have focused on clarifying which isoforms of P450 are involved and the importance of these isoforms in the biphasic response (Lewandowski *et al.*, 1990; Adams *et al.*, 1993a,b; Ryu *et al.*, 1995).

5. INTERACTIONS WITH CYTOCHROME P450

Perry and Bucknor (1970) and Matthews and Casida (1970) using house flies, and Matthews *et al.* (1970) using mice, first demonstrated that MDP compounds

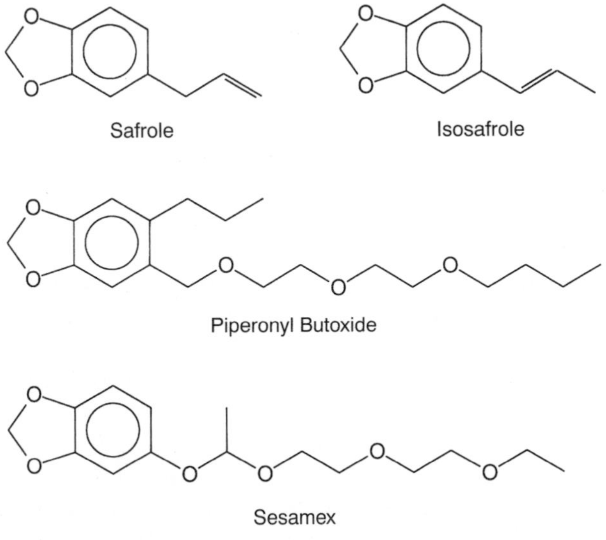

Safrole

Isosafrole

Piperonyl Butoxide

Sesamex

Figure 3.1.   Structures of some MDP compounds.

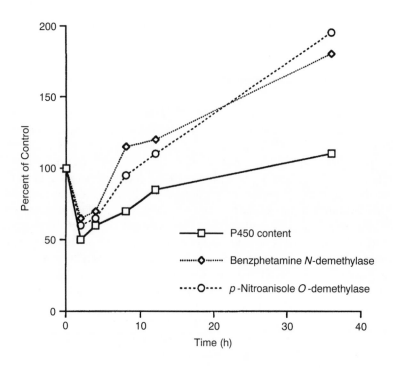

Figure 3.2. Biphasic effect of piperonyl butoxide on mouse hepatic cytochrome P450 activities (from Kinsler et al., 1990, with permission).

interact with the P450 mono-oxygenase oxidase system in a manner not explained by a typical enzyme–substrate relationship. These studies demonstrated an apparent reduction in P450 levels in microsomes prepared from animals treated *in vivo* with PBO. A similar reduction in the cytochrome P450 level caused by PBO was subsequently achieved with an *in vitro* system by Philpot and Hodgson (1971). Incubation of mouse hepatic microsomes with PBO in the presence of NADPH was found to result in a spectrally observable complex that blocked CO binding to the cytochrome. This provided an explanation for the apparent loss of cytochrome P450 since the combination of CO with the reduced cytochrome is employed in the assay of this pigment (Omura and Sato, 1964). The complex thus formed gives rise to a P450 difference spectrum with Soret peaks at 455 and 427 nm (Fig. 3.3) which exists in pH-dependent equilibrium, referred to as a 'type III' P450 spectrum (Philpot and Hodgson, 1971). In the absence of NADPH, PBO forms a typical type I substrate difference spectrum; this spectrum was originally described by Matthews *et al.* (1970) using mouse liver microsomes and by Philpot and Hodgson (1970) using abdominal microsomes from resistant house flies. Inhibition of CO binding by PBO is noncompetitive and corresponds to the type III interaction. Similarities between the complex formed by the addition of PBO to microsomes and that formed by ethyl isocyanide (Omura and Sato, 1964; Schenkman *et al.*, 1967) were noted in these studies (Philpot and Hodgson, 1971; Franklin, 1971;

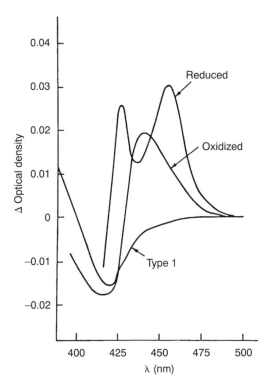

Figure 3.3. Type III optical difference spectra caused by the incubation of piperonyl butoxide with mouse liver microsomes in the presence of NADPH. NADPH is present in both reference and sample cuvettes (from Hodgson and Philpot, 1974, with permission).

Elcombe *et al.*, 1976). The inhibition of CO binding by ethyl isocyanide is competitive, however, in contrast to the noncompetitive inhibition caused by PBO.

The possibility that the type III complex formed by PBO and P450 in the presence of NADPH is related to the synergistic action of MDP compounds was strengthened when the same interaction was obtained by *in vivo* administration of PBO to mice (Philpot and Hodgson, 1971/72; Franklin, 1972, 1976; Elcombe *et al.*, 1976). In microsomes from treated mice the type III complex is detected in an oxidized state which can be converted to the double-Soret reduced spectrum by the addition of either NADPH or dithionite.

The demonstration that *in vivo* treatment of mice and insects (Philpot and Hodgson, 1971, 1971/72) with PBO results in a tightly bound cytochrome P450 complex that cannot be displaced by CO and withstands preparative procedures led Philpot and Hodgson (1972a) to suggest that this complex accounts for the apparent reduction in cytochrome P450 levels observed when animals are treated with MDP compounds (Philpot and Hodgson, 1971/72; Skrinjaric-Spoljar *et al.*, 1971; Perry and Bucknor, 1970; Matthews and Casida, 1970; Matthews *et al.*, 1970). The fact that the K_s for PBO in the formation of the type

III complex and the K_i for PBO in the inhibition of CO binding are the same confirms this hypothesis (Philpot and Hodgson, 1972b).

Examination of the difference spectra formed by MDP compounds and P450 from insecticide-susceptible and -resistant strains of house flies (A.P. Kulkarni and E. Hodgson; unpublished work, 1973; Kulkarni *et al.*, 1974) reveals that the same type of spectra are formed although some differences may be noted.

As inhibitors of P450 activity, MDP compounds such as PBO and sesamex have been used extensively as synergists with the pyrethroid and carbamate pesticides, groups of pesticides whose detoxifying metabolism in insects is P450-mediated (Haley, 1978). Dahl and Hodgson (1979) proposed that the inhibition of P450 activity *in vivo* results from the formation of a complex between the haem iron of P450 and the carbene formed when water is cleaved from the hydroxylated methylene carbon of the MDP compound (Fig. 3.4). Although the carbene, a short-lived reactive intermediate, has never been demonstrated directly, this hypothesis still appears to be theoretically sound, and no evidence has been put forward to shed doubt on its validity. In any event, the P450–MDP metabolite complex is extremely stable, since even in subcellular microsomal preparations the enzyme remains inhibited, and the effects of *in vivo* interaction can be demonstrated *in vitro*.

Little is known of the P450 isozyme specificity for either PBO inhibition, metabolite-inhibitory complex formation or type III spectrum formation. Studies involving five isoforms purified from mouse liver (Levi and Hodgson, 1985; Beumel *et al.*, 1985) showed that epoxidation of aldrin by any of these isoforms was inhibited by PBO and that all isoforms appeared to form a stable inhibitory complex. However, the form of the type III spectrum was variable, with little correlation between the extent of inhibition and the nature of the spectrum. More recently, Pappas and Franklin (1996) have provided indirect evidence that

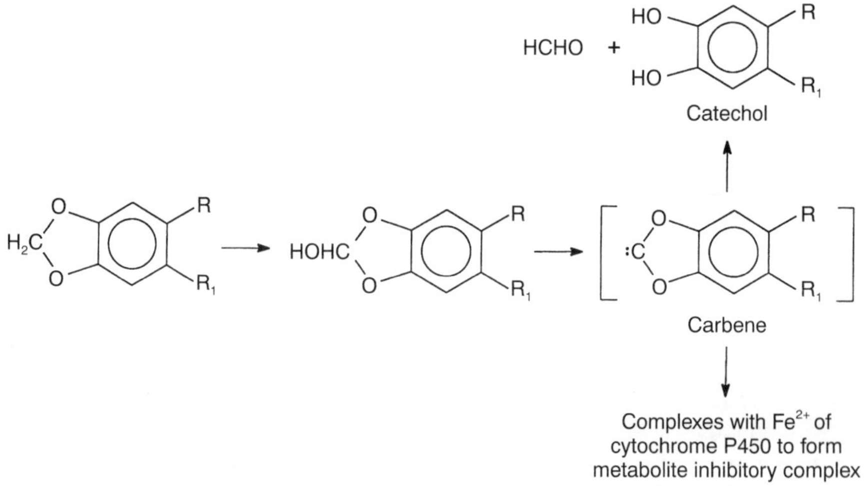

Figure 3.4.   Oxidation of MDP compounds by cytochrome P450.

in the rat PBO preferentially enhances the formation of the metabolite-inhibitory complex with members of the P450 3A subfamily.

It is also clear that the reactions leading to the formation of the metabolite-inhibitory complex are not the only metabolic reactions undergone by PBO. Early studies (Fishbein *et al.*, 1969) demonstrated the formation of at least 28 metabolites *in vivo* in the rat. Subsequently, Casida *et al.* (1966) and Kamienski and Casida (1970) showed that in the house fly and in rats and mice α-methylene-[^{14}C]PBO was metabolized to CO_2, with metabolism to CO_2 being much faster in the mammals than in the insect.

6. INDUCTION OF CYTOCHROME P450

Brown *et al.* (1954) first observed the stimulation of liver microsomal enzyme activity by xenobiotics and since that time many examples of such stimulation have been noted. The early investigations were the subject of a landmark review by Conney (1967) and it was clear by that time that all microsomal enzyme inducers did not induce the same enzyme activities. It is now known (Okey, 1990; Hodgson, 1994) that inducers of P450 may be highly isoform-specific, some acting through a cytosolic receptor (the Ah receptor) and others via a variety of alternative mechanisms.

While short-term (less than 24 hours) treatment with PBO causes inhibition of many monooxygenase activities, early studies (summarized in Hodgson and Philpot, 1974) demonstrated induction of a number of enzyme activities as well as increased protein levels when a large time interval elapsed between treatment of the animal and assay of enzyme activity. In the mouse, rabbit and rat an increase in the metabolism of a variety of substrates, including biphenyl, parathion, hexobarbital, aniline, *p*-nitroanisole, dimethyl-*p*-nitrophenyl carbamate, phenylphosphonothioic acid *O*-ethyl *O*-(4-nitrophenyl) ester (EPN) and *p*-nitrobenzoic acid occurred through several different reactions including hydroxylation, oxidative ester cleavage, *O*-demethylation, *N*-demethylation and nitroreduction. Studies were also carried out to determine the effect of such induction on cytochrome P450. In the light of our current knowledge of P450 isoforms, these latter studies are no longer important; however, they did demonstrate spectral changes and confirmed the biphasic response at the cytochrome level. The appearance of novel haemoproteins (Fennel *et al.*, 1980; Murray *et al.*, 1983a, b; Ohyama *et al.*, 1984) was also proposed, although this was controversial. Recent studies have not confirmed the induction of novel haemoproteins and none of the recent induction results discussed below require novel haemoproteins for their interpretation. Induction in nonmammalian species such as the Southern armyworm, *Spodoptera eridania,* has also been demonstrated (Marcus *et al.*, 1986). Several studies using fish have shown an increase in P450 activities, especially those associated with induction of CYP 1A1, after treatment with MDP compounds (Vodicnik *et al.*, 1981; Erickson *et al.*, 1988). For the most part, however, the identity of the P450 species induced have not been defined.

In a series of papers, Cook and Hodgson (1983, 1984, 1985, 1986) demon-strated that P450 induction by isosafrole was not dependent on the Ah receptor, even though some of the enzyme activities induced were characteristic of those induced by Ah-dependent P450 inducers. In a structure–activity study, it was shown using 1-t-butyl-3,4-benzodioxole that substitution of the methylene car-bon with two methyl groups prevented the induction of P450, although other proteins, including epoxide hydrolase and NADPH-cytochrome P450 reductase, were induced. Of particular interest was the observation that isosafrole, while not acting as a ligand for the Ah receptor, caused induction of the Ah receptor.

In a more recent series of communications, Adams *et al.* (1993a, b) examined the regulation of three different P450 isozymes, 1A1, 1A2 and 2B10, in the liver of C57 (Ah-responsive) and DBA (Ah-nonresponsive) mice by safrole, isosaf-role, PBO and sesamex. P450s 1A1 and 1A2 are known to be associated with induction by ligand binding to the Ah receptor, while 2B10 induction is not mediated by this mechanism. Levels of mRNA, protein and enzyme activities for these three P450 isozymes were determined. In addition to an increase in 2B10 in both strains, an increase in P450 1A2 was also seen in both strains, sug-gesting that this isozyme was induced by an Ah-independent mechanism. On the other hand, 1A1 induction was seen only at the higher doses, only with PBO and sesamex, and only in C57 mice, suggesting that, even at high doses, these two compounds are only marginally effective as Ah-dependent inducers.

This pattern of induction has been confirmed with C57 mice using three other benzodioxole compounds with four-carbon side chains, 5-n-butyl-, 5-t-butyl- and 5-(3-oxobutyl)-1,3-benzodioxole (Ryu *et al.*, 1995). Western and northern blots indicated the induction of P450 1A2 by all three compounds, while there was no induction of 1A1. In addition the t-butyl compound also caused induc-tion of 2B10. *In vitro*, all three compounds inhibited ethoxyresorufin *O*-de-ethylation and acetanilide hydroxylation, both 1A2-associated activities in mice, with the n-butyl analogue being the most effective. Pentoxyresorufin *O*-dealkylation, a 2B10 activity, was inhibited equally by all three compounds. Thus the structure–activity relationships for induction and inhibition were not the same, suggesting that the isozyme specificity for induction and inhibition is different.

The induction patterns by MDP compounds in the rat are somewhat different from those described above for the mouse (Wagstaff and Short, 1971; Goldstein *et al.*, 1973; Murray *et al.*, 1983a, b; Marcus *et al.*, 1990a, b). While PBO induces P450s from the 1A and 2B families in both rats and mice, it is only at high doses that 1A1 is induced in mice. Other MDP compounds, such as isosafrole, that have been shown not to induce P450 1A1 in mice, will induce 1A1-associated activi-ties in the rat. The reason for this difference – whether resulting from metabolism and/or differences in binding to the Ah receptor – is not known.

7. EFFECTS ON MULTIENZYME SYSTEMS

The metabolism of xenobiotics is usually complex, frequently involving more than one enzyme or enzyme system. One example is the oxidation of the organophosphorus pesticide phorate by the flavin-containing monooxygenase (FMO) and P450 systems (Levi and Hodgson, 1988, 1989) (Fig. 3.5). Both FMO and P450 readily catalyse the initial sulfoxidation of the thioether moiety of phorate to form the sulfoxide. Subsequent oxidation reactions, however, such as sulfone formation or oxidative desulfuration, are catalysed entirely by P450. Although both FMO and P450 form phorate sulfoxide, the products are stereochemically different, with FMO forming the (−) isomer, and P450 forming either the (+) isomer or a racemic mixture. Since the preferred substrate for further oxidation by P450 to either more or less toxic metabolites is the (+) sulfoxide, changes in the relative contributions of the two enzymes could have toxicological significance.

To study the relative contributions of these two enzymes with common substrates, we developed methods to measure each activity separately in the same microsomal preparations. The most useful of these techniques is the use of an antibody to NADPH-P450 reductase that inhibits P450 activity. In this situation oxidation results primarily from FMO activity. A second procedure, heat treatment of microsomal preparations (50°C for 1 minute), inactivates hepatic FMO, allowing measurement of only P450 activity. Both of these methods are performed at the same time in order to generate more accurate results. These methods work well with liver microsomes, since the total activity of the two determinations described above agrees closely with the value for untreated microsomes. They do not work as well with lung microsomes since one of the FMO forms, FMO2, expressed in lung (but not liver), is known to be more heat stable than other FMO isozymes.

Using these methods Kinsler *et al.* (1990) investigated the relative contributions of the two enzyme systems to the oxidation of phorate following treatment of mice with various agents, including PBO. As noted earlier, after PBO treatment there is an initial inhibition of P450 followed by induction of P450 activity (Fig. 3.2). In the case of phorate oxidation the initial inhibition of P450 at 2 hours resulted in a decrease in both the overall rate of oxidation and in the percentage contribution by P450 (from 76% to 58%). After 4 hours both the

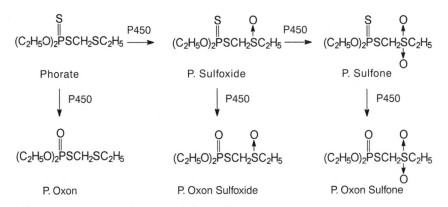

Figure 3.5.   Oxidation of phorate by P450 and the flavin-containing monooxygenase.

overall rate of oxidation and the relative contributions were similar to control levels. After 12 hours the rate of oxidation was increased, as was the per cent contribution of P450 (to 89%). Such interactions are complex and the implications, especially when toxicities of enzyme products are different, are difficult to predict.

8. CONCLUSION

Numerous studies conducted over the last 35 years have established that PBO is a substrate, an inhibitor and an inducer of P450 in mammals and in insects. Earlier studies demonstrated the nature of the metabolite-inhibitory complex and its spectral manifestation and a mechanism was proposed for the inhibitory reaction with P450. Similarly, early studies demonstrated that PBO could function as a P450 inducer and that, because of the relatively fast rate of P450 inhibition in comparison with P450 induction, the time course of effects in mammals is biphasic. Recent and ongoing studies are concerned primarily with the isozyme specificity of the induction process in mammals and have demonstrated that in mice PBO not only induces P450 2B10, but also induces P450 1A2 by an Ah-receptor-independent mechanism and, at high doses, 1A1 by an Ah-receptor-dependent mechanism. In the rat, however, PBO appears to induce 2B1/2 as well as 1A1/2, the latter by Ah-receptor-dependent mechanisms.

PBO is not only an effective insecticide synergist because of its interaction with P450 but, owing to the same interaction, is also useful as a diagnostic tool in pesticide toxicology in the determination of whether or not P450-dependent mono-oxygenation is involved in the metabolism of a particular pesticide and whether or not P450 mono-oxygenation is involved in resistance when resistant populations are discovered.

REFERENCES

Adams, N.H., Levi, P.E. and Hodgson, E. (1993a). Regulation of cytochrome P450 isozymes by methylenedioxyphenyl compounds. *Chem.–Biol. Interact.* **86**, 255–274.

Adams, N.H., Levi, P.E. and Hodgson, E. (1993b). Differences in induction of three P450 isozymes by piperonyl butoxide, sesamex and isosafrole. *Pestic. Biochem. Physiol.* **46**, 15–26.

Adams, N.H., Levi, P.E. and Hodgson, E. (1995). Regulation of cytochrome P450 isozymes by methylenedioxyphenyl compounds – an updated review of the literature. Toxicology Communications Inc., Raleigh, NC. *Rev. Biochem. Toxicol.* **11**, 205–222.

Beumel, G.A., Levi, P.E. and Hodgson, E. (1985). Spectral interactions of piperonyl butoxide and isocyanides with purified hepatic cytochrome P-450 from uninduced mice. *Gen. Pharmacol.* **16**, 193–197.

Brown, R.R., Miller, J.A. and Miller, E.C. (1954). The metabolism of methylated aminoazo dyes. 4. Dietary factors enhancing demethylation *in vitro*. *J. Biol. Chem.* **209**, 211–221.

Casida, J.E. (1970). Mixed-function oxidase involvement in the biochemistry of insecticide synergists. *J. Agric. Food Chem.* **18**, 753–759.

Casida, J.E., Engel, J.L., Esaac, E.G., Kamienski, F.X. and Kuwatsuka, S. (1966). Methylene-^{14}C-dioxyphenyl compounds: metabolism in relation to their synergistic action. *Science* **153**, 1130–1133.

Conney, A.H. (1967). Pharmacological implications of microsomal induction. *Pharmacol. Rev.* **19**, 317–366.

Cook, J.C. and Hodgson, E. (1983). Induction of cytochrome P-450 by methylene-dioxyphenyl compounds: importance of the methylene carbon. *Toxicol. Appl. Pharmacol.* **68**, 131–139.

Cook, J.C. and Hodgson, E. (1984). 2,2-Dimethyl-5-t-butyl-1,3-benzodioxole: an unusual inducer of microsomal enzymes. *Biochem. Pharmacol.* **33**, 3941–3946.

Cook, J.C. and Hodgson, E. (1985). The induction of cytochrome P-450 by isosafrole and related methylenedioxyphenyl compounds. *Chem.–Biol. Interact.* **54**, 299–315.

Cook, J.C. and Hodgson, E. (1986). Induction of cytochrome P-450 in congenic C57B/6J mice by isosafrole: lack of correlation with the Ah locus. *Chem.–Biol. Interact.* **58**, 233–240.

Dahl, A.R. and Hodgson, E. (1979). The interaction of aliphatic analogs of methylene-dioxyphenyl compounds with cytochromes P-450 and P-420. *Chem.–Biol. Interact.* **27**, 163–175.

Elcombe, C.R., Bridges, J.W. and Nimmo-Smith, R.H. (1976). Substrate-elicited disso-ciation of a complex of cytochrome P-450 with a methylenedioxyphenyl metabolite. *Biochem. Biophys. Res. Commun.* **71**, 915–918.

Erickson, D.A., Goodrich, M.S. and Lech, J.L. (1988). The effect of piperonyl butoxide on hepatic cytochrome P450-dependent monooxygenase activities in rainbow trout (*Salmo gairdneri*). *Toxicol. Appl. Pharmacol.* **94**, 1–10.

Fennel, T.R., Sweatman, B.C. and Bridges, J.W. (1980). The induction of hepatic cytochrome P450 in C57 BL/10 and DBA/2 mice by isosafrole and piperonyl butoxide. *Chem.–Biol. Interact.* **31**, 189–201.

Fishbein, L., Falk, H.L., Fawkes, J., Jordan, S. and Corbett, B. (1969). The metabolism of PBO in the rat with ^{14}C in the methylenedioxy or 2-methylene group. *J. Chromatogr.* **41**, 61–79.

Franklin, M.R. (1971). The enzymic formation of a methylenedioxyphenyl derivative exhibiting an isocyanide-like spectrum with reduced cytochrome P-450 in hepatic microsomes. *Xenobiotica* **1**, 581–591.

Franklin, M.R. (1972). Inhibition of hepatic oxidative metabolism by piperonyl butoxide. *Biochem. Pharmacol.* **21**, 3287–3293.

Franklin, M.R. (1976). Methylenedioxyphenyl insecticide synergists as potential human health hazards. *Environ. Health Perspect.* **14**, 29–32.

Goldstein, J.A., Hickman, P. and Kimbrough, R.D. (1973). Effects of purified and tech-nical piperonyl butoxide on drug-metabolizing enzymes and ultrastructure of rat liver. *Toxicol. Appl. Pharmacol.* **26**, 444–458.

Gonzalez, F.J. and Gelboin, H.V. (1994). Role of human cytochromes P450 in the metabolic activation of chemical carcinogens and toxins. *Drug Metabol. Rev.* **26**, 165–183.

Guengerich, F.P. (1992). Metabolic activation of carcinogens. *Pharmacol. Ther.* **54**, 17–61.

Guengerich, F.P. (1993). Bioactivation and detoxication of toxic and carcinogenic chemicals. *Drug Metabol. Disp.* **21**, 1–6.

Guengerich, F.P. (1994). Catalytic selectivity of human cytochrome P450 enzymes – relevance to drug metabolism and toxicity. *Toxicol. Lett.* **70**, 133–138.

Haley, T.J. (1978). Piperonyl butoxide, α[2-(2-butoxyethoxy)ethoxy]-4,5-methylene-dioxy-2-propyltoluene: a review of the literature. *Ecotoxicol. Environ. Safety* **2**, 9–31.

Hodgson, E. (1994). Chemical and environmental factors affecting metabolism of xeno-biotics. In: *Introduction to Biochemical Toxicology* (Hodgson, E. and Levi, P. E., eds), ch. 7, pp. 153–175. Appleton and Lange, Norwalk, CT.

Hodgson, E. and Casida, J.E. (1960). Biological oxidation of *N,N*-dialkyl carbamates. *Biochim. Biophys. Acta* **42**, 184–186.

Hodgson, E. and Casida, J.E. (1961). Metabolism of *N,N*-dialkyl carbamates and related compounds by rat liver. *Biochem. Pharmacol.* **8**, 179–191.

Hodgson, E. and Philpot, R.M. (1974). Interactions of methylenedioxyphenyl (1,3-benzodioxole) compounds with enzymes and their effects on mammals. *Drug Metab. Rev.* **2**, 231–301.

Ioannides, C., Delaforge, M. and Parke, D.V. (1981). Safrole: its metabolism, carcinogenicity and interactions with cytochrome P-450. *Food Cosmet. Toxicol.* **19**, 657–666.

Kamienski, F.X. and Casida, J.E. (1970). Importance of demethylation in the metabolism *in vivo* and *in vitro* of methylenedioxyphenyl synergists and related compounds in mammals. *Biochem. Pharmacol.* **19**, 91–112.

Kinsler, S., Levi, P.E. and Hodgson, E. (1990). Relative contributions of the cytochrome P450 and flavin-containing monooxygenase to the microsomal oxidation of phorate following treatment of mice with phenobarbital, hydrocortisone, acetone and piperonyl butoxide. *Pestic. Biochem. Physiol.* **37**, 174–181.

Kulkarni, A.P., Mailman, R.B., Baker, R.C. and Hodgson, E. (1974). Cytochrome P-450 difference spectra: Type II interactions in insecticide-resistant and -susceptible houseflies. *Drug Metabol. Disp.* **2**, 309–320.

Levi, P.E. (1994). Reactive metabolites and toxicity. In: *Introduction to Biochemical Toxicology, 2nd edn.* (Hodgson, E. and Levi, P.E., eds), pp. 219–239.) Appleton and Lange, Norwalk, CT.

Levi, J.E. and Hodgson, E. (1985). Oxidation of pesticides by purified cytochrome P-450 isozymes from mouse liver. *Toxicol. Lett.* **24**, 221–228.

Levi, P.E. and Hodgson, E. (1988). Stereospecificity in the oxidation of phorate and phorate sulfoxide by purified FAD-containing monooxygenase and cytochrome P450 isozymes. *Xenobiotica* **18**, 29–39.

Levi, P.E. and Hodgson, E. (1989). Monooxygenations: interactions and expression of toxicity. In: *Insecticide Action: From Molecule to Organism* (Narahashi, T. and Chambers, J.E., eds), pp. 233–244. Plenum Press, New York.

Lewandowski, M., Chui, Y.C., Levi, P.E. and Hodgson, E. (1990). Differences in induction of hepatic cytochrome P450 isozymes in mice by eight methylenedioxyphenyl compounds. *J. Biochem. Toxicol.* **5**, 47–55.

Marcus, C.B., Murray, M., Wang, C. and Wilkinson, C.F. (1986). Methylenedioxyphenyl compounds as inducers of cytochrome P450 and monooxygenase activity in the Southern armyworm (*Spodoptera eridania*) and the rat. *Pestic. Biochem. Physiol.* **26**, 310–322.

Marcus, C.B., Wilson, N.M., Jefcoate, C.R., Wilkinson, C.F. and Omiecinski, C.J. (1990a). Selective induction of cytochrome P450 isozymes in rat liver by 4-n-alkyl-methylenedioxybenzenes. *Arch. Biochem. Biophys.* **277**, 8–16.

Marcus, C.B., Wilson, N.M., Keith, I.M., Jefcoate, C.R. and Omiecinski, C.J. (1990b). *Arch. Biochem. Biophys.* **277**, 17–25.

Matthews, H.B. and Casida, J.E. (1970). Properties of microsomal cytochromes in relation to sex, strain, substrate specificity and apparent inhibition by synergist and insecticide chemicals. *Life Sci.* **9**, 989–1001.

Matthews, H.B., Skrinjaric-Spoljar, M. and Casida, J.E. (1970). Insecticide synergist interactions with cytochrome P450 in mouse liver microsomes. *Life Sci.* **9**, 1039–1048.

Miller, E.C. and Miller, J.A. (1985). Some historical perspectives on the metabolism of xenobiotic chemicals to electrophiles. In: *Bioactivation of Foreign Compounds* (Anders, M.W., ed.), pp. 3–28. Academic Press, New York.

Monks, T.J. and Lau, S.S. (1988). Reactive intermediates and their toxicological significance. *Toxicology* **52**, 1–53.

Murray, M., Wilkinson, C.F., Marcus, C. and Dube, C.E. (1983a). Structure–activity relationships in the interactions of alkoxymethylenedioxybenzene derivatives with rat hepatic microsomal mixed-function oxidases *in vivo*. *Mol. Pharmacol.* **24**, 129–136.

Murray, M., Wilkinson, C.F. and Dube, C.E. (1983b). Effects of dihydrosafrole on cytochromes P-450 and drug oxidation in hepatic microsomes from control and induced rats. *Toxicol. Appl. Pharmacol.* **68**, 66–76.

Nelson, D.R., Kamataki, T., Waxman, D.J., Guengerich, F.P., Estabrook, R.W., Feyereisen, R., Gonzalez, F.J., Coon, M.J., Gunsalus, I.C., Gotoh, O., Okuda, K.

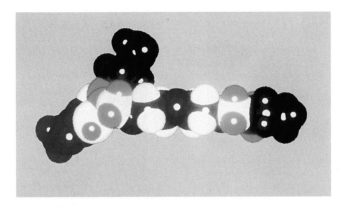

Plate 1. Lipophilicity of piperonyl butoxide. The atoms are coloured according to their lipophilicity. Hydrophilic atoms are water-soluble, while lipophilic atoms are lipid-soluble. Dark blue, most lipophilic; light blue, somewhat lipophilic; white, neutral/intermediate; light bright red, somewhat hydrophilic; dark red, most hydrophilic.

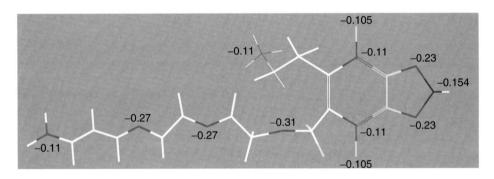

Plate 2. Calculated major residue partial surface charges on carbon and oxygen atoms in piperonyl butoxide. Colours indicate the degree of charge as follows: dark blue, most positive; light blue, somewhat positive; white, neutral; light red, somewhat negative; dark red, most negative.

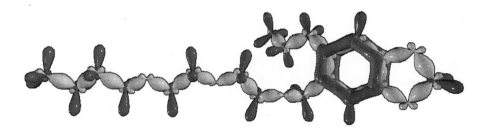

Plate 3. Atomic orbitals for minimized piperonyl butoxide showing σ and π orbitals. The π orbitals are red toroids on both sides of the ring of carbon atoms. The σ orbitals are coloured yellow for carbon and oxygen and green for hydrogen.

Plates 4 & 5. Applying a measured drop to a test insect such as the housefly. This apparatus allows for the very rapid treatment of large batches.

Plate 6. The Aerosol Test Room permits the proper evaluation of Aerosol Insecticides under standard conditions.

Plate 7. The measured drop equipment can also be used to dose roaches.

and Nebert, D.W. (1993). The P450 superfamily: update on new sequences, gene mapping, accession numbers, early trivial names of enzymes, and nomenclature. *DNA and Cell Biol.* **12**, 1–51.

Ohyama, T., Nebert, D.W. and Negishi, M. (1984). Isosafrole-induced cytochrome P2-450 in DBA/2N mouse liver. *J. Biol. Chem.* **259**, 2675–2682.

Okey, A.B. (1990) Enzyme induction in the cytochrome P450 system. *Pharmacol. Ther.* **45**, 241–298.

Omura, T. and Sato, R. (1964). The carbon monoxide-binding pigment of liver microsomes. *J. Biol. Chem.* **239**, 2370–2378.

Pappas, J. and Franklin, MR. (1996). Enhanced cytochrome P450 metabolic-intermediate complex formation from methylenedioxybenzene derivatives in CYP3A enriched rat liver microsomes. *Toxicologist* **30**, 274 (abstract number 1405).

Perry, A.S. and Bucknor, A.J. (1970). Studies on microsomal cytochrome P-450 in resistant and susceptible houseflies. *Life Sci.* **9**, 335–350.

Philpot, R.M. and Hodgson, E. (1970). Unpublished work cited in Hodgson, E. and Plapp, F. W. (1970). Biochemical characteristics of insect microsomes, *J. Agric. Food Chem.* **18**, 1048–1055.

Philpot, R.M. and Hodgson, E. (1971). A cytochrome P-450–piperonyl butoxide spectrum similar to that produced by ethyl isocyanide. *Life Sci.* **10**, 503–512.

Philpot, R.M. and Hodgson, E. (1971/72). The production and modification of cytochrome P-450 difference spectra by *in vivo* administration of methylene-dioxyphenyl compounds *Chem.–Biol. Interact.* **4**, 185–194.

Philpot, R.M. and Hodgson, E. (1972a). The effect of piperonyl butoxide concentration on the formation of cytochrome P-450 difference spectra in hepatic microsomes from mice. *Mol. Pharmacol.* **8**, 204–214.

Philpot, R.M. and Hodgson, E. (1972b). Differences in the cytochrome P-450s from resistant and susceptible houseflies. *Chem.–Biol. Interact.* **4**, 399–408.

Ryu, D.Y., Levi, P.E. and Hodgson, E. (1995). Regulation of cytochrome P450 isozymes CYP1A1, CYP1A2 and CYP2B10 by three benzodioxole compounds. *Chem.–Biol. Interact.* **96**, 235–247.

Schenkman, J.B., Remmer, H. and Esterbrook, R.W. (1967). Spectral studies of drug interaction with hepatic microsomes. *Mol. Pharmacol.* **3**, 113–123.

Skrinjaric-Spoljar, M., Matthews, H.B. and Casida, J.E. (1971). Response of hepatic microsomal mixed-function oxidases to various types of insecticide chemical synergists administered to mice. *Biochem. Pharmacol.* **20**, 1607–1618.

Smith, C.A.D., Smith, G. and Wolf, C.R. (1994). Genetic polymorphisms in xenobiotic metabolism. *Eur. J. Cancer* **30A**, 1935–1941.

Sun, Y.P. and Johnson, E.R. (1960). Synergistic and antagonistic action of insecticide–synergist combinations and their mode of action, *J. Agric. Food Chem.* **8**, 261–268.

Vodicnik, M.J., Elcombe, C.R. and Lech, J.J. (1981). The effect of various types of inducing agents on hepatic microsomal monooxygenase activity in rainbow trout. *Toxicol. Appl. Pharmacol.* **59**, 364–374.

Wagstaff, D.J. and Short, C.R. (1971). Induction of hepatic microsomal hydroxylating enzymes by technical piperonyl butoxide and some of its analogs. *Toxicol. Appl. Pharmacol.* **19**, 54–61.

4

A Review of the Chemistry of Piperonyl Butoxide

GIOVANNA DI BLASI

1. INTRODUCTION

Piperonyl butoxide (PBO) was patented by Herman Wachs (1949) in the USA, although early production occurred from 1946 onwards. PBO is now manufactured as a speciality chemical in five principal sites using essentially the same process and input raw materials as described by the inventor. Although safrole, which provides over half the weight of the PBO molecule, has been synthesized, no cost-effective process is known which can compete with natural sources, e.g. sassafras oil.

This chapter presents an overview of the principal physical and chemical characteristics of PBO and gives descriptions of several analytical methods each applicable to differing use patterns.

2. NOMENCLATURE OF PBO

The common chemical names, identity numbers and formulae for PBO are:

ISO	PBO
IUPAC	2-(2-butoxyethoxy)ethyl 6-propylpiperonyl ether
CA (9th CIPs)*	5-[[2-(2-butoxyethoxy)ethoxy]methyl]-6-propyl-1,3-benzodioxole
Others	(butylcarbityl) (6-propyl-piperonyl) ether
Commercial names	PBO, PB, Butacide® (AgrEvo), Prentox® (Prentiss Inc.), Butoxide, S1
Empirical formula	$C_{19}H_{30}O_5$
Structural formula	

* CA; Chemical Abstract Name is the uninverted form of the name corresponding to that used by the Chemical Abstract Service (CAS) during the 9th and/or subsequent Collective Index Periods (CIPs).

PIPERONYL BUTOXIDE
ISBN 0-12-286975-3

Relative molecular mass   338.43
CAS number   51-03-6
EEC number   200-076-7

3. MAIN PHYSICAL AND CHEMICAL CHARACTERISTICS

The following values summarize some significant data developed on purified PBO by various laboratories.

Refractive index   1.4985   at 20°C
Density $1.057\,g\,mL^{-1}$ at 25°C
Viscosity   40 mPa s   at 25°C
Boiling point   180°C   at 1 mm Hg
    195°C   at 2 mm Hg
Vapour pressure $< 1 \times 10^{-7}\,mm\,Hg$ at 25°C
Flash point (ASTM D93)   140°C   (Pensky-Martens closed-cup)
Solubility in water $14.3\,g\,L^{-1}$ at 25°C
Solubility in organic solvents   Highly soluble in most organic solvents
Partition coefficient (log P_{OW})   4.75

4. SPECTROSCOPIC CHARACTERIZATION

Infrared (IR), ultraviolet (UV) and nuclear magnetic resonance (NMR) characteristic spectra are shown in Figs 4.1, 4.2 and 4.3.

4.1. Infrared Spectrum

The IR spectrum shown in Fig. 4.1 has been recorded with a Perkin–Elmer FT-IR instrument without any sample preparation, utilizing the potassium bromide technique.

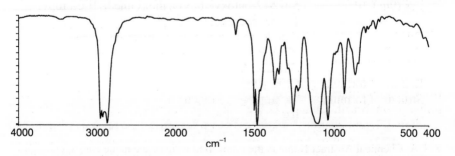

Figure 4.1.   Infrared spectrum of PBO.

4.2. UV Spectrum

The following UV–visible spectrum (Fig. 4.2) has been recorded with a Varian instrument on a hexane solution (0.00206% w/v) and a 1 cm optical path quartz cell. Two maximum absorption bands are observed at 237 nm and 290 nm.

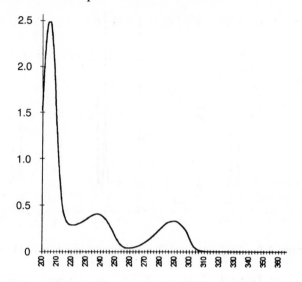

Figure 4.2. Ultraviolet spectrum of PBO.

4.3. NMR Spectrum

The proton NMR spectrum shown in Fig. 4.3 has been recorded with a Varian continuous wave instrument at 60 MHz on a deuterized chloroform solution of PBO.

5. CHARACTERISTICS OF THE COMMERCIAL PRODUCT

5.1. Influence of the Input Raw Materials Used in Manufacture

PBO is an aromatic-aliphatic polyether produced from the condensation of the sodium salt of 2-(2-butoxyethoxy)ethanol and the chloromethyl derivative of hydrogenated safrole.

Safrole is the major component of sassafras essential oil, which is distilled from the wood of a few species of trees growing in different geographical areas, primarily in Brazil, China and Vietnam. The various sources of sassafras oil result in different qualities and quantities of impurities, some of which can lead to the origination of by-products during the manufacturing process. In order to minimize this effect, the sassafras oil is normally distilled before hydrogenation.

Industrial grade 2-(2-butoxyethoxy)ethanol may contain the lower molecular

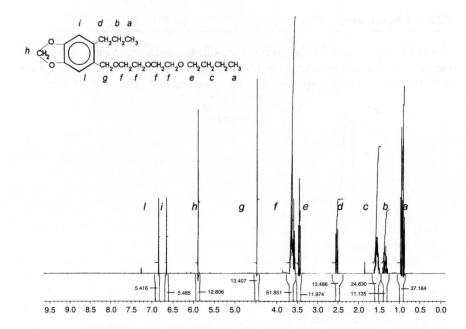

Figure 4.3.   Nuclear magnetic resonance spectrum of PBO.

weight ethoxylated butanol, 2-butoxyethanol, as a typical impurity. This impurity has the same opportunity to react with chloromethyldihydrosafrole as the 2-(2-butoxyethoxy)ethanol, and therefore traces may appear as lower molecular weight harmonics, or homologues, of the PBO products.

5.2. Purity, Specifications and Related Compounds

Wachs (1947) first described PBO by claiming a purity of 80%. The components of the remaining 20% were not identified, even though they were considered to be insecticidally active. Later Miller *et al.* (1952) fractionally distilled a commercial sample of PBO and tested each of the 10 fractions obtained in order to characterize them physically, chemically and biologically.

   The first official technical specification for PBO was published by the World Health Organization (WHO, 1954) and a minimum content of 80% was once again reported. Velenowskj (1963) described a typical technical PBO containing a minimum of 80% of pure compound and stated that '… the remainder consists of related insecticidally active materials'. Brown (1970) published a report in which specifications for a commercial PBO and identification of the allied compounds were reported. Details are given in Table 4.1. He also confirmed that, with the exception of 2-(2-butoxyethoxy)ethanol, all the chemicals present had some synergistic properties with pyrethrins. Albro *et al.* (1972) characterized and identified the minor compounds following their chromatographic separation of technical PBO and subsequent NMR, mass spectrometry (MS) and IR analyses. The Food and Agriculture Organization (FAO) specifications in 1971

Table 4.1.   Typical composition of PBO manufactured in UK in 1970

Amount (%)	Chemical structure	Name
86–89	$R-CH_2OC_2H_4OC_2H_4OC_4H_9$	PBO
1–3	$HOC_2H_4OC_2H_4OC_4H_9$	2-(2-butoxyethoxy)ethanol
1	$R-CH_2OC_2H_4OH$	
3	$R-CH_2OC_2H_4OC_4H_9$	
2	$R-CH_2OC_2H_4OC_2H_4OH$	
2–4	$R-CH_2-R$	Dipiperonyl methane
1–3	$R-CH_2OCH_2-R$	Dipiperonyl ether

$R = H_2C$ ... $CH_2CH_2CH_3$

reported a purity value of 90% with a maximum tolerance of ±2% (from which comes the minimum 88% purity, still requested by some users).

Many improvements in both the raw material's quality and the manufacturing process have been developed since PBO was first produced and the average worldwide production is presently at a minimum assay level of 92%. The chemical and physical characteristics of a commercial product as it is currently manufactured are listed in Table 4.2.

Figure 4.4 shows a recent gas chromatography–mass spectrometry (GC-MS) chromatogram in which the main PBO-related compounds are identified. The corresponding chemical structures are listed in Table 4.3. 2-(2-butoxyethoxy)ethanol, CAS no. 112-34-5, was not revealed in this assay. Figure 4.5 shows a typical high-resolution gas chromatography (HRGC) chromatogram in which the main impurity peaks are also indicated.

5.3. Notes on Other Related Compounds

Early productions of PBO did contain limited amounts of safrole and dihydrosafrole (DHS). Systematic improvements in the distillation techniques have resulted in the elimination of safrole and in the reduction of DHS to a level normally below the detection limit of the HRGC analytical method (about 40 ppm).

Pastor *et al.* (1989) described a high-performance liquid chromatography (HPLC) method for the detection of DHS in PBO which has a 3 ppm detection limit.

Table 4.2.   Typical physical and chemical characteristics of commercial PBO

Characteristic	Value
Refractive index	1.497–1.512 at 20°C
Density	1.05–1.07 g mL^{-1} at 25°C
Purity	92–95%
Odour	Mild and pleasant
Colour	2–5 Gardner scale (ASTM 1544–68)
Acidity number	Less than 0.1 mg KOH g^{-1}
Moisture	Less than 0.1% (K. Fisher method)

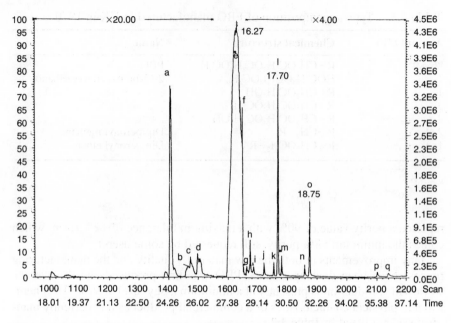

Figure 4.4. GC-MS chromatogram of a PBO composite sample from various manufacturers.

Table 4.3. Identification of the peaks listed in Fig. 4.4

Name of compound	Source	CAS no.	Peak identification
Methyl eugenol derivative	Raw material impurity	34827-25-3	F
Dipiperonyl methane	Reaction by-product	34827-26-4	L
Dipiperonyl ethane	Reaction by-product	37773-74-3	O
Monobutyl glycol-derivative	Raw material impurity	68547-67-1	A
6-Propyl piperonyl alcohol	Reaction by-product	21809-60-9	B
Geometric ring isomers of PBO	Reaction by-products	Unknown	C–D
Disubstituted PBO	Reaction by-product	68850-28-2	J
Disubstituted PBO	Reaction by-product	68850-29-3	I
Higher diglycol ether dimers	Reaction by-products	Unknown	P–Q

5.4. Notes on Acidity

Although PBO is a neutral molecule without any significant ionization constant, a residual acidity may be detected in the technical product. This acidity seems to be caused by some low boiling point compounds that are not completely removed during distillation of PBO and which remain at trace level in the technical product. The acidity number can be determined by titration with an alcoholic KOH solution. This value does not generally exceed 0.07 (expressed as mg of KOH per g of fresh product), but can increase slightly during storage.

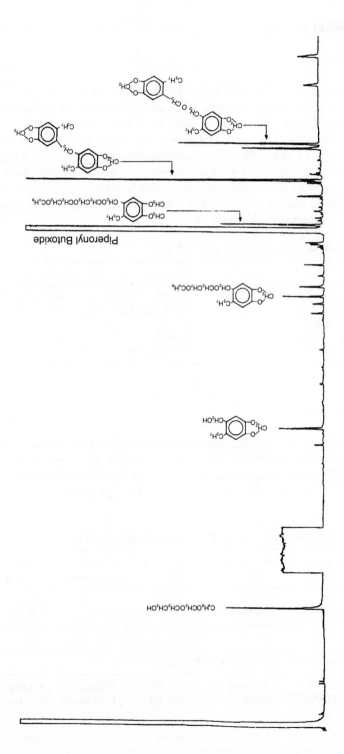

Figure 4.5. Typical HRGC chromatogram of a technical PBO.

6. PBO STABILITY

6.1. Storage Stability

Under normal conditions, PBO can be stored for 2 years at room temperature without any significant degradation and, when the product is kept in a suitable internally lacquered drum, it can remain stable for a much longer period.

A demonstration of this long-term stability was obtained when a drum of PBO was recovered after 18 years. The manufacturing date was 1977. At that time, the minimum purity guaranteed for PBO was 88% by gas–liquid chromatography (GLC). The product was reanalysed with a more sensitive analytical method and the new HRGC analysis showed no significant degradation. The relevant analytical data are summarized in Table 4.4.

Table 4.4.   Analysis of an '18-year-old' PBO

Characteristic	Value
Purity	85.4%
Acidity number	$0.1 \, \text{mg KOH g}^{-1}$
Colour (Gardner scale)	2
Refractive index	1.5012
Density	$1.060 \, \text{mg mL}^{-1}$ at 25°C

6.2. Thermal Stability

The data in Tables 4.5 and 4.6 summarize the results of thermal stability tests which were conducted in Bologna (Italy) on PBO stored in a dark glass and sealed bottle at 40°C for 6 months and at 54°C for 15 days. A slight increase in colour was noted, while the PBO assay did not change significantly.

Table 4.5.   Thermal stability test on PBO after storage at 40°C

Duration (days)	Assay % by HRGC	Refractive index	Relative density	Colour (Gardner)	Acidity number
0	90.90	1.5050	1.060	3–3.5	0.07
180	90.51	1.4990	1.060	5.5–6	0.08

Table 4.6.   Thermal stability test after storage at 54°C

Duration (days)	Assay % by HRGC	Refractive index	Relative density	Colour (Gardner)	Acidity number
0	90.90	1.5050	1.060	3–3.5	0.07
15	90.52	1.4990	1.060	6.5–7	0.20

7. ANALYTICAL METHODS FOR TECHNICAL GRADE PBO

The 'green test' proposed by Labat (1933) was the first qualitative test for PBO analysis, based on the colorimetric method used for compounds containing the methylenedioxyphenyl group. The green colour was developed when the substance was heated with concentrated sulfuric and gallic acids. Unfortunately, this test cannot be used for quantitative determinations because of the charring effect of the sulfuric acid and the consequent rapid fading of the colour.

The method of Jones *et al.* (1952), also known as the 'PBO blue test', was an improvement of the above test and could be used quantitatively for both quality control analyses and research studies. The reaction was specific for PBO and some compounds closely related to it, but the colour was not obtained with many other compounds containing the methylenedioxyphenyl group.

The Association of Official Agricultural Chemists (AOAC) in 1952 and the World Health Organization in 1954 adopted this test as their official method for PBO assay. In order to minimize the effect of the allied compounds on the colour, Blazejewicz (1966) modified the WHO test by introducing a thin-layer chromatographic purification of technical PBO prior to the colorimetric reaction.

Brown (1970) mentioned the 'blue test' and commented that it can be considered a useful indicator of the biological activity of PBO, since the allied compounds, believed to have synergistic properties, contribute to the colorimetric response of the commercial product. In 1971, the Collaborative International Pesticides Analytical Council (CIPAC) also adopted the colorimetric test as its official method for PBO analysis.

Gas-chromatographic analyses were widely adopted soon after the technique was evolved, first employing packed columns and then, with improved separation, by means of capillary columns. A comparative analytical study on the effect of the column's polarity in the chromatographic separation of 3,4-methylenedioxyphenyl derivatives, including PBO, was carried out by Zielinski and Fishbein (1966).

Gas–liquid chromatography is probably the most frequently encountered method in PBO assays and quantitative analyses are normally carried out using an internal standard (IS) method. The substances most commonly used as internal standards are dioctylphthalate, eluting after the PBO peak, or dibutyl-phthalate, eluting before it.

The AOAC–CIPAC method (1983) adopts a 1.22 m × 4 mm i.d. glass column with 5% OV-101 or OV-1 on 80/100 mesh Chromosorb W (HP), using dicyclo-hexyl phthalate as IS. The 1993 edition of the *British Pharmacopoeia* lists a GLC method using a 1.0 m × 4 mm glass column, packed with 3% of phenyl-methyl-silicone fluid (OV-17) on an acid-washed, silanized diatomaceous support (100/120 mesh), with tetraphenylethylene 0.2% as IS. A typical GLC chromatogram is shown in Fig. 4.6.

Whilst GLC is a very suitable analytical method for all routine assays, HRGC, which uses capillary columns, may be preferred when more detailed information is needed. The current method, currently adopted by the Italian manufacturer for the production quality control assay, uses a 25 m long

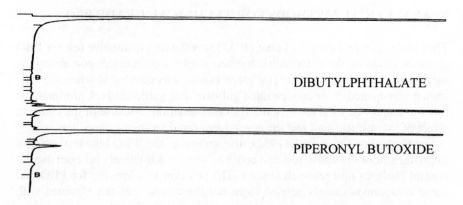

Figure 4.6.   Typical GLC chromatogram of PBO using dibutylphthalate as IS.

capillary fused silica column of 0.32 mm internal diameter. A thin 0.45 µm bonded film of commercial methyl-silicone phase OV-101 is used with the following thermal profile.

Initial isotherm	70°C per 2 minutes
Gradient	5°C per minute
Final isotherm	285°C per 10 minutes
Detector temperature	310°C

A typical chromatogram obtained with the above conditions is shown in Fig. 4.5. Even though the HRGC method requires more than 1 hour, it is still considered a suitable technique for routine analyses, since it allows the determination of PBO purity as well as having a good evaluation of the quantity of major allied compounds.

The HPLC technique, because of its optimal separation characteristics, is widely used in residue and formulation analyses, but can also be successfully applied to quality control operations. This analysis is usually carried out with a C_{18} reversed-phase (RP) column such as Lichrocart RP-18 or equivalent. Suggested mobile-phase and gradient elution conditions for HPLC analysis are given in Table 4.7. An HPLC chromatogram obtained under the conditions mentioned is shown in Figure 4.7.

Table 4.7.   Mobile phase and gradient elution used in HPLC analysis of PBO

	Mobile phase	
Time	Methyl alcohol (%)	Water (%)
0	60	40
25	98	2
30	98	2
[a]10	60	40

[a] Column restoring time.
Readings are normally made with a UV detector set at 240 nm.

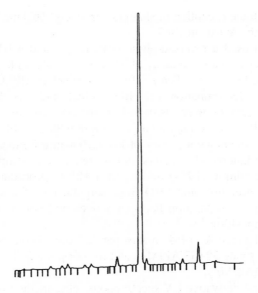

Figure 4.7.   HPLC chromatogram of PBO.

8. ANALYSIS OF FORMULATIONS CONTAINING PBO

The following is a summary of selected published methods for analysing PBO in various formulated products. Suggestions for avoiding the most common interferences by the other components are often listed.

- The Jones colorimetric method (1952) can be used, after necessary modifications, for the PBO assay in formulations containing pyrethrins. It should be emphasized that other pyrethrum synergists do not produce the blue colour.
- In 1960 Stahl suggested the separation of PBO from pyrethrins by means of a silica gel thin-layer chromatography, followed by treatment with antimony trichloride, which produces a violet spot in the presence of PBO (Zweig and Sherma, 1973).
- The AOAC 960.11 method (1961) is based on the Jones colorimetric reaction with a photoelectric detector set at the wavelength range of 625–635 nm.
- A selective liquid chromatography analysis of emulsifiable concentrates and aerosol formulations of pyrethroids, utilizing an IR spectrophotometer as a detector, was described by Papadopoulou-Mourkidou *et al.* (1981). The mobile phase consisted of various combinations of solvents, depending on the pyrethroids under study.
- Meinen and Kassera (1982) published a collaborative study on GLC determination of pyrethrins and PBO in various formulations. The method uses dicyclohexylphthalate as IS and a 4 mm i.d. × 122 cm glass column packed with 5% OV-101 or OV-1, on 80–100 mesh Chromosorb W. It is applicable to most formulations containing more than 0.1% pyrethrins and PBO, including

aerosols, with the exception of shampoo products. This method was adopted officially by the AOAC in 1983.
- Perez (1982) used a reversed-phase (RP) HPLC and a UV detector set at 240 nm to perform a rapid assay of PBO in aerosols containing fenitrothion and bioresmethrin. Later, Perez (1983) described an HPLC method for the simultaneous determination of PBO, Folpet and pyrethrins in aerosol formulations containing emulsifiers. Here the UV detector was set at 254 nm.
- Sura *et al.* (1984) described a GC technique with a highly polar stationary phase for the simultaneous assay of PBO, Pynamin-Forte and resmethrin in aerosol formulations. This method did not require any sample pre-treatment.
- Nehmer and Dimov (1984) developed a HPLC procedure for separating Kadethrin, permethrin and PBO from components of a spray insecticide, obtaining a good separation between actives and coformulants, even in a perfume composition.
- The CIPAC handbook (1985) makes parallel suggestions for PBO assay in formulations (code 32+33/AL/M/-) and for sampling and analysing aerosols (code 32+33/AE/M/-).
- A rapid second-derivative UV spectrophotometric method was developed by Jimena Garciá *et al.* (1992). This method provided satisfactory results in the analysis of a binary mixture of PBO with permethrin, tetramethrin or fenitrothion.

9. ANALYSIS OF RESIDUES

Owing to its widespread use as a synergist for natural pyrethrum, the assay of PBO/pyrethrin mixtures has been studied intensively. The following key publications are mentioned.

9.1. PBO Residue Detection in Grains, Cereals and Derivatives

The AOAC 29.161 method (1965) is applicable for determining PBO colorimetrically in Alaska peas, barley, hulled rice, oats, pinto beans and wheat. The extracted PBO is first treated in order to develop a colorimetric reaction with chromotropic acid. The absorbance is then measured at 575 nm and compared against a blank sample.

A spectrophotofluorimetric method, more sensitive than the colorimetric assay, based on the fluorescence produced by the methylenedioxybenzene group, was proposed for rice, flour and cornmeal by Bowman and Beroza (1967). Secreast and Cail (1971) described a low-residue detection method in flour by means of an extraction with pentane, followed by a chromatography on a column of Florisil with two successive eluting solutions of ethylacetate in pentane.

Cereals can be treated with petroleum ether, after which the residue is analysed by GC (10% SE 30 (a methyl silicone) on Chromosorb W) equipped with a flame ionization detector (FID), as described by Mestres and Susilo (1980). Residues of PBO on grain or cereals can be determined by HPLC of a concentrated and cleaned-up extract. Different techniques utilizing various extraction solvents or mobile phases are suggested by various authors, among which are Argauer (1980) and Molinari *et al.* (1987).

Cave (1981) described the simultaneous detection of bioresmethrin and PBO by a GLC technique with chemical ionization mass spectroscopy on whole-grain wheat. Simonaitis (1983) extracted the residual PBO from bread prepared with cornmeal and wheat flour. The extract was cleaned up and chromatographed on a column of Florex XXS (60–90 mesh).

Several synthetic pyrethroids and PBO can be detected on paddy rice by RP-HPLC with detection at 225 nm, as described by Haddad *et al.* (1989). This method gave good recoveries and linear calibration plots. Detection limits were in the order of 0.05 mg mL^{-1}.

9.2. PBO Residue Detection in Products of Animal Origin

Bruce (1967) developed a method based on a gas chromatograph equipped with a specially constructed electron capture cell. This detector cell was claimed to allow the determination of extremely low quantities (picogram levels) of organophosphorous and pyrethroid insecticides, as well as synergists in food, feeds and water. Moore (1972) analysed PBO residues in different foodstuffs (oils and fats, potato chips, bacon, chicken and eggs, dried codfish) by means of the Bruce gas chromatographic method.

Hill (1979) published some recommendations for the analysis of PBO and pyrethrin residues in milk, fish, fat and meat of several animals using GLC and TLC, while Nijhuis *et al.* (1982, 1984) published an HPLC method for detection in milk. The latter method described the clean-up and concentration treatment on Florisil or Sep-Pak C_{18} and the final determination of PBO with a C_{18} column and methanol/water eluent.

Marti-Mestres *et al.* (1995) described the determination of deltamethrin and PBO residue in poultry fed with cereal treated with the mentioned active agents. The PBO analyses were performed with HPLC and a UV detector after solvent extraction and clean-up.

9.3. PBO Residue Detection in Vegetables and Plant Materials

A method applicable to the analysis of several pesticide residues on plant material was developed by Thier (1972), by previous bromination of the chemicals (among which PBO), followed by gas chromatography with an electron capture detector. This very sensitive technique allows the detection of residues at the nanogram level.

Krause (1980) developed a method for determining methyl carbamate insecticides in fruit and vegetables using the HPLC technique. Krause and August (1983) later applied this method to the detection of other pesticides and synergists, including PBO.

9.4. PBO Detection in the Environment

Debon and Segalen (1989) described how to detect pyrethrins and PBO at trace levels (down to 0.1 mg L^{-1}) in tap water using a solid-phase extraction cartridge to isolate and preconcentrate the compounds. Elution with small quantities of methanol provides a solution which can be analysed directly by RP-HPLC.

10. ANALYTICAL STANDARDS

All accurate assays require an analytical standard, preferably supplied by an independent source. Brown (1970) suggested the preparation of an analytical standard PBO by redistilling technical PBO at 0.1 mm Hg and rejecting the first and final 20% of the distillate. The resulting sample was 96% pure.

A PBO analytical standard was prepared by the National Physical Laboratory, UK, using a single chromatographic separation on a Florisil (60–100 mesh) column. The chromatographic phase was activated (pH 8.5) at 650°C and hydrated with 5% water. The column (12 mm) was packed with hexane and the PBO was loaded with the same solvent (15 mg PBO per g of Florisil max.). Five mixtures of solvents were used to develop the column: n-hexane, benzene/hexane 50 : 50, benzene/hexane 90 : 10, benzene/diethylether 95 : 5, methylene chloride/glacial acetic acid 95 : 5. The material collected with fractions 3–5 resulted as a 99% pure PBO.

Albro *et al.* (1972) suggested that a 99% pure PBO can be obtained from technical material after a single chromatographic separation on Florisil.

11. ACKNOWLEDGEMENTS

My gratitude goes particularly to Denys Glynne Jones, who constantly encouraged me, and to Dr Alberto Zilli, Product and Quality Manager of Ciba Speciality Chemicals, for his valuable assistance in writing this review.

REFERENCES

Albro, P.W., Fishbein, L. and Fawkes, J. (1972). Purification and characterisation of pesticidal synergists piperonyl butoxide. *J. Chromatogr.* **65**, 521–532.

AOAC Official Methods of Analysis (1961). PBO in pesticidal formulations, colorimetric method. Code 960.11. *J. Assoc. Offic. Anal. Chem.* (1990) (XV edn) **43**, 350.

AOAC Official Methods of Analysis (1965). Official final action. Code 29.161. *J. Assoc. Offic. Anal. Chem.* (X edn) **35**.

Argauer, R.J. (1980). Fluorescence and ultraviolet absorbance of pesticides and naturally occurring chemicals in agricultural products after HPLC separation on a bonded-CN polar phase. *Am. Chem. Soc. Symp. Ser.* 1980-Vol. *Pest. Anal. Methodol.* **136**, 103–126.

Blazejewicz, L. (1966). Beitrang zur Kenntnis von technischem Piperonylbutoxid-Dünnschichtchromatographish-colorimetrische Gehaltsbestimmung von 2-Propyl-

3, 4-methylendioxybenzyl-butoxyäthoxyäthyläther in technischem Piperonylbutoxid. *Z. Anal. Chem.* **3**, 327–341.

Bowman, M.C. and Beroza, M. (1967). Spectra and analyses of insecticide synergists and related compounds containing the methylenedioxyphenyl group by spectro-photofluorometry (SPF) and spectrophotophosphorimetry (SPP). *Residue Rev.* **17**, 1–22.

British Pharmacopoeia Veterinary. (1993).

Brown, N.C. (1970). A Review of the Toxicology of Piperonyl Butoxide. R&D Report No. A28/52 from the Wellcome Foundation Ltd., Berkhamsted, Herts, UK.

Bruce, W.N. (1967). Detector cell for measuring picogram quantities of organo-phosphorus insecticides, pyrethrins synergists and other compounds by gas-chromatography. *Agric. Food Chem.* **15**, 178–181.

Cave, S.J. (1981). Simultaneous estimation of Bioresmethrin and PBO by gas-liquid chromatography with chemical ionisation mass spectrometry. *Pestic. Sci.* **12**, 156–160.

CIPAC Handbook (1980). *Analysis of Technical and Formulated Pesticides*, Vol. 1A (addendum to CIPAC 1). CIPAC, Harpenden, Herts, UK.

CIPAC Handbook (1985). *Analysis of Technical and Formulated Pesticides*, Vol. 1C (addendum to CIPAC 1, 1A & 1B). CIPAC, Harpenden, Herts, UK.

Debon, A. and Segalen, J.L. (1989). Trace analysis of pyrethrins and piperonyl butoxide in water by HPLC. *Pyreth. Post* 17, 43–46.

FAO (1971). Specifications for plant protection products, pyrethrum and piperonyl butoxide AGP:CP/39, Draft specification Code 33/1/S/4.

Haddad, P.R., Brayan, J.G., Sharp, G.J., Dilli, S. and Desmarchelier, J.M. (1989). Determination of pyrethroid residue on paddy rice by reversed phase HPLC. *J. Chromatogr.* **461**, 337–346.

Haley, T.J. (1978). Piperonyl butoxide: a review of the literature. *Ecotoxicol. Environ. Safety* 2, 9–31.

Hill, K.R. (1979). Recommended methods for the determination of residues of pyrethrins and piperonyl butoxide. *Pure Appl. Chem.* **51**, 1615–1623.

Jimena Garciá, J.A., Giménez Plaza, J. and Cano Pavón, J.M. (1992). Determination of active components in insecticide formulations by derivative ultraviolet spectro-photometry. *Anal. Chem. Acta.* **268**, 153–157.

Jones, H.A., Ackermann, H.J. and Webster, M.E. (1952). The colorimetric determination of piperonyl butoxide. *J. Assoc. Offic. Anal. Chem.* **35**, 771–780.

Krause, R.T. (1980). Multi residue method for determining *N*-methylcarbamate insecti-cides in crops, using HPLC. *J. Assoc. Offic. Anal. Chem.* **63**, 1114–1124.

Krause, R.T. and August, E.M. (1983). Applicability of a carbamate insecticide multiresidue method for determining additional types of pesticides in fruits and vegetables. *J. Assoc. Offic. Anal. Chem.* **66**, 234–240.

Labat, J.A. (1933). Sur une réaction de la fonction éther méthylenique dans la série aromatique. *Bull. Soc. Chim. Biol.* **14**, 1344–1345.

Marti-Mestres, G.N., Cooper, J.-F.M., Mestres, J.-P., M., De Wilde, G. and Wynn, N.R. (1995). Effects of a supplemented Deltamethrin and Piperonyl Butoxide diet on levels of residues in products of animal origin. 2: Feeding study in poultry. *J. Agric. Food Chem.* **43**, 1039–1043.

Meinen, V.J. and Kassera, D.C. (1982). Gas–liquid chromatographic determination of pyrethrins and piperonyl butoxide: collaborative study. *J. Assoc. Offic. Anal. Chem.* **65**, 249–255.

Mestres, G. and Susilo, H. (1980). Study and determination of PNO residues in the cereal products. *Travaux Soc. Pharmacol. Montpellier* **40**, 1–18.

Miller, A.C., Pellegrini, J.P. Jr, Pozewsky, A. and Tomlinson, J.R. (1952). Synergist action of piperonyl butoxide fractions and observations refuting a pyrethrins–butox-ide complex. *J. Econ. Entomol.* **45**, 94–97.

Molinari, G.P., Battini, M.L., Fontana, E. and Giunchi, P. (1987). Ricerche sul comporta-mento dei residui di Deltametrina e piperonyl butoxide durante la molitura del grano. Atti del 4° Simposio 'La difesa antiparassitaria nelle industrie alimentari e la protezione degli alimenti' Piacenza 23–25 Settembre 1987. *Ed.Camera di Commercio Industria Artigianato e Agricoltura di Piacenza.* 415–427.

Moore, J.B. (1972). Terminal residues of pyrethrin-type insecticides and their synergists in foodstuffs. *Pyreth. Post* **11**, 106–110.

Nehmer, U. and Dimov, N. (1984). HPLC determination of Kadethrin, Permethrin and Piperonyl Butoxide in spray solutions. *J. Chromatogr.* **288**, 227–229.

Nijhuis, H., Heeschen, W. and Hahne, K.H. (1982). Zur Bestimmung von pyrethrum und piperonyl butoxide in milch mit der Hochleistungs-Flussigkeits-Chromatographie (HPLC). *Milchwissenschaft* **37**, 97–100.

Nijhuis, H., Heeschen, W. and Hahne, K.H. (1984). Determination of pyrethrum and piperonyl butoxide in milk by high performance liquid chromatography (HPLC). *Pyreth. Post* **16**, 14–17.

Pastor, J., Pauli, A.M. and Schreiber-Deturmeny, E. (1989). Dosage du dihydrosafrole dans le piperonyl butoxide par HPLC. *J. Chromatogr.* **462**, 435–441.

Papadopoulou-Mourkidou, E., Iwata, Y. and Gunter, F.A. (1981). Utilisation of infrared detector for selective LC analysis. 2: Formulation analysis of some pyrethroid insecticides. Allethrin, Decamethrin, Cypermethrin, Phenothrin and Tetramethrin. *J. Agric. Food Chem.* **29**, 1105–1111.

Perez, R.L. (1982). Determination of Fenitrothion, Bioresmethrin and Piperonyl Butoxide in aerosol concentrates by HPLC. *J. Chromatogr.* **243**, 178–182.

Perez R.L. (1983). Simultaneous determination of Folpet, Piperonyl Butoxide and pyrethrins in aerosol formulations by HPLC. *J. Assoc. Offic. Anal. Chem.* **66**, 789–792.

Secreast, M.F. and Cail, R.S. (1971). A chromatographic colorimetric method for determining low residues of piperonyl butoxide in flour. *J. Agric. Food Chem.* **19**, 192–193.

Simonaitis, R.A. (1983). Recovery of piperonyl butoxide residues from bread made from cornmeal and wheat flour. *Pyreth. Post* **15**, 66–70.

Sura, D.J., Simon, R.H. and Beyerlein, F.H. (1984). Simultaneous determination of Pynamin-Forte, Resmethrin and Piperonyl Butoxide in insecticide formulations by GC. *J. Chromatogr.* **314**, 471–475.

Thier, H.P. (1972). Analysengang zur Ermittlung von Pestizid Rueckstaenden in Pflanzenmaterial. *Deutsche Lebensmittel Rundsch.* **68**, 397–401.

Velenowskj, J.J. (1963). Analytical method for pesticides and plant growth regulators and food additives. In: *Piperonyl Butoxide* (Zweig, G., ed.), vol. II **34**, 393–398. Academic Press, London.

Wachs, H. (1947). Synergistic insecticides. *Science* **105**, no. 2733, 530–531.

Wachs, H. (1949). Methylenedioxyphenyl derivatives and method for the production thereof. US Patent 2 485 681.

WHO (1954) Tentative Specification no. WHO/SAC/3: approved 11 September 1954.

Zielinski, W.L. Jr and Fishbein, L. (1966). Gas chromatography of 3,4-methylendioxyphenyl derivatives. *Anal. Chem.* **38**, 41.

Zielinski, W.L. Jr and Fishbein L. (1967). Gas chromatography of 3,4-methylendioxyphenyl derivatives. *Pyreth. Post* **9**, 6–9.

Zweig, G. and Sherma, J. (1973). Analytical method for pesticides and plant growth regulators. In: Vol. VII, 61–62. Academic Press, San Diego, CA.

5

A Note on Molecular Modelling of Piperonyl Butoxide Using a Computational Chemistry Program for the IBM Personal Computer PC*

DONALD R. MACIVER

1. INTRODUCTION

During the development of data for the US Environmental Protection Agency (EPA), a number of studies of residue fate and animal metabolism have been conducted with technical piperonyl butoxide (PBO). The theoretical pathways in which the residues formed, both as a result of reaction with air and sunlight and also by plant or animal tissue, could be deduced by working back from fragment identification and applying existing structural knowledge. We were interested in examining a computer model (Molecular Modeling Pro™, 1992–95) of the pure substance since this might be used to supplement and reinforce the proposed pathways for the breakdown of this chemical.

2. DATA

The data with molecular structures are presented in the Appendices.

3. DISCUSSION

Most of the fragments encountered during metabolism and degradation studies pointed to a progressive chain shortening of the polyethoxy side chain. The usual pattern of this chain shortening would be the formation of a preoxidative state (epoxide or peroxide), followed by cleavage and formation of a hydroxyl

* IBM ® Personal Computer PC ® are registered names of International Business Machines Inc.

PIPERONYL BUTOXIDE
ISBN 0-12-286975-3

group. Following this, a progressive cascade oxidation of the hydroxy carbon through aldehyde to a final carboxylic acid would follow. Some partial oxidation of the n-propyl side chain could also be detected in certain studies. In no studies were less polar substances such as dihydrosafrole formed.

Ethers with the oxygen adjacent to a methylene group are prone to peroxidation at the adjacent carbon and are then likely to undergo fission, particularly in the presence of light. The computer simulation showed partial charges calculated as a modified Del Re charge (Del Re, 1958) on all the ether oxygens in the range −0.23 to −0.27, which is in line with their electronegative character. A slightly higher partial charge of −0.31 was noted on the first oxygen interruption from the ring (computer carbon numbering '19').

The computer calculated some physical characteristics which require explanation. It is useful to remember that the computer model is of a 'pure' 100% molecule, not technical material, so any calculated figures should differ from those determined practically on technical PBO. Quite surprisingly, the molecule has only a small calculated dipole moment, 2.70, but shows a Griffen's HLB number of over 12. This HLB value is close to the range found in some commercial surfactants.

The computer simulation also coloured the atoms by lipophilicity and hydrophilicity (see Plate 1). It was evident that the two main hydrophilic groups (the alkoxy and the 1,3-dioxole group) were separated from the lipophilic groups by the aromatic ring. The lipophilic group (n-propyl- and terminal n-butyl-) were also separated from each other by the aromatic ring. The relatively small dipole moment is explained by this balance of hydrophilic groups – the twin ethoxy groups separated by a relatively hydrophobic ring and n-propyl group from the other hydrophilic group, the 1,3-dioxole ring (see Fig. 5.1). The HLB number is a summation of the hydrophile versus lipophile groups *irrespective* of their position or geometry. In this case it is not strictly relevant, but nonetheless of certain interest. In other words, the balance of the two dissimilar group types calls for a small dipole moment and a nonemulsifier character (Fig. 5.1). This unique balance explains the special very broad range of solvent properties exhibited by PBO.

The expected chain shortening by oxidation would be expected to yield the most stable specimens of chain shortening, as shown in the theoretical scheme (Fig. 5.2). In aerobic aquatic metabolism studies, the degradative pathway

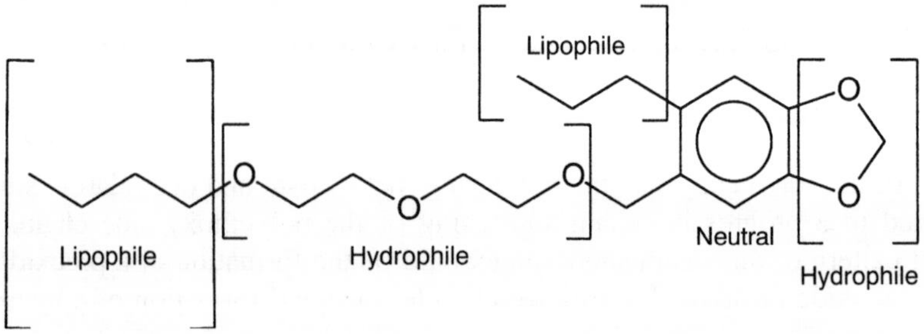

Figure 5.1.   Balance of lipophile and hydrophile groups in PBO.

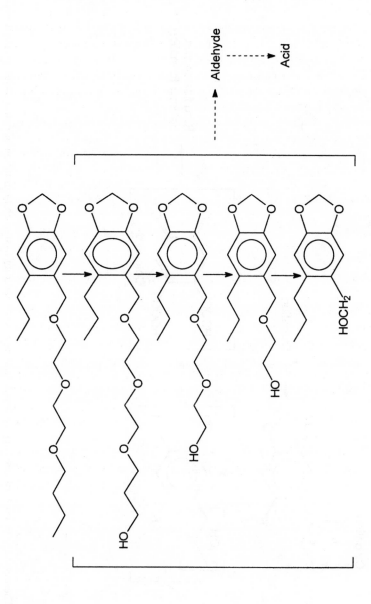

Figure 5.2. Theoretical expected chain shortening yielding alcohols → aldehydes → acids progressively by attack on activated/available chain carbons.

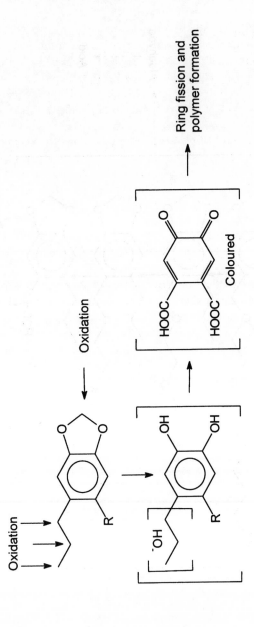

Figure 5.3 Theoretical sites for further oxidative attack leading to eventual ring fission and formation of colours and oxidized polymers. R, progressively oxidized chain.

proposed showed the last alcohol, 3,4-methylenedioxy-6-propylbenzyl alcohol, being oxidized to the corresponding aldehyde and thence to the acid. Indications of fuller progressive chain shortening are evident in residues detected in leaf lettuce; these show all of the above alcohols conjugated with glucose or similar hydrophilic group.

Computer analysis indicates that there is a small electronegativity in the terminal carbon atoms of the butoxy- and n-propyl groups and progressive oxidation of the n-propyl group would also be expected to occur. In addition, there is the opportunity for the methylene group in the 1,3-dioxole group to oxidize and open theoretically to a substituted catechol, which would then form a substituted orthoquinone with very evident brown colour, ring fission and polymer formation (Fig. 5.3).

The relative ease in which PBO is oxidized may account for its known action as a mixed function oxidase inhibitor. In more stable analogues of PBO the alkoxy side chain is replaced with a more oxidatively stable n-alkyl side chain of equal length, which greatly diminishes the synergistic effect. The same is true of other synergists such as Tropital (Maciver, 1966).

4. CONCLUSIONS

The computer-simulated model for PBO indicates an unusual separation of lipophilic and hydrophilic groups. Overall this would indicate that the molecule has a detergent potential; however on closer inspection the groups balance one another and the total dipole moment is small. Indications would favour chain shortening in the alkoxy group following a sequence of alcohol → aldehyde → acid. The eventual environmental fate of PBO is dissolution via ring fission through possible orthoquininoid structures.

REFERENCES

Chemsite™ for Windows™ (1995) The 3d Molecular Graphics Studio.© (1995) Pyramid Learning, LLC, P.O. Box 5817, Stamford, CA 94309, USA.
Del Re, G. (1958). A simple MO-LCAO method for the calculation of charge distributions in saturated organic molecules. *J. Chem. Soc.* 4031–4040.
Del Re, G. (1963). Electronic structure of the α-amino acids of proteins I. Charge distributions and proton chemical shifts. *Biochem. Biophys. Acta* **75**, 153–182.
Maciver, D.R. (1966). The development of Tropital, a polyalkoxy acetal of piperonaldehyde as a potential synergist for pyrethrins. *Pyrethrum Post* **8**, 3–6.
Molecular Modelling Pro™, (1992–95). Molecular Modelling Pro™ Computational Chemistry Programme. Revision 1.44. Window Chem Software™ Inc. Fairfield, California, USA. Manual by James A. Quinn,© (1992–94) NorGwyn Montgomery Software™ Inc. c/o James A Quinn, 216 Lower Valley Riad, North Wales, PA 19454, USA.

APPENDIX A

Calculated Values for Physical Properties of Pure PBO Developed from Computer Modelling

Table A.1   Physical properties of $C_{19}H_{30}O_5$

Molecular weight	338.443	Molecular volume	211.311
Surface area	27.8108	Molecular length	22.8066
Molecular width	9.0314	Molecular depth	10.676
Lop *P*	3.4786	HLB	12.4208
Solubility parameter	21.4589	Dispersion	18.6284
Polarity	3.27228	Hydrogen bonding	10.1371
Percent hydrophilic service	48.2446		
H bond acceptor	0.576298	H bond donor	0.0107583
Dipole moment	2.70011	Connectivity 0	47.0373
Connectivity 1	11.8469	Connectivity 2	9.20196
Connectivity 3	7.28209	Valence index 0	44.9741
Valence index 1	8.99367	Valence index 2	5.85506
Valence index 3	3.85796	Kappa shape index 2	12.3689

Table A.2   Solubility parameters (25°C) for PBO

Log *P*	3.4756
Griffin's HLB	12.4208
Volumetric HLB	10.3522
Solubility parameter	21.4589
Hansen's 3-D solubility parameters (delta/sqr(MPa)):	
Dispersion	18.6284
Polarity	3.27228
Hydrogen bonding	10.1371
Solubility parameter	21.4589
van Krevelen and Hoftyzer's 3-D solubility parameters (delta/sqr(MPa)):	
Dispersion	18.8756
Polarity	3.05129
Hydrogen bonding	7.14894
Molar volume ($cm^3\,mol^{-1}$)	293.5
Solubility parameter	20.4134
Hydration number	5
Hydrophilic surface area	13.4172
% Hydrophilic surface area	48.2446
Surface tension ($\times 10^{-2}\,N\,m^{-1}$)	35.8086
Log water solubility ($mol\,L^{-1}$) of piperonyl butoxide	−4.44517
Water solubility ($mol\,L^{-1}$)	0.0000358781
Water solubility ($g\,L^{-1}$)	0.0121427
Water solubility (ppm)	12.1427
Molecular volume	211.311
Log water solubility ($mol\,L^{-1}$) from log *P*	−3.37302

APPENDIX B

Complete Del Re Modified Partial Charges on all Atoms in PBO (see Numbering Diagram in Fig. B1 of Major Carbon and Oxygen Atoms)

Table B.1   Del Re charge summary for PBO

C 1	8.8839E-02	H 28	3.8877E-02
C 2	−11.6098E-02	C 29	3.5484E-02
C 3	8.9232E-02	H 30	5.2669E-02
O 4	−23.4372E-02	H 31	5.2669E-02
C 5	−6.5930E-02	O 32	−27.0686E-02
H 6	10.5205E-02	H 33	5.1121E-02
C 7	−11.1993E-02	H 34	5.1121E-02
O 8	−23.4291E-02	C 35	3.3741E-02
C 9	15.4555E-02	C 36	3.3699E-02
C 10	−3.4901E-02	H 37	5.0877E-02
C 11	−1.5621E-02	H 38	5.0877E-02
H 12	10.5554E-02	O 39	−27.3410E-02
H 13	6.4023E-02	H 40	5.0836E-02
H 14	6.4023E-02	H 41	5.6836E-02
C 15	3.3510E-02	C 42	2.2413E-02
C 16	−6.5918E-02	C 43	−6.3173E-02
H 17	4.5513E-02	H 44	4.9286E-02
H 18	4.5513E-02	H 45	4.9286E-02
O 19	−31.3539E-02	C 46	−7.2330E-02
H 20	8.1365E-02	H 47	3.9117E-02
H 21	8.1365E-02	H 48	3.9117E-02
C 22	−11.6855E-02	C 49	−1.7621E-02
H 23	3.8770E-02	H 50	3.7906E-02
H 24	3.8770E-02	H 51	3.7906E-02
C 25	4.8706E-02	H 52	3.8733E-02
H 26	3.8877E-02	H 53	3.8733E-02
H 27	3.8877E-02	H 54	2.8733E-02

Total charge −5.551115E-17

Table B.2   Dipole moment

x component	−7.7262E-02
y component	3.1331E-02
z component	−55.6309E-02
Total	9.007552E-30 C m
Hydrogen bond acceptor	57.6299E-02
Hydrogen bond donor	1.0758E-02

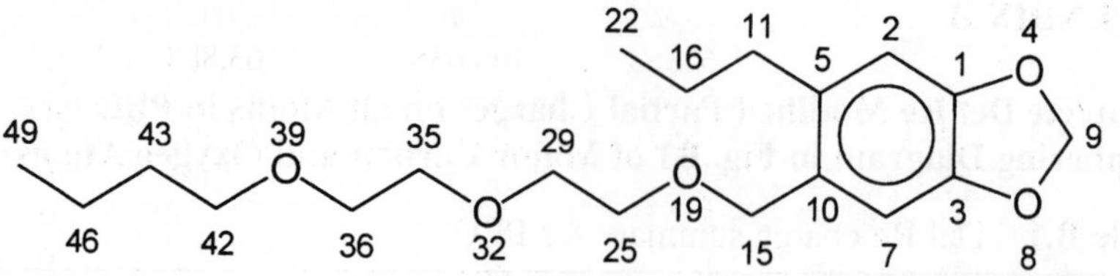

Figure B.1.   Computer numbering diagram for carbon and oxygen atoms in PBO.

APPENDIX C

Table C.1   Bond dimensions and angles in PBO (see numbering diagram in Fig. B.1 of major carbon and oxygen atoms)

PBO atom number	Atomic symbol	Bond length (Å)	Bond angle (°)	Torsional angle (°)	Reference atom
1	C				2
2	C	1.47494			1
3	C	1.38672	120.77		1
4	O	1.45054	131.386	180.25	1
5	C	1.61783	123.321	1.70201	2
6	H	1.11752	118.391	181.366	2
7	C	1.40757	120.319	16.7687	3
8	O	1.51087	112.073	186.154	3
9	C	1.56963	166.169	166.145	4
10	C	1.51969	116.001	347.209	5
11	C	1.77786	123.576	163.979	5
12	H	1.09078	118.566	176.952	7
13	H	1.15873	113.242	254.714	9
14	H	1.1377	111.718	135.554	9
15	C	1.61644	127.498	192.617	10
16	C	1.73281	118.448	176.7	11
17	H	1.69138	108.656	51.8513	11
18	H	1.1543	104.484	295.636	11
19	O	1.50576	111.521	78.9817	15
20	H	1.07648	108.647	316.3	15
21	H	1.12385	108.238	200.334	15
22	C	1.67162	116.321	169.798	16
23	H	1.12894	169.349	54.7478	16
24	H	1.16366	110.003	294.306	16
25	C	1.40698	119.755	211.325	19
26	H	1.1432	108.334	177.991	22
27	H	1.11151	111.697	57.2356	22
28	H	1.13713	168.511	294.882	22
29	C	1.64364	107.307	139.731	25
30	H	1.11365	113.645	17.3084	25
31	H	1.11219	109.375	255.653	25
32	O	1.46695	102.526	171.751	29
33	H	1.12936	111.445	53.2263	29
34	H	1.13565	110.055	289.586	29

35	C	1.42943	119.496	199.047	32
36	C	1.62863	108.638	163.888	35
37	H	1.13636	111.314	42.7022	35
38	H	1.16571	108.841	281.845	35
39	O	1.47333	166.616	176.98	36
40	H	1.14633	110.639	54.8398	36
41	H	1.12483	168.843	294.975	36
42	C	1.42435	115.144	187.749	39
43	C	1.67277	116.18	176.601	42
44	H	1.1288	167.88	57.9324	42
45	H	1.11912	110.742	299.784	42
46	C	1.67802	102.724	177.583	43
47	H	1.13473	110.697	58.2715	43
48	H	1.14013	116.548	295.541	43
49	C	1.5892	167.824	180.678	46
50	H	1.14525	110.588	58.1561	46
51	H	1.12359	109.673	297.837	46
52	H	1.12966	116.464	183.318	49
53	H	1.13393	112.247	66.6891	49
54	H	1.12496	108.29	300.935	49

APPENDIX D

Plates 2 and 3 are computer-generated coloured images showing the major residue partial surface charges on the carbon and oxygen atoms in PBO, and the atomic orbitals for PBO, respectively.

6
Photolytic Degradation of Piperonyl Butoxide

CHRISTOPHER A.J. HARBACH, ROBERT LARGE,
KARL ROPKINS, DENYS GLYNNE JONES and
GIOVANNA DI BLASI

1. RATIONALE FOR STUDY

In the USA, the Environmental Protection Agency (EPA) now requires very detailed studies on the effects of pesticides on the environment. In addition to providing data on the rates of degradation of the particular pesticide in soil, water and when exposed to sunlight, the exact composition of the major degradates has to be determined so that an opinion can be expressed as to their safety to humans when compared with the parent compound. In particular, as safrole is the starting material for the manufacture of piperonyl butoxide (PBO), in which it is hydrogenated to dihydrosafrole, special concern has, inevitably, been expressed as to whether either of these known mild rodent carcinogens is formed during degradation.

Initial environmental studies with PBO labelled with ^{14}C in the butyl carbitol side chain proved unacceptable to the EPA. This work was repeated at the Huntingdon Life Science Centre in the UK with all carbon atoms in the ring labelled.

Freidman and Epstein (1970) exposed films and vials of 70% PBO to fluorescent light and reported photolytic stability. Fishbein and Gaibel (1970) also reported photolytic stability of vials of PBO under a variety of extreme conditions of irradiation. However they reported the formation of a dark brown compound (3% by weight) after exposure in sunlight for 7 days. Following these publications, the generally held view has been that PBO was stable in sunlight.

Maciver (personal communication) reported routine laboratory tests on commercial formulations which clearly indicated that PBO packaged in plastic containers was affected by exposure to sunlight. Further experiments suggested that very thin films of PBO were unstable in strong sunlight.

In order to provide bulk quantities of degradates for analysis and possible isolation, large samples of technical PBO were exposed to natural sunlight, UV light, tungsten lamps and infrared radiation. The resulting samples were analysed

PIPERONYL BUTOXIDE
ISBN 0-12-286975-3

by various mass spectrometric methods and the initial data suggested further experimentation on the degradative processes.

2. DEGRADATION PROCEDURES

2.1. Exposure to Sunlight

2.1.1. Menorcan Aerobic Sample

Twenty Whatman No. 1 filter papers, each 15 cm in diameter, were washed in dichloromethane and dried in sunlight. Each paper was treated with 1.5 g of technical PBO and placed in open glass vessels resting on aluminium foil on a rooftop site in Menorca (latitude 40° N). Exposure was during early June 1994, with the papers receiving 5 to 6 hours of strong sunlight per day up to a total exposure time of 35 hours. When not exposed, the vessels were covered to reduce contamination from dust.

After 4 hours exposure, the papers developed a yellow colour which became deeper and finally showed intense colours (deep yellow to red). A distinct odour change was noticeable after 3 hours of irradiation.

2.1.2. UK Aerobic Sample

One gram of the same PBO batch was pipetted on to each of twelve Whatman No. 1 filter papers prewashed with dichloromethane. The filters were held by steel pins 10 mm above aluminium foil prewashed in dichloromethane. These filter papers were exposed to sunlight in the UK for 47 hours during July 1994. Similar changes were observed to those seen after exposure to Menorcan sunlight.

2.1.3. South African Anaerobic Sample

Samples of PBO were sealed into ordinary glass tubes of 5 mm internal diameter and 150 mm length, with pure nitrogen in the headspace. These tubes were exposed to sunlight near Durban, South Africa (latitude 39° 82′ S), for 20 days. Analysis of a dichloromethane solution of the PBO by gas chromatography–electron ionization mass spectrometry (GC-EIMS) showed no evidence for photolytic degradation.

2.2. Exposure to UV Light

2.2.1. Aerobic Sample

Glass fibre filter circles (75 mm diameter) were extracted with dichloromethane, dried and coated with 2.5–3.0 g of PBO. These filter circles were exposed to the UV light emanating from a Philips HB 311/A lamp with six tubes, each of 40 W, encased in a metal reflector with the tubes 0.5 m above the irradiated PBO. Three exposures were carried out: 2, 5 and 12 days.

The filters were reweighed after 24 hours of exposure, indicating a weight gain of *c.* 15%. A distinct odour was noticed during the irradiation procedure.

2.2.2. Anaerobic Samples

Samples of PBO were sealed into ordinary glass tubes of 5 mm internal diameter and 150 mm length, with pure nitrogen in the head-space. These tubes were exposed to the UV lamp for 15 days. Analysis of a dichloromethane solution of the PBO by GC-EIMS showed no evidence for photolytic degradation.

2.3. Infrared and Tungsten Light (aerobic)

Filter papers coated with PBO were exposed to IR radiation from a 60 W Philips IF bulb at a distance of 100 mm for 48 hours. This experiment was repeated using a 60 W tungsten household bulb.

In neither instance was there any colour change and no odour was detected.

3. ANALYTICAL PROCEDURES

3.1. GC-MS

All of the gas chromatography–mass spectrometry (GC-MS) work, except the headspace analysis, was carried out using a Carlo-Erba 8030 GC interfaced to a VG Analytical AutoSpecE double-focusing mass spectrometer operating at 8 kV. Samples were dissolved in or diluted with dichloromethane (DCM) or other suitable solvent and aliquots (*c.* 1 μL) were injected via a cold on-column injector onto a J&W DB-5MS capillary GC column (30 m, 0.32 mm i.d., 0.25 μm film thickness). During all of the GC-MS experiments, the GC column was held at 40°C for 1 minute and then heated at 8°C min^{-1} to 300°C. The majority of the experiments were carried out in the electron ionization (EI) mode at an electron energy of 70 eV and at low resolution (between 1500 and 2000) with the mass spectrometer scanning from 500 to 30 amu at 1 scan per second. Perfluorokerosene (PFK) was used as the external mass calibrant in all the GC-EIMS analyses.

3.2. Screening for Safrole, Isosafrole and Dihydrosafrole

Authentic samples of safrole, isosafrole and dihydrosafrole were supplied by Endura Spa. These samples and a sample of photolytically degraded PBO were analysed by GC-EIMS. The retention time data and mass spectra from the three authentic compounds were used to screen the PBO data for the presence of safrole, isosafrole and dihydrosafrole.

3.3. GC-CIMS: Confirmation of Molecular Weights

Photolytically degraded PBO was examined by GC-CIMS. The GC conditions were identical to those described in Section 3.1. The VG AutoSpecE mass spectrometer was operated in the chemical ionization (CI) mode at low resolution, using ammonia as the CI reagent gas. Chemical ionization is a lower energy process than EI and, with ammonia as reagent gas, generally gives strong $[M + H]^+$ and/or $[M + NH_4]^+$ ions which confirm molecular weight.

3.4. GC-HRMS: Confirmation of Empirical Formulae

Photolytically degraded PBO was re-examined by GC-EIMS at a higher mass spectrometer resolution (5000) with PFK in the mass spectrometer source as an internal mass calibrant. This procedure permits the accurate mass measurement of molecular and key fragment ions and hence the determination of their elemental compositions and fragmentation pathways. This information, along with the molecular weight information (GC-CIMS) and the low-resolution GC-EIMS spectrum, allows a structure to be proposed for each of the degradation compounds observed.

3.5. Confirmation of Structural Assignments with Authentic Reference Standards

Authentic samples of 3,4-methylenedioxy-6-propylbenzyl alcohol (MDB alcohol) and 3,4-methylenedioxy-6-propylbenzoic acid (MDB acid) were provided by Endura SpA, Italy. An authentic sample of 3,4-methylenedioxy-6-propyl-benzaldehyde (MDB aldehyde) was provided by Biological Test Centre (BTC), California. These three samples were analysed by GC-EIMS and retention data and mass spectra used to investigate the presence of these compounds in photolytically degraded PBO.

3.6. Derivatization/GC-EIMS

A portion of the photolytically degraded PBO was treated with diazomethane to methylate any acid functions present. Half of this methylated sample was treated with MSTFA (*N*-methyl-*N*-trimethylsilyl-trifluoroacetamide) to trimethylsilylate any alcohol groups. The two derivatized samples were examined under the standard GC-EIMS conditions used for the photolytically degraded PBO and the data obtained were compared with the data from the photolytically degraded PBO.

3.7. Thin-Layer Chromatography

A concentrated DCM solution of photolytically degraded PBO was spotted onto a silica thin-layer chromatography (TLC) plate and chromatographed using 20% ethyl acetate in DCM. The most intense coloured band (red/pink; close to the origin) was scraped from the plate and eluted from the silica with methanol.

After concentration, this extract was examined by EIMS using a heated solids probe. The probe was heated from ambient to 500°C at 50°C min^{-1}, whilst the VG AutoSpecE mass spectrometer was scanned from 950 to 25 amu at 5 s per decade in EI mode at low resolution. The sample was also analysed using the standard GC-EIMS conditions. The TLC plate was left on the bench for 2 days, at which point all of the spots had changed colour to match the darkest spot.

3.8. Headspace/TD-GC-MS: Odour Characterization

Small circles of filter paper were soaked in PBO and placed into 20 mL headspace vials. The vials were sealed and exposed to UV light for *c*. 3 hours. The PBO on the filter papers developed a pale yellow colour, indicating that photolytic degradation had occurred. An aliquot (1 mL) of the headspace organic vapours in the vial was extracted with a gas-tight syringe and injected onto a Chrompack thermal desorption (TD) system. The organic vapours in the headspace were cryotrapped (-130°C) before being flash-vaporized onto a thick-film (3 μm) DB-5 equivalent column for subsequent analysis by GC-EIMS.

4. DISCUSSION OF ANALYTICAL RESULTS

4.1. PBO

The GC-EIMS chromatogram from a blended composite PBO sample is shown in Fig. 6.1.

PBO

The dominant peak at *c*. 28 minutes is PBO itself, whilst the other peaks are assumed to be impurities present from the manufacturing process. Many of these other peaks have been identified by the mass spectrometric techniques employed in these studies, but the data are not considered relevant to the current discussion. The EI mass spectrum of PBO (Fig. 6.2) is similar to the published library spectrum. The fragmentation pattern of PBO, and hence many of its degradation products, is unusual. Whilst the EI spectrum shows a clear molecular ion (*m/z* 338), the spectrum is dominated by the fragment ion at *m/z* 176, which has been attributed by previous workers (Williams and Williamson, 1991) to the rearrangement ion:

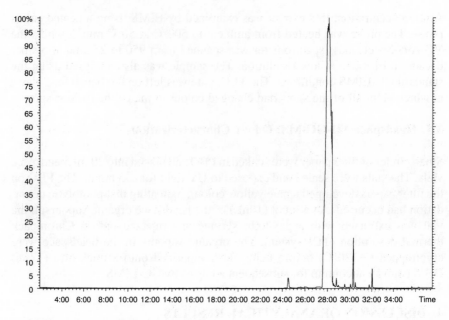

Figure 6.1.   GC-EIMS chromatogram of composite sample of undegraded PBO.

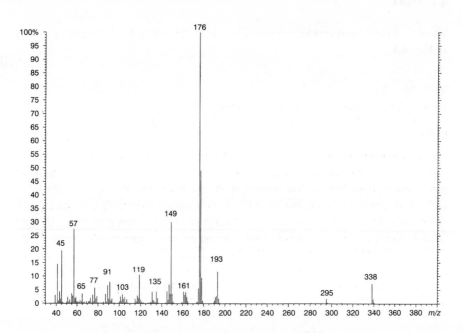

Figure 6.2.   Electron ionization mass spectrum of PBO.

A quinonoid-type structure, however, is also possible:

EI rearrangement ions of this type appear diagnostic of (alkoxy) substituted 3,4-methylenedioxy-benzo compounds. In the case of PBO the ion is formed by the loss of neutral diethyleneglycolmonobutylether from the molecular ion.

4.2. Screening for Safrole, Isosafrole and Dihydrosafrole

The authentic reference samples of safrole and dihydrosafrole showed single peaks by GC-EIMS, whilst isosafrole gave two peaks, assumed to be the *cis*- and *trans*- isomers. The mass spectra of the two isosafrole isomers were similar to that of safrole, but the three retention times were different. GC-EIMS mass chromatography is, therefore, an appropriate screening method for the presence of safrole, isosafrole and dihydrosafrole in photolytically degraded PBO. There was no evidence for the presence of safrole or isosafrole. However, trace quantities (< *c.* 10 ppm) of dihydrosafrole were detected.

4.3. Photolytic Degradation Products

The GC-EIMS chromatogram from PBO photolytically degraded in sunlight in Menorca is shown in Fig. 6.3. A number of relatively major chromatographic peaks are observed which are absent or much smaller in the composite undegraded PBO sample (Fig. 6.1). These peaks have been labelled as A–H on Fig. 6.3. The GC-CIMS chromatogram from the same photolytically degraded PBO sample (Fig. 6.3) is closely similar to the GC-EIMS trace, allowing correlation of the compounds observed. Chemical ionization with ammonia as reagent gas generally gives strong $[M + H]^+$ and/or $[M + NH_4]^+$ ions which confirm the relative molecular mass of the compounds seen.

The GC-EIMS chromatograms from the methylated and methylated/trimethysilylated photolytically degraded PBO are shown in Fig. 6.4. As well as assisting in the GC-EIMS analysis and identification of some of the more polar components (see below), derivatization assisted in the identification of components A–H by highlighting free hydroxyl groups.

The GC-HRMS analysis of the photolytically degraded PBO provided accurate masses and empirical formulae for 23 ions from compounds C–H. The data for these high-resolution measurements are summarized in Table 6.1.

The identities of components A–H and other compounds detected will now be discussed.

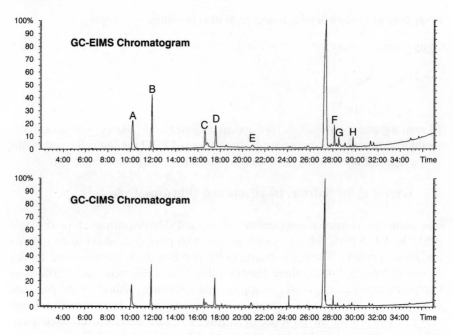

Figure 6.3.   GC-EIMS and GC-CIMS chromatograms of photolytically degraded PBO.

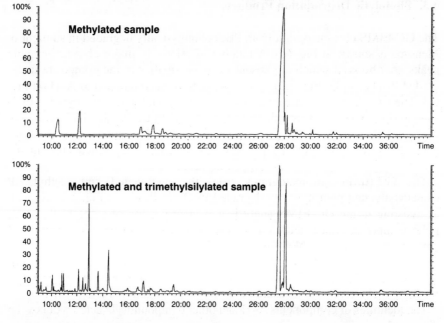

Figure 6.4.   GC-EIMS chromatograms of derivatized photolytically degraded PBO.

Table 6.1 GC-HRMS analysis of photolytically degraded PBO

Component	Measured mass (Da)	Empirical formula	Theoretical mass (Da)	Error (mDa)	Assignment
C	208.0721[a]	$C_{11}H_{12}O_4$	208.0736	1.5	M
	180.0779[a]	$C_{10}H_{12}O_3$	180.0786	0.7	(M–CO)
	151.0398	$C_8H_7O_3$	151.0395	0.3	(M–CO–C_2H_5)
D	192.0790	$C_{11}H_{12}O_3$	192.0786	0.4	M
	191.0715	$C_{11}H_{11}O_3$	191.0708	0.7	(M–H)
	177.0551	$C_{10}H_9O_3$	177.0552	0.1	(M–CH_3)
	163.0395	$C_9H_7O_3$	163.0395	0.0	(M–C_2H_5)
	149.0604	$C_9H_9O_2$	149.0603	0.1	(M–CH_3–CO)
	135.0450	$C_8H_7O_2$	135.0446	0.4	(M–C_2H_5–CO)
E	206.0576[a]	$C_{11}H_{10}O_4$	206.0579	0.3	M
	177.0187	$C_9H_5O_4$	177.0188	0.1	(M–C_2H_5)
F	352.1902[a]	$C_{19}H_{28}O_6$	352.1886	1.6	M
	191.0717	$C_{11}H_{11}O_3$	191.0708	0.9	
	190.0651	$C_{11}H_{10}O_3$	190.0630	2.1	
	179.0341	$C_9H_7O_4$	179.0344	0.3	
	175.0390	$C_{10}H_7O_3$	175.0395	0.5	(190–CH_3)
	135.0449	$C_8H_7O_2$	135.0446	0.3	
G	191.0707	$C_{11}H_{11}O_3$	191.0708	0.1	
	190.0644	$C_{11}H_{10}O_3$	190.0630	1.4	
H	340.1674[a]	$C_{21}H_{24}O_4$	340.1675	0.1	M
	176.0833	$C_{11}H_{12}O_2$	176.0837	0.4	

[a] Voltage scanning; remainder of measurements by magnet scanning.

4.3.1. Component A

Component A, a major degradation product, gives dominant MH^+ and $[M + NH_4]^+$ ions under CIMS conditions (Fig. 6.5). Its EI mass spectrum gave a good library match (96%) as diethyleneglycolmonobutylether (2-(2-butoxy-ethoxy)ethanol; butyl carbitol; relative molecular mass 162). Derivatization/GC-EIMS analysis showed that component A can be trimethylsilylated, underlining the presence of a free alcohol group.

$C_4H_9OCH_2CH_2OCH_2CH_2OH$

4.3.2. Component B

Component B, another major degradation product, gives dominant MH^+ and $[M + NH_4]^+$ ions under CIMS conditions (Fig. 6.6), suggesting a relative molecular mass of 190. It was unchanged by the derivatization process and, hence, contains no free hydroxyl groups. Its EI spectrum suggests that it is a homologue of component A and it has been tentatively assigned as diethyleneglycolbutylethylether:

$C_4H_9OCH_2CH_2OCH_2CH_2OC_2H_5$

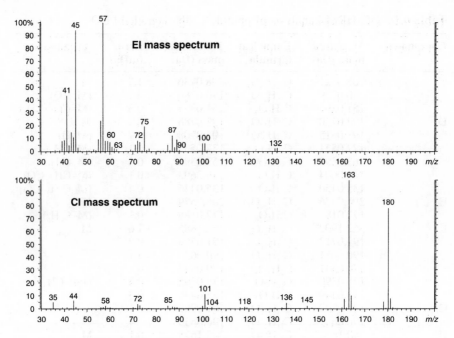

Figure 6.5. EI and CI mass spectra of component A.

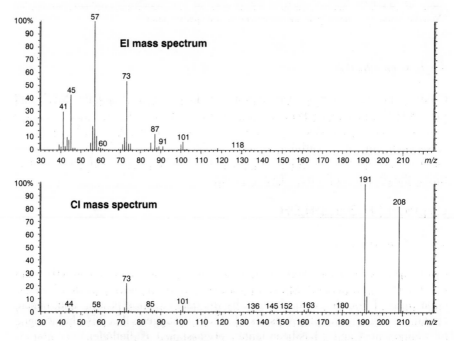

Figure 6.6. EI and CI mass spectra of component B.

4.3.3. Component C

Although component C (Fig. 6.7) gives many fragment ions under CIMS conditions, a strong ion at m/z 226 ($[M + NH_4]^+$) is observed, confirming the relative molecular mass as 208. Accurate mass measurement (Table 6.1) suggests an empirical formula of $C_{11}H_{12}O_4$, with m/z 180 being (M–CO) and m/z 151 (M–CO–C_2H_5). Component C is unchanged by derivatization (Fig. 6.4), indicating no free hydroxyl groups. A *hypothetical* structure for this relatively major degradation product is shown below:

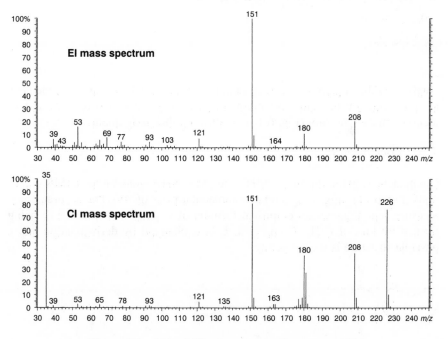

4.3.4. Component D

The CI mass spectrum of major degradation product D (Fig. 6.8) is dominated by the MH$^+$ ion at m/z 193, with a weaker $[M + NH_4]^+$ ion at m/z 210. Accurate mass measurement of the m/z 192 molecular ion (Table 6.1) suggests an empirical formula of $C_{11}H_{12}O_3$, with some of the key fragment ions being due to loss of alkyl groups followed by loss of CO. Component D is unchanged by

Figure 6.7. EI and CI mass spectra of component C.

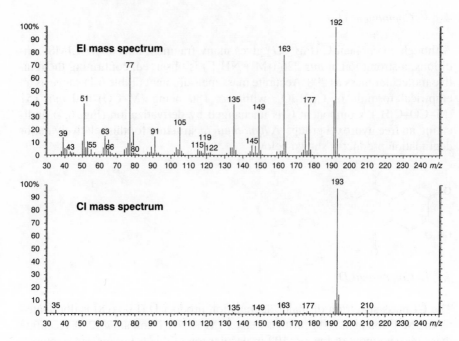

Figure 6.8. EI and CI mass spectra of component D.

derivatization, indicating no free hydroxyl groups, and could be 3,4-methylene-dioxy-6-propylbenzaldehyde (MDB aldehyde):

Analysis of the authentic reference sample of MDB aldehyde gave a single peak at approximately the same retention time as component D (Fig. 6.9) and an identical mass spectrum (Fig. 6.10), confirming this assignment.

4.3.5. Component E

Component E gives dominant MH^+ and $[M + NH_4]^+$ ions under CIMS conditions (Fig. 6.11) suggesting a relative molecular mass of 206. The accurate mass measurement suggests an empirical formula of $C_{11}H_{10}O_4$, with m/z 177 being formed by loss of C_2H_5. Component E is unchanged by derivatization and a possible structure is shown below:

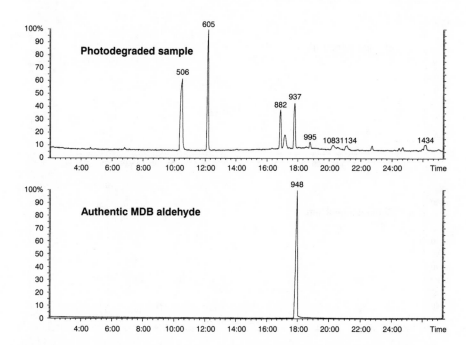

Figure 6.9. GC-EIMS chromatograms of photolytically degraded PBO and authentic MDB aldehyde.

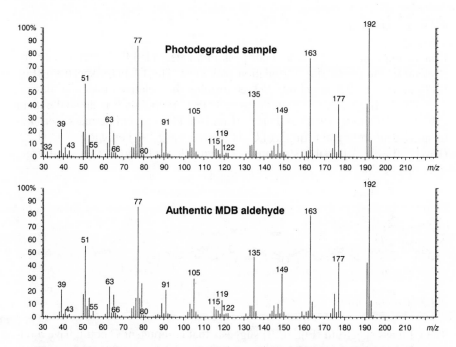

Figure 6.10. EI mass spectra of component D and authentic MDB aldehyde.

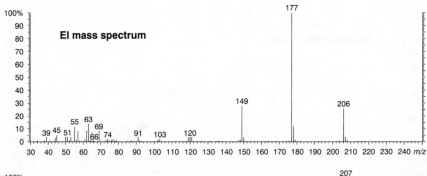

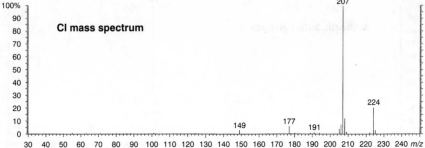

Figure 6.11.   EI and CI mass spectra of component E.

4.3.6. Component F

Component F appears to be a minor peak in nondegraded PBO and may not be relevant to the photolytic degradation pathways. The CI mass spectrum (Fig. 6.12) shows an MH$^+$ ion at *m/z* 353, confirming the relative molecular mass as 352. The EI fragmentation pattern suggests that compound F is an analogue of PBO (*m/z* 190 under EIMS and *m/z* 191 under CIMS). Accurate mass measurement gave an empirical formula of $C_{19}H_{28}O_6$, with *m/z* 190 being loss of $C_8H_{18}O_3$ (butyl carbitol). A possible structure is shown below, although other valid structures can be drawn.

4.3.7. Component G

The CI mass spectrum (Fig. 6.13) suggests that component G could have a relative molecular mass of 350. The accurate mass measurement of the *m/z* 190 fragment ion gave an empirical formula of $C_{11}H_{10}O_3$, which is the same as that

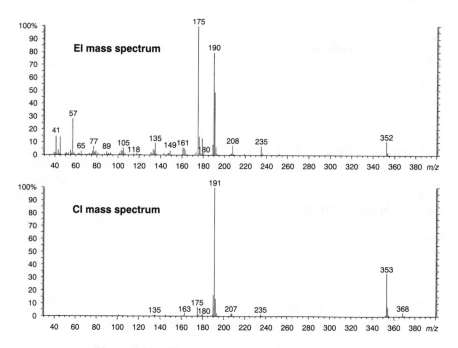

Figure 6.12. EI and CI mass spectra of component F.

for the *m/z* 190 fragment in component F. A possible structure is shown below, where R has a mass of 159:

4.3.8. Component H

Component H appears to be a significant peak in nondegraded PBO and may not be relevant to the photolytic degradation pathways. The mass spectra (Fig. 6.14) suggested a relative molecular mass of 340 for component H and accurate mass measurement gave an empirical formula of $C_{21}H_{24}O_4$. The base peaks in both the EI and CI mass spectra are the same as those seen for PBO itself. These data are consistent with the structure shown below:

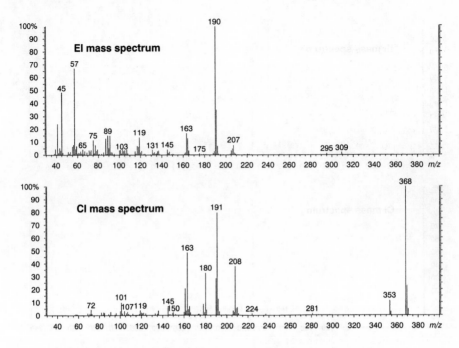

Figure 6.13. EI and CI mass spectra of component G.

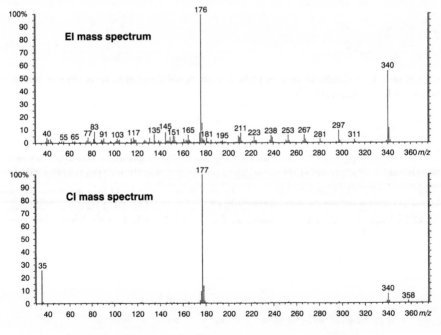

Figure 6.14. EI and CI mass spectra of component H.

4.3.9. MDB Alcohol

Analysis of the authentic reference sample of MDB alcohol gave a single chromatographic peak (Fig. 6.15) with a characteristic spectrum (Fig. 6.16). Inspection of the data from the photolytically degraded PBO showed a minor peak at
similar retention time with an identical spectrum (Fig. 6.16), confirming the
presence of MDB alcohol in the photolytically degraded PBO. Inspection of the
data from the doubly derivatized sample showed a compound of relative molecular mass 266, the spectrum of which is consistent with the TMS (trimethyl-
silyl) ether of MDB alcohol. This component (MDB alcohol) will be referred to
as component I; its structure is shown below:

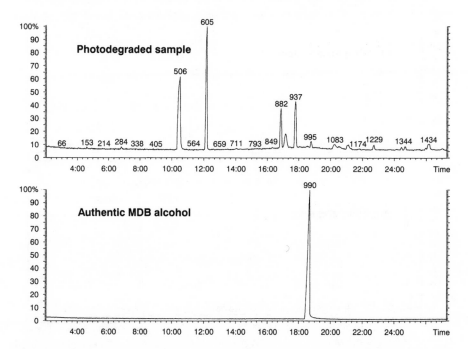

4.3.10. MDB Acid

Analysis of a partially derivatized sample of the authentic reference sample of
MDB acid gave a peak doublet (Fig. 6.17). The second, sharper, peak is the
TMS ester of the acid. The spectrum of the authentic MDB acid is shown in
Fig. 6.18. Inspection of the data from the photolytically degraded PBO showed

Figure 6.15. GC-EIMS chromatograms of photolytically degraded PBO and authentic
MDB alcohol.

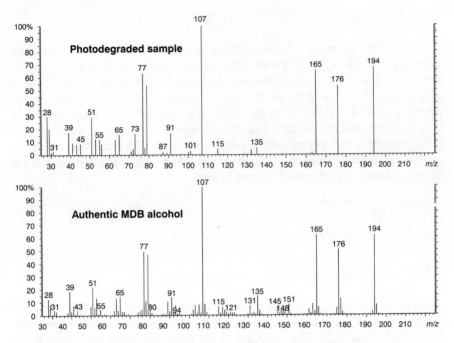

Figure 6.16.   EI mass spectra of component I and authentic MDB alcohol.

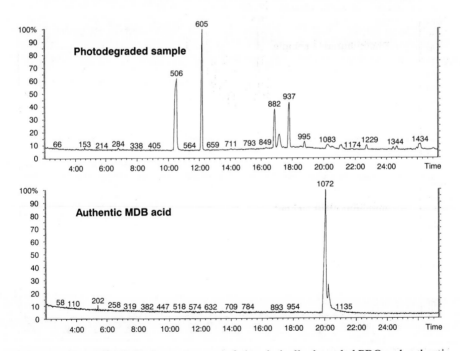

Figure 6.17.   GC-EIMS chromatograms of photolytically degraded PBO and authentic MDB acid.

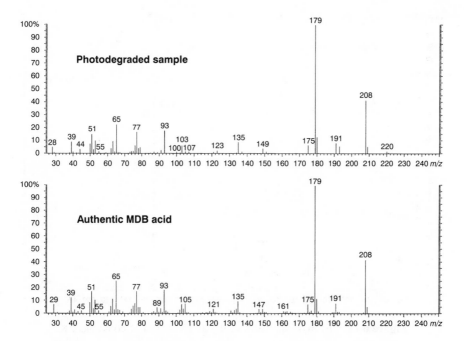

Figure 6.18. EI mass spectra of component J and authentic MDB acid.

a minor peak at similar retention time with an identical spectrum (Fig. 6.18), confirming the presence of MDB acid in the photolytically degraded PBO. Inspection of the data from the methylated sample showed a compound of relative molecular mass 222, the spectrum of which is consistent with the methyl ester of MDB acid. This component (MDB acid) will be referred to as component J; its structure is shown below:

4.3.12. Major New Component in the Doubly Derivatized Sample

The GC-EIMS chromatogram of the doubly derivatized sample (Fig. 6.4) shows a major peak at *c.* 28 minutes, which is absent from the underivatized sample. The EI mass spectrum of this peak (Fig. 6.19) is consistent with a di-TMS ether of a diphenolic analogue of PBO. The underivatized component will be referred to as component K:

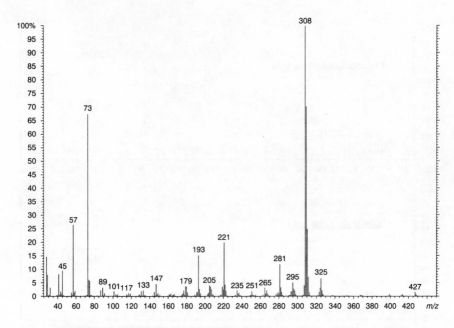

Figure 6.19.   EI mass spectrum of the di-TMS ester of component K.

4.3.13. *TLC Results*

The extract of the pink/red coloured spot from the TLC plate was examined under the standard GC-EIMS conditions and by probe EIMS. The fraction proved to be a very complex mixture of components. However, no compounds were observed which could be responsible for the colour. At least seven components were identified by mass spectrometry, including butyl carbitol, a phenol (component L) and two alcohols (components M and N).

L

M

N

4.3.14. Odour Characterization

The TD-GC-MS chromatograms from the headspace organic vapours above PBO exposed to UV light and from a blank air sample are shown in Fig. 6.20. Two main components were seen in the sample chromatogram which were absent from the blank. They have been identified as n-butanal (*c.* 14½ minutes) and n-butanol (*c.* 18¼ minutes). The butanol presumably arises from cleavage of the side chain of PBO at the oxygen furthest from the ring. The butanal is, presumably, an oxidation product of the butanol. The approximate concentrations and published odour thresholds (Devos *et al.*, 1990) are shown in Table 6.2. Both compounds would therefore be expected to contribute to the odour.

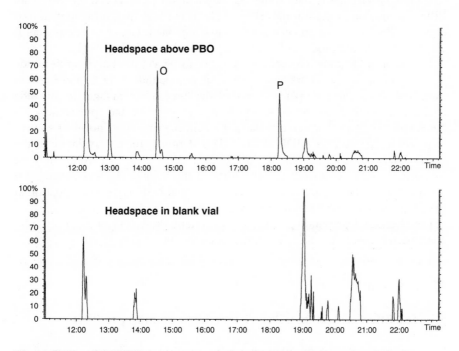

Figure 6.20. GC-EIMS chromatograms of the headspaces above photolytically degraded PBO and a blank.

Table 6.2 Approximate concentrations and published odour thresholds for components O and P (Devos *et al.*, 1990)

Compound	Approximate concentration ($\mu g\ L^{-1}$)	Published odour threshold ($\mu g\ L^{-1}$)	Component
n-Butanal	9.3	0.028	O
n-Butanol	5.0	1.5	P

4.4. Degradation Pathways

It is clear that PBO degradation is facilitated by light, oxygen and moisture. The wavelength of the light is important, since degradation appears faster under UV light than under sunlight and almost nonexistent under IR light. Under anaerobic conditions, with the exclusion of moisture, PBO degradation is negligible. At least three main degradation routes appear possible:

- Degradation of the butyl carbitol side chain
- Oxidation of the propyl group
- Degradation of the methylenedioxy group

The mass spectrometric data show that degradation can proceed directly from PBO via any one of these routes (Fig. 6.21). It is likely that, once started, oxidation of PBO does not proceed solely down any one route, since component E can be an oxidation product of both components D and F (Fig. 6.21).

The primary degradation pathway appears to be via the butyl carbitol side chain, as degradation occurs here to produce compounds A, D, O and P. However, hydrolysis can take place at any of the three ether linkages in this side chain, producing components A and I, N and P or M and, presumably, 2-butoxyethanol. The cleavage closest to the benzene ring does, however, appear to be favoured, with the MDB alcohol (I) undergoing further oxidation to the MDB aldehyde (D) and then the MDB acid (J). The butyl carbitol may be hydrolysed and/or oxidized further, although the butanol (P) and butanal (O) observed in the headspace analysis may arise from hydrolysis and oxidation at the ether linkage furthest from the ring.

The origin of degradation product B is uncertain. Product G has not been fully characterized; its origin therefore remains uncertain.

5. ADDENDUM

5.1. The Effect of Adding a Proprietary Light Stabilizer to PBO

With most use patterns the photoinstability of PBO described in the previous pages is a distinct advantage in reducing the residues remaining in the environment. However, certain uses, such as the insect proofing of timber, require both the insecticide and the synergist to be as persistent as possible in light.

Tinuvin 292, a liquid hindered amine light stabilizer manufactured by CIBA Specialty Chemicals, found to extend significantly the life of various paints, was tested as an additive at 5% w/w.

Whatman No. 1 filter papers (150 mm diameter) were treated with 1 mL of either a standard sample of PBO or the same PBO containing the Tinuvin 292. One set of control and treated papers were subjected to UV light for 20 hours and another set to alternating sunlight and UV. Semiquantitative assays were performed by solvent extraction of the papers and analysis of the extracts by

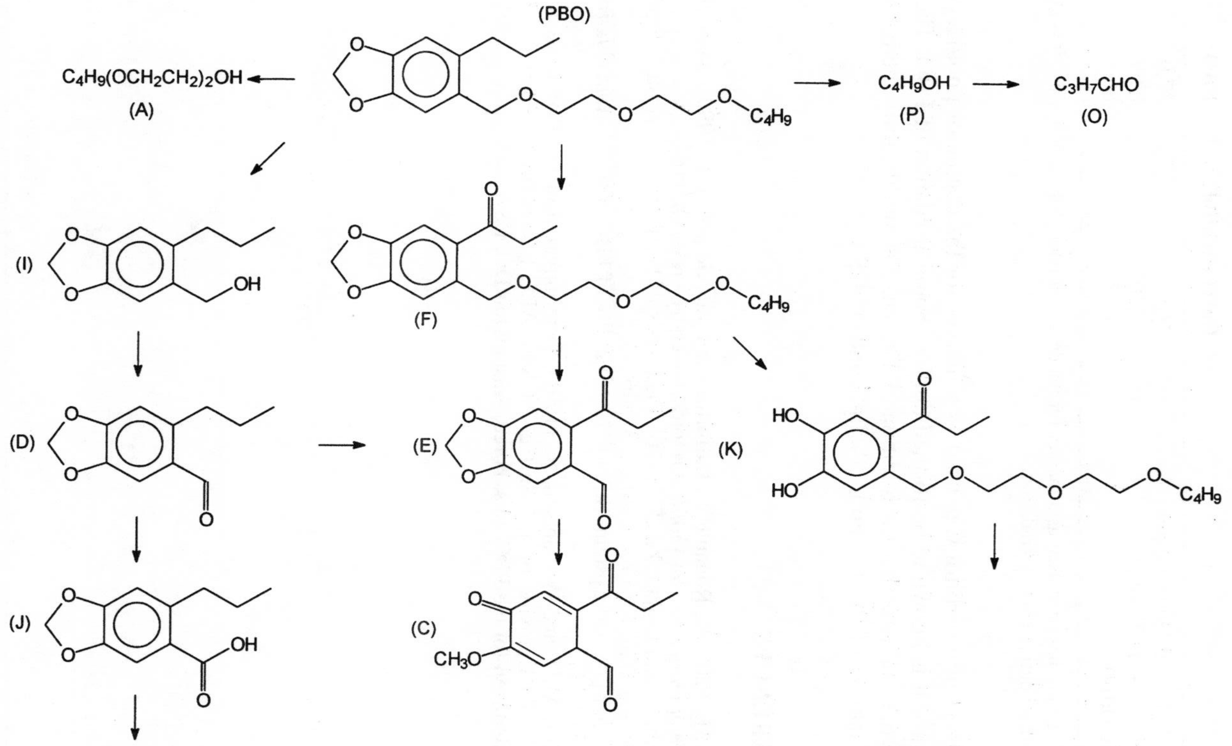

Figure 6.21. Proposed degradation pathways for PBO.

Table 6.3   Reduction in photodegradation of PBO in the presence of a UV stabilizer

Treatment	GC-FID peak area (%)	
	Butyl carbitol[a]	PBO
UV/sunlight with no stabilizer	21.7	37.0
UV/sunlight with Tinuvin	3.20	83.9
None (control)	1.03	86.1

[a] Butyl carbitol has been used as a marker for the degree of photodegradation, since it is known to be formed during photodegradation.

GC-FID. Butyl carbitol was used as a marker for the degree of photodegradation since it is known to be formed during photodegradation of PBO. The data (Table 6.3) indicate that the addition of Tinuvin 292 significantly increased the photostability of PBO in both UV light and sunlight.

REFERENCES

Devos M., Patte, F., Rouault, J., Laffort, P. and Van Gemert, L.J (1990). *Standardised Human Olfactory Thresholds*. Oxford University Press, Oxford.

Fishbein, L. and Gaibel, Z.L.F. (1970). Photolysis of Pesticidal Synergists I. Piperonyl Butoxide. *Bull. Environ. Contam.* **5**, 546–552.

Freidman, M.A. and Epstein, S.S. (1970). Stability of PBO. *Toxicol. Appl. Pharmacol.* **17**, 810–812.

Williams, M.D. and Williamson, K. (1991). Determination of the Photolysis Rate of Piperonyl Butoxide on the Surface of Soil. ABC Laboratories Report no. 38687. Undertaken for the PBO Task Force, Washington DC, USA.

7

The Fate and Behaviour of Piperonyl Butoxide in the Environment

DAVID J. ARNOLD

1. INTRODUCTION

During agricultural and outdoor environmental health applications piperonyl butoxide (PBO) may either directly or indirectly reach soil and water. It may also have a small potential to volatilise.

Whilst the compound has not given rise to any environmental concerns, despite its long-term use, PBO has now undergone a series of environmental studies to meet current regulatory standards. To provide both quantitative and qualitative data all studies were carried out under laboratory conditions using ^{14}C-radio-labelled compound. Additional *ad hoc* evaluations, using nonlabelled PBO, but involving mass spectral analysis of samples have provided qualitative data that assisted in the interpretation of factors influencing the breakdown of PBO.

All 'core' studies used PBO ^{14}C-labelled in the benzyl ring (Fig. 7.1), as this was the most stable moiety. One of the shortcomings of earlier studies has been the lack of authenticated reference standards against which to compare isolated radiolabelled degradates or metabolites. A range of nonradiolabelled reference standards was therefore synthesized for use in chromatography and mass spectral analysis to aid the characterization of potential radiolabelled degradates. These included safrole, isosafrole and dihydrosafrole, because of their reported toxicological significance in rats. However, in none of the environmental studies was any of these compounds detected. They can therefore be excluded as potential environmental contaminants.

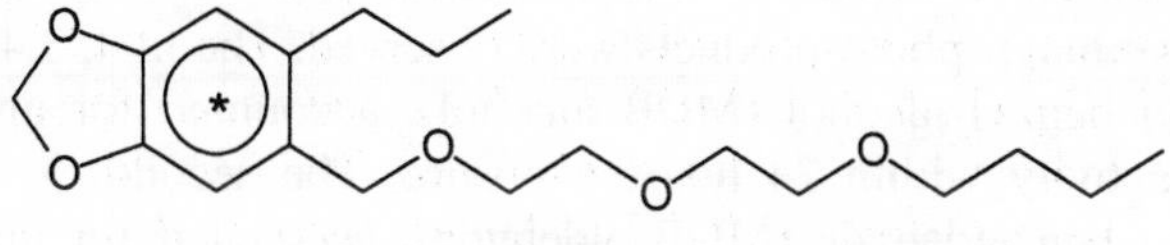

Figure 7.1. ^{14}C ring labelled PBO.

PIPERONYL BUTOXIDE
ISBN 0-12-286975-3

A description of the tests themselves and the results obtained is presented in three parts: the aqueous environment, soil and air. This is followed by a discussion on metabolic pathways. The chapter concludes with a reference to the fate and behaviour of PBO in field situations.

2. ANALYTICAL PROCEDURES

To remove PBO and its degradates from substrates such as sediment and soil, sequential extraction was carried out, generally at ambient temperatures, using solvents of differing polarity. Extraction procedures were designed to remove as much of the radioactive residue as possible without affecting the stability of PBO. Stability checks were employed where appropriate. Procedural recovery checks were also made to ensure that no losses of radioactivity occurred during 'work up' procedures.

Quantitative analysis of radioactivity remaining unextracted from soil was carried out by oxidation of the residue (combustion) and recovery of the evolved $^{14}CO_2$. The majority of analyses of aqueous samples and solvent extracts were carried out by gradient high-performance liquid chromatography (HPLC). Co-chromatography of radio peaks with authentic reference standards was used for initial characterization and quantification.

Because of the similar polarity of certain degradates on HPLC, thin-layer chromatography (TLC) systems were developed to provide improved separation and enable quantitation of components that could not be separated by HPLC. Recovery checks were also made comparing the quantity of radioactivity recovered from the TLC plate prior to and after elution in a solvent system.

Finally, samples were analysed by mass spectrometry to confirm the nature of the degradates isolated.

3. FATE AND BEHAVIOUR IN THE AQUEOUS ENVIRONMENT

PBO is stable in sterile aqueous environments in the absence of light. In a laboratory study, carried out in sterile buffer solution at pH 5, 7 and 9 at 25°C in darkness, no significant degradation of PBO had occurred in a 30-day incubation period. The extrapolated half-life was greater than 1 year (Kirkpatrick, 1995a).

In contrast, PBO was rapidly degraded in sterile aqueous environments exposed to sunlight. The stability of ^{14}C-radiolabelled PBO was investigated in 10 mmol L^{-1} HEPES buffer solution (at pH 7) exposed to natural sunlight for up to 84 hours at 25°C. The parent compound was rapidly degraded with a half-life of 8.4 hours. The distribution of radiolabelled products over time is shown in Fig. 7.2. Two major photo-products were observed. The first, 3,4-methylene-dioxy-6-propyl benzyl alcohol (MDB alcohol), accounted for up to 50% of applied radioactivity within 24 hours exposure. The second, 3,4-methylene-dioxy-6-propyl benzaldehyde (MDB aldehyde), accounted for up to 10% of applied radioactivity in the same time period (Selim, 1995).

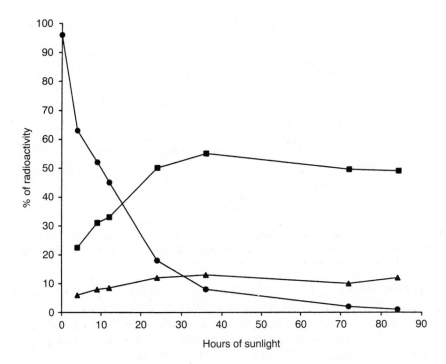

Figure 7.2. Concentrations of PBO MDB alcohol and MDB aldehyde versus time of exposure to sunlight. ━●━, PBO; ━■━, MDB alcohol; ━▲━, MDB aldehyde.

The degradation of [^{14}C]PBO applied to glass fibre filter papers and exposed to UV light has also been examined in a non-GLP study (Harbach and Large, 1995a). Results, whilst qualitative, indicated a similar route of PBO degradation via MDB alcohol, through MDB aldehyde, to the corresponding acid (3,4-methylenedioxy-6-propyl benzoic acid).

In nonsterile aqueous systems, simulating ponds or other static water with a bottom sediment, but in the absence of light, the rate of PBO degradation was influenced by the presence of oxygen.

When applied as a spray, PBO may come into contact with the surface of water bodies either indirectly through spray drift or by direct application, for example in insect control in rice paddies. This situation can be simulated in the laboratory using small-scale sediment and water 'microcosms' (Fig. 7.3). Such systems are useful in estimating the rate of loss of the applied compound from the water phase to the sediment and any subsequent redistribution of parent compound or metabolites.

The degradation of [^{14}C]PBO, at a concentration of 10 mg L^{-1}, was investigated in a water/sediment system (using a sandy loam soil, 2.5 cm deep) incubated under aerobic conditions in darkness for 30 days (Elsom, 1995a). Following application to the surface water, PBO partitioned into both sediment (60%) and water (40%), where it was further degraded. The calculated DT$_{50}$ (loss of 50% of the initial residue) in the system was 213 days. Both MDB

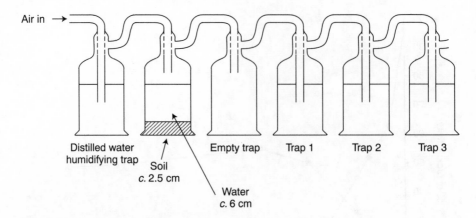

Figure 7.3. Test apparatus for incubation of sediment (soil)/water. Trap 1 contains ethyl digol, trap 2 contains 1 mol L^{-1} potassium hydroxide with phenolphthalein indicator, and trap 3 contains 1 mol L^{-1} potassium hydroxide.

alcohol and MDB acid were formed as minor degradates (5% of applied radioactivity). Negligible mineralization to $^{14}CO_2$ (*c.* 1%) occurred.

In the same experimental system, but one in which anaerobic conditions were established by replacing air with nitrogen gas throughout the incubation period, negligible degradation of PBO took place (Mayo, 1995a).

4. FATE AND BEHAVIOUR IN SOIL

PBO is rapidly degraded in soils under aerobic conditions both in the presence and absence of sunlight. Different moisture levels and treatment methods influence the rate at which PBO is degraded and, as a consequence, the rates of formation and decline of metabolites.

4.1. Soil Photodegradation

The degradation of [^{14}C]PBO was evaluated on thin (2 mm) layers of a sandy loam soil exposed to artificial sunlight (Kirkpatrick, 1995b). The compound was applied to soil plates at a rate equivalent to 10 kg active ingredient (a.i.) ha^{-1}. Soil moisture levels were adjusted to 75% water-holding capacity at $\frac{1}{3}$ bar at intervals during the study period. Soil plates were continuously exposed to a xenon arc artificial sunlight source (Fig. 7.4) for a period of 15 days, equating to 41 days exposure to natural sunlight.

The spectral energy distribution of the lamp source and natural sunlight were measured using a spectrad system spectrometer (Glen Spectra UK). Light intensity measurements were made at each soil plate position at the start and end of the study. The average light intensity for each position was used to calculate the equivalent time of irradiation by natural sunlight at latitude 40° N in summer.

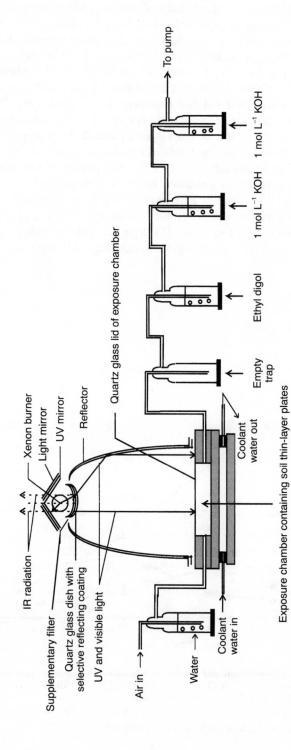

Figure 7.4. Diagrammatic representation of the irradiation apparatus.

A comparative treatment was incubated in darkness. PBO was degraded both in the presence and absence of light with a half-life of approximately 1 day.

A similar degradation pathway was observed to that under aqueous photolysis (Selim, 1995). Principle degradates were MDB alcohol, MDB aldehyde and MDB acid, accounting for 44%, 8% and 10%, respectively, of applied radioactivity. In the dark control, MDB acid accounted for up to 49% of applied radioactivity. In both irradiated and dark treatment, a non-polar minor ^{14}C product was formed (up to 5% applied radioactivity) which was subsequently characterized by mass spectroscopy as bis(3,4-methylene dioxy-6-propyl benzyl) ether. Benzyl ring cleavage was observed through the formation of a significant amount (up to 28% in irradiated soil) of $^{14}CO_2$. The pattern of formation and decline of major degradates in irradiated soil is shown in Fig. 7.5.

In additional non-GLP* studies, PBO was shown to be degraded on filter papers exposed to sunlight under aerobic conditions (Large and Harbach, 1994), whereas PBO exposed to either UV light or sunlight under anaerobic conditions (N_2 gas) underwent negligible breakdown (Harbach and Large, 1995b).

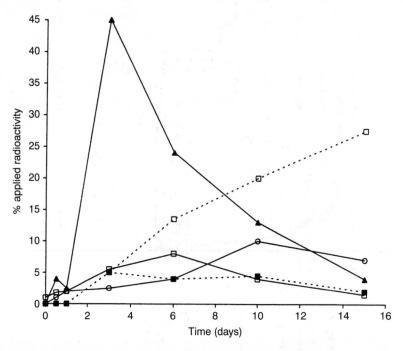

Figure 7.5.   Pattern of formation and decline of the major degradation products of [^{14}C]PBO on irradiated soils. ⋯▫⋯, CO_2; ⋯■⋯, bis(MDB)ether; —○—, MDB acid; —□—, MDB aldehyde; —▲—, MDB alcohol.

* Study not performed to present good laboratory practise.

4.2. Aerobic Soil Degradation

The degradation of $[^{14}C]$PBO was determined in soil incubated under aerobic conditions for up to 8 months (Mayo, 1995b).

The compound was applied to 50 g aliquots of sandy loam soil, held in glass conical flasks, at a concentration of 10 mg kg^{-1}. Treated soils were incubated at 75% water-holding capacity at $\frac{1}{3}$ bar at 25°C in darkness. A stream of CO_2- free air was continuously drawn through the flasks and effluent air was passed through trapping solutions to collect volatile radioactivity, primarily $^{14}CO_2$.

PBO was rapidly degraded with a DT_{50} (loss of 50% of applied residue) of approximately 14 days. More than 90% of the compound had been degraded after 50 days and more than half the applied PBO had been mineralized to $^{14}CO_2$ by the end of the study period (Fig. 7.6).

A large number of degradates (more than 16) were isolated by HPLC from solvent extracts of soil. The majority accounted for less than 1% of applied radioactivity. Those of 3% or more were worked up for further characterization. The principal soil degradate (MDB acid) accounted for up to 17% applied radioactivity after 30 days and was further degraded with time (Fig. 7.6). Other degradates accounting for >3% applied radioactivity were characterized as:

(i) 5[[2-(2-butoxyethoxy)-ethoxy]-hydroxymethyl]6-carboxy-1,3-benzodiox-ole (9% after 30 days).

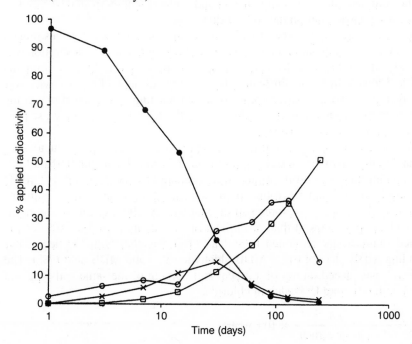

Figure 7.6. Distribution of $[^{14}C]$PBO and major components in aerobic soil. ─•─, PBO; ─×─, MDB acid; ─▫─, $^{14}CO_2$; ─○─, unextracted (bound).

(ii)   3,4-methylenedioxy–6-(prop-1-one)benzaldehyde (6% after 7 days).
(iii) 5-[[2-(2-butoxyethoxy)ethoxymethyl]-1,3-benzodioxol-6-yl]propan-1-one
       (3% after 30 days).

Compounds (ii) and (iii) above were also confirmed as ^{14}C degradates in another study using [^{14}C]PBO applied to filter papers and exposed to natural sunlight (Large and Harbach, 1994). Nonextractable soil-bound residues increased to a maximum of approximately 37% after 128 days but were themselves further degraded, with a corresponding increase in $^{14}CO_2$.

4.3. Adsorption and Mobility

Whilst practical outdoor uses of PBO are unlikely to lead to significant amounts of the chemical reaching soil, its potential for leaching has been assessed in a soil batch adsorption equilibrium study and soil column leaching studies using nonaged and aged PBO residues.

In the adsorption study (Elsom, 1995b), four concentrations of [^{14}C]PBO (0.4, 2.0, 3.0 and 4.0 mg per litre of 0.01 mol L^{-1} calcium chloride) were equilibrated for 24 hours at 25°C in darkness in four soils (sand, sandy loam, clay loam and silt loam) using a soil : solution ratio of 1 : 10. Following the adsorption phase, the soil and solution were separated and two sequential desorption steps, involving the complete removal of supernatant and replacement by fresh 0.01 mol L^{-1} $CaCl_2$, were carried out on the soil residue.

PBO was moderately sorbed to all soils except the sand. K_{OC}* values ranged from 399 to 830, indicating moderate to low leaching potential, but sorption was probably also influenced by clay content. Desorption was highest from the sand and lowest from the silt loam where *c.* 25% of applied radioactivity was desorbed after two desorption steps. Nonaged leaching was carried out in the same four soil types referred to above whilst the aged leaching study used the sandy loam only (Elsom, 1995c).

For nonaged leaching, [^{14}C]PBO was applied at a rate equivalent to 5 kg a.i. ha^{-1} to the top of 30 cm soil columns (Fig. 7.7) and eluted with 0.01 mol L^{-1} calcium chloride. Aged soil residues were prepared by incubation of [^{14}C]PBO in sandy loam soil under aerobic conditions for 18 days, under the conditions previously described for aerobic soil metabolism. Analysis of one of the replicate soil samples showed that, at the end of the ageing period, *c.* 40% of the applied radioactivity remained in soil as PBO, together with ^{14}C degradates including MDB alcohol (6%), MDB aldehyde (3%) and MDB acid (7%). The aged soil was placed on top of a 30 cm soil column of the same soil type and eluted with 0.01 mol L^{-1} calcium chloride.

$$* \ K_{OC} = \frac{Kd}{\% \ \text{organic carbon}} \times 100, \ \text{where}$$

$$Kd = \frac{\text{conc. of pesticide in soil at equilibrium (mg/L)}}{\text{conc. of pesticide in solution at equilibrium (mg/L)}}$$

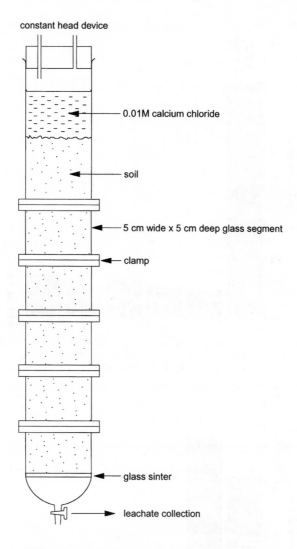

Figure 7.7.   Soil column leaching apparatus.

Nonaged PBO was not readily leached in any of the loam soils (0.2–1.3% of applied ^{14}C), whereas 74% of applied ^{14}C was leached from the sand.

A mean of approximately 9% of applied radioactivity was leached from soil columns treated with aged soils. Most of the radioactivity was retained in the aged soil sample layer applied to the soil column. The leachate radioactivity comprised MDB alcohol (4.7%), aldehyde (1.6%) and acid (7.4%), all of which are more mobile than the parent compound. A comparison of the leaching pattern of nonaged and aged PBO in sandy loam soil is shown in Figure 7.8.

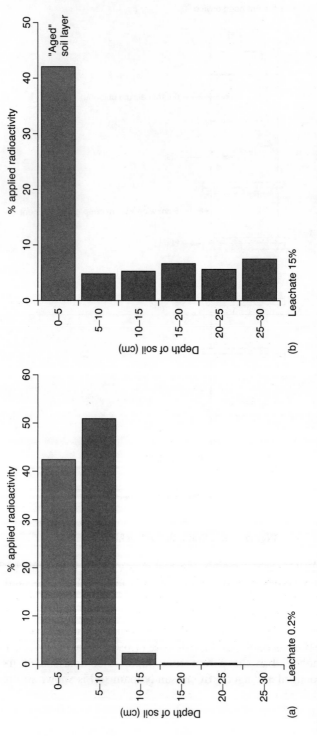

Figure 7.8.   Distribution of radioactivity from (a) nonaged and (b) aged PBO applied to columns of sandy loam soil.

4.4. Losses to Air

It is generally recognized that compounds with vapour pressures above 10^{-5} Pa are potentially volatile. Whilst PBO falls into this category, with a vapour pressure of 10^{-4} Pa, losses to air are influenced by a range of factors, not least formulation type, climate and use pattern.

With reference to insecticidal activity, volatility could be seen as either a positive or a negative property. Losses from surfaces could reduce efficacy whereas vapour activity could be beneficial. In terms of environmental fate, loss to air is not considered to be a major issue other than for highly volatile and/or stable molecules. Nevertheless, the volatility of $[^{14}C]$PBO from plant leaf surfaces was assessed in a laboratory system. PBO as an 80% emulsifiable concentrate was applied to french bean leaves, 0.25 µg a.i. cm^{-2}, which equated to a field rate of 250 g a.i. ha^{-1}, and moist air passed over the surface at a rate of 1–2 m s^{-1} for 24 hours. Volatile ^{14}C was trapped and analysed. Approximately 9% of applied radioactivity, all as $[^{14}C]$PBO, was volatilized in 24 hours which, under the test conditions, was comparatively low.

4.5. Degradation Pathways

All environmental studies show the formation of the same key products of PBO degradation. It is clear that PBO degradation is facilitated by sunlight, moisture and soil microbial activity. Under strictly anaerobic conditions, with the total exclusion of oxygen, PBO degradation is negligible. Under these conditions, such as are found in deep soil and sediment layers, PBO would be only slowly degraded.

The principal route of PBO degradation proceeds via an initial oxidation and hydrolysis at the position of the ethoxy ethoxybutoxy side chain leading to the formation of MDB alcohol, also referred to as piperonyl alcohol. This undergoes further oxidation, through MDB aldehyde, to MDB acid. This acid, also referred to as piperonylic acid (Kamienski and Casida, 1970), was detected as a metabolite in the urine of mice, either as the free acid or conjugated. It has also been confirmed as a metabolite in the rat (Lin and Selim, 1991; Selim, 1991), where it accounted for c. 3% of the applied dose in both urine and faeces.

At least two other routes of degradation in soil are apparent from the nature of the degradates formed. An initial propyl side chain oxidation of PBO leads to the formation of prop-1-one PBO. This then either undergoes carboxylic acid substitution at the 6-propyl position and hydroxylation to give 5[2-(2-butoxyethoxy)-ethoxy-hydroxymethyl]-6-carboxy-1,3-benzodioxole, or loss of the butoxy ethoxy ethoxy methyl side chain to form the corresponding prop-1-one benzaldehyde (Fig. 7.9).

Whilst the dimer, bis(3,4-methylenedioxy-6-propyl benzene)ether, was observed in soil photodegradation studies, it occurred in both irradiated soil and the dark control. The product was not seen in other soil studies, so if it were formed it must have been very rapidly degraded to undetectable levels.

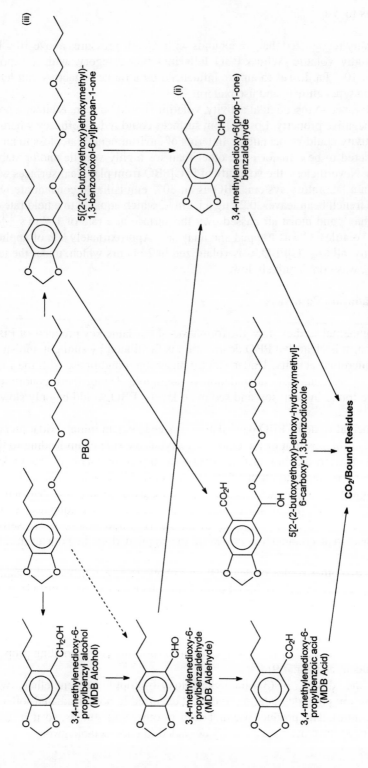

Figure 7.9. Proposed routes of PBO degradation in soil and water.

All degradation products are transient and ultimately undergo ring cleavage to form bound residues and carbon dioxide.

4.6. Relating Laboratory Data to the Field

Two series of field experiments were carried out to examine the behaviour of PBO in soil and water. The data from these field studies confirm the rapid dissipation of PBO in the environment.

The soil dissipation rate of PBO was measured at three sites in the USA following spray treatment of Pyrenone (pyrethrum (0.46 lb acre^{-1}) + PBO (4.6 lb acre^{-1})) (Hatterman, 1994a). The half-life of PBO in the 0–6 inches 'soil horizon' was a maximum of 4.3 days and residues reached nondetectable levels after a maximum of 30 days. No residues above the limit of determination were detected at lower soil depths up to 36 inches (90 cm).

Field aquatic dissipation studies at three locations in the USA showed that following Pyrenone application to water (pyrethrum 0.46 lb acre^{-1} + PBO 4.6 lb acre^{-1}) PBO was rapidly dissipated from the water phase with half-lives of 0.55–1.64 days, depending upon the clay/silt content of the sediment. In sediments the maximum half-life was 24 days and negligible residues were detected after 120 days (Hatterman, 1994b).

The behaviour of PBO in the environment therefore shows that residues of the parent compound are extremely short-lived. As a consequence rapid dissipation of any major degradate would also have occurred. Because levels of PBO declined in soil to below the limit of determination so rapidly, the potential for residues to leach or to run off into surface water is negligible. Similarly in aquatic systems residues will rapidly decline below levels of concern.

5. DISCUSSION

PBO has a long history of use in a wide variety of pesticide treatments, involving both indoor and outdoor applications, for environmental health and agricultural purposes. Whilst there has been no suggestion that PBO has ever led to any environmental concern, it is only right that the synergist should be subjected to scrutiny in terms of its fate, behaviour and environmental impact, in the form of regulatory tests. Hitherto, the limited amount of physicochemical data available has suggested that PBO was stable to both hydrolysis and UV radiation and is lipophilic. The more recent studies referred to in this chapter have shown that the stability of PBO is highly influenced by a combination of environmental parameters. The compound is readily degraded when exposed to sunlight and UV light and in the presence of oxygen. Environmental studies have shown that the residues of PBO are dissipated to levels that have no environmental impact.

In those situations requiring residual activity, persistence of the synergist and the insecticide is desirable. For wood treatments, particularly below soil level where oxygen and daylight are excluded, there is likely to be little breakdown of PBO and effective synergistic potential would persist (expressed as prolonged activity of the insecticide).

PBO is also effective as a means of insect control in grain stores over long periods (Desmarchelier *et al.*, 1979; Bengston *et al.*, 1983). This stability is the result of the absence of daylight and low metabolism in dry grain. Whilst there is no direct evidence, volatility studies indicate that the effectiveness of pest control may also be due to PBO activity in the vapour phase as a consequence of a limited (but sufficient) degree of volatilization from treated seed surfaces.

PBO therefore provides a double benefit of prolonged and effective synergism of insecticidal activity where stability is needed and rapid dissipation in the environment where the compound undergoes decomposition to inactive moieties, and ultimately mineralization to carbon dioxide.

REFERENCES

Bengston, M., Davies R.A.H., Desmarchelier, J.M., Henning, R., Murray, W., Simpson, B.W., Snelson, J.T., Sticka, R. and Wallbank, B.E. (1983). Organophosphorothioates and synergised synthetic pyrethroids as grain protectants on bulk wheat. *Pesti. Sci.* **14**, 373–384.

Desmarchelier, J.M., Bengston, M. and Sticka, R. (1979). Stability and efficacy of pyrethrins on grain in storage. *Pyreth. Post* **15**, 3–8.

Elsom, L.F. (1995a). Piperonyl Butoxide Aerobic Aquatic Metabolism. Unpublished report no. PBT9/951114. Huntingdon Life Sciences, PO Box 2, Huntingdon, Cambridge PE18 6ES. Undertaken for the PBO Task Force, Washington DC, USA.

Elsom, L.F. (1995b). Piperonyl Butoxide Adsorption/Desorption on Soil. Unpublished report no. PBT10A/950775. Huntingdon Life Sciences, PO Box 2, Huntingdon, Cambridge PE18 6ES. Undertaken for the PBO Task Force, Washington DC, USA.

Elsom, L.F. (1995c). Piperonyl Butoxide: Soil Column Leaching of Non-aged and Aged Residues of ^{14}C-piperonyl butoxide. Unpublished report no. PBT10B/950899. Huntingdon Life Sciences, PO Box 2, Huntingdon, Cambridge PE18 6ES. Undertaken for the PBO Task Force, Washington DC, USA.

Harbach, C.A.J. and Large, R. (1995a). GC/EIMS Analysis no. 9506/7804. M-Scan Limited, Sunninghill, Berkshire, UK. Undertaken for the PBO Task Force, Washington DC, USA.

Harbach, C.A.J. and Large, R. (1995b). GC/EIMS Analysis of Piperonyl Butoxide Samples Exposed to Light under Nitrogen. Certificate of analysis 9501/7550. M-Scan Limited, Sunninghill, Berkshire, UK. Undertaken for the PBO Task Force, Washington DC, USA.

Hatterman, D.R. (1994a). Piperonyl Butoxide Analytical Phase for Pyrenone Crop Spray (Pyrethrin + Piperonyl Butoxide) Field Dissipation – Terrestrial Trial Applied to Bare Ground in California, Georgia and Michigan. Landis International Inc., PO Box 5126, 3025 Madison Highway, Valdosta, GA 31603-5126, USA.

Hatterman, D.R. (1994b). Piperonyl Butoxide Analytical Phase for Pyrenone Crop Spray (Pyrethrin + Piperonyl Butoxide) Aquatic Dissipation Study Applied Post Emergence to Bare Water in California, Arkansas and Mississippi. Landis International Inc., PO Box 5126, 3025 Madison Highway, Valdosta, GA 31603-5126, USA.

Kamienski, F.X. and Casida, J.E. (1970). Importance of demethylenation in the metabolism *in vivo* and *in vitro* of methylenedioxyphenyl synergists and related compounds in mammals. *Biochem. Pharmacol.* **19**, 91–112.

Kirkpatrick, D. (1995a). Piperonyl Butoxide Hydrolysis as a Function of pH at 25°C. Unpublished report no. PBT4/943285. Huntingdon Life Sciences, PO Box 2, Huntingdon, Cambridge PE18 6ES. Undertaken for the PBO Task Force, Washington DC, USA.

Kirkpatrick, D. (1995b). Piperonyl Butoxide Photodegradation on Soil. Unpublished report no. PBT6/950382. Huntingdon Life Sciences, PO Box 2, Huntingdon, Cambridge PE18 6ES. Undertaken for the PBO Task Force, Washington DC, USA.

Large, R. and Harbach, C.A.J. (1994). Photolytic Degradation of Piperonyl Butoxide: GC/HRMS Analysis of Menorca Sunlight Filter. Certificate of analysis 9409/7228. M-Scan Limited, Sunninghill, Berkshire, UK. Undertaken for the PBO Task Force, Washington DC, USA.

Lin, P. and Selim S. (1991). Absorption, Distribution, Metabolism and Excretion (ADME) Studies of Piperonyl Butoxide in the Rat. Unpublished report no. A490-0530. Biological Test Centre, PO Box 19791, 2525 McGaw Avenue, Irvine, CA 92713-9791, USA. Undertaken for the PBO Task Force, Washington DC, USA.

Mayo, B.C. (1995a). ^{14}C Piperonyl Butoxide Anaerobic Aquatic Metabolism. Unpublished report no. PBT8/951877. Huntingdon Life Sciences, PO Box 2, Huntingdon, Cambridge PE18 6ES. Undertaken for the PBO Task Force, Washington DC, USA.

Mayo, B.C. (1995b). Piperonyl Butoxide Aerobic Soil Metabolism. Unpublished report no. PBT7/951484. Huntingdon Life Sciences, PO Box 2, Huntingdon, Cambridge PE18 6ES. Undertaken for the PBO Task Force, Washington DC, USA.

Selim, S. (1991). Addendum to Report Entitled Absorption, Distribution, Metabolism and Excretion (ADME) Studies of Piperonyl Butoxide in the Rat. Unpublished report no. A491-0214. Biological Test Centre, PO Box 19791, 2525 McGaw Avenue, Irvine, CA 92713-9791, USA. Undertaken for the PBO Task Force, Washington DC, USA.

Selim, S. (1995). Isolation and Identification of Major Degradates of Piperonyl Butoxide (PBO) Following Aqueous Photolysis. Unpublished report no. 95-0306.btc. Biological Test Centre, 2525 McGaw Avenue, PO Box 19791, Irvine, CA 92713-9791, USA. Undertaken for the PBO Task Force, Washington DC, USA.

8

An Ecological Risk Assessment of Piperonyl Butoxide

THOMAS G. OSIMITZ and JAMES F. HOBSON

1. INTRODUCTION

The purpose of this chapter is to assess the likelihood that use of piperonyl butoxide (PBO) according to the label directions could potentially result in unacceptable risks to the environment. The assessment of ecological risk requires the characterization of the inherent toxicity (i.e. hazard or effects assessment) of PBO as well as the estimate of the potential exposure (i.e. exposure assessment) that may occur to organisms in the environment. The final step in the assessment of risk is the risk characterization which integrates the results of the effects and exposure assessments. Risk characterization is often expressed as the ratio of effects over exposure. This chapter will present these steps in an assessment of risk to aquatic and terrestrial ecosystems, based on the methodologies used for Ecological Risk Assessment of pesticides by the United States Environmental Protection Agency (USEPA).

2. EFFECTS ASSESSMENT

2.1. Aquatic Effects

The acute toxicity of PBO has been evaluated in several species of freshwater and estuarine/marine fish and invertebrates. The acute LC_{50} (or EC_{50}) values and the no observed effect concentrations (NOECs) from these studies are presented in Table 8.1 (the LC_{50} is the concentration which is lethal to or affects 50% of the tested group in a set time period, e.g. 48 or 96 hours). According to the EPA's comparative toxicity categories for acute aquatic toxicity studies (US Environmental Protection Agency, 1985), PBO is 'moderately' toxic to fish species, with reported LC_{50} values ranging from 3.94 to 6.12 mg L^{-1} (Table 8.1). Based on the data in Table 8.1, PBO is 'highly' acutely toxic to invertebrate species. Invertebrates are consistently more sensitive than fish species, with $L(E)C_{50}$ values

Table 8.1.   Acute aquatic toxicity of PBO

	LC/EC$_{50}$ (mg L^{-1})	NOEC (mg L^{-1})	Toxicity category[a]
Freshwater			
Rainbow trout (Holmes and Smith, 1992d)	6.12	1.19	Moderate
Bluegill (Holmes and Smith, 1992c)	5.37	1.46	Moderate
Daphnia magna (Holmes and Smith, 1992b)	0.51	0.15	High
Estuarine/marine			
Sheepshead minnow (Holmes and Smith, 1992e)	3.94	0.92	Moderate
Mysid shrimp (Holmes and Smith, 1992f)	0.49	0.085	High
Eastern oyster (Holmes and Smith, 1992g)	0.23	–	High

[a] According to EPA's comparative toxicity categories for acute aquatic toxicity studies (US Environmental Protection Agency, 1985).

ranging from 0.23 to 0.51 mg L^{-1}. Two acute studies were run on Mysid shrimp; the results are reported in Table 8.1. Finally, there are no marked differences in the sensitivity of freshwater and estuarine/marine species, although the latter are slightly more sensitive for the species evaluated.

Chronic toxicity of PBO has been evaluated in two *Daphnia magna* 21-day life cycle studies (Holmes and Smith, 1992a; Putt, 1994) and a fathead minnow early life stage study (Machado, 1994). The resulting LOECs (lowest observed effect concentration), NOECs and MATCs (maximum acceptable toxicant concentration) are presented in Table 8.2. (The MATC is the geometric mean of the LOEC and NOEC and is a commonly used expression of chronic aquatic toxicity.) The most sensitive indicators of effect observed in the *D. magna* and fathead minnow studies were reproduction, and hatching and larval growth, respectively.

Two consistent patterns can be observed in the aquatic toxicity data presented in Tables 8.1 and 8.2. First, fish are approximately an order of magnitude less sensitive than invertebrates, for both acute and chronic toxicity. Second, chronic toxicity values are approximately three to five fold lower than acute data for the same species (*D. magna*) or phylogenetic group (fish). The ACR (acute to chronic ratio) of 3–5 for PBO is also consistent for different phylogenetic groups, i.e. fish versus invertebrates. The ACR toxicity value reflects the underlying toxicology of the compound and an ACR of 10 (one order of magnitude) is typical of most organic compounds (Rand, 1995). This ratio indicates that the acute and chronic toxicity may occur through similar mechanisms of action and is consistent for the fish and invertebrates tested. This is supported by the results of the bioconcentration study, discussed below. Using the low ACR, only moderate additional incremental adjustment factors would be justified in any extrapolation from acute to chronic risk (see Section 4).

Table 8.2. Chronic aquatic toxicity of PBO

Study	LOEC (mg L^{-1})	NOEC (mg L^{-1})	MATC (mg L^{-1})
Daphnia magna – 21-day life cycle I (Holmes and Smith, 1992a)	0.047	0.030	0.038
Daphnia magna – 21-day life cycle II (Putt, 1994)	0.012	0.066	0.089
Fathead minnow – 35-day early life stage (Machado, 1994)	–	0.18	0.27

A flow-through bioconcentration study using bluegill sunfish exposed to PBO produced bioconcentration factors (BCFs) of 260×, 91× and 380× for whole fish, edible tissue and nonedible tissue, respectively (Sved *et al.*, 1992). In this context nonedible tissue includes the head, viscera and fins; edible tissue includes the muscles, skin, scales, bones etc. (J. Swigert, personal communication). Compounds for which the BCF is less than 1000× are considered to have low potential for bioconcentration. Concentration factors were approx. 4× higher in the nonedible tissue than in the edible tissue.

Steady-state concentration was reached rapidly. The 90% steady-state concentrations were reached after 2.2, 5.2 and 4.2 days, for edible, nonedible and whole body compartments, respectively. Depuration of residues also occurred rapidly when the fish were placed in clean water. The 50% clearance levels of PBO in edible and nonedible tissues and whole fish were 0.67, 1.6 and 1.3 days, respectively. Even under this extreme scenario of continuous exposure for 28 days, the BCFs were well below 1000× and depuration was relatively rapid, i.e. 90% clearance within a few days.

2.2. Avian Effects

Avian toxicity has been evaluated in five studies: one acute oral study (Campbell *et al.*, 1991), two acute dietary feeding studies (Grimes *et al.*, 1991a, b) and two reproduction studies (Rodgers, 1995a, b). The results are listed below in Table 8.3. Evaluations of avian acute toxicity show that PBO is 'practically nontoxic' to avian species in single dose oral or dietary exposures. The acute median lethal concentrations were all greater than the respective 'limit dose' of 2250 mg per kg body weight and limit exposure level of 5620 mg per kg feed, for oral and dietary exposures, respectively. (In a toxicology study the limit dose or limit exposure level is the maximum level of treatment or exposure beyond which testing is not required.) In all three acute avian toxicity studies, median lethal (LD$_{50}$ or LC$_{50}$) values are above the respective limit of required testing (i.e. limit dose/ exposure levels). In the Northern Bobwhite Quail chronic reproduction study, dietary administration of up to 3000 mg PBO per kg diet had no effect on reproductive performance (Table 8.3). A significant effect on adult body weight was observed, however, at both 1500 and 3000 mg PBO per kg diet. No effects were observed at 300 mg PBO per kg diet. For mallard ducks, dietary administration

Table 8.3. Avian toxicity of PBO

Species	Oral administration (mg per kg bodyweight)		Dietary administration (ppm in diet)[a]		Toxicity category
	LD$_{50}$	NOEC	LD$_{50}$	NOEC	
Acute toxicity					
Northern Bobwhite Quail (Campbell *et al.*, 1991; Grimes *et al.*, 1991a)	>2250	486	>5620	1000	Practically non-toxic
Mallard Duck (Grimes *et al.*, 1991b)	–	–	>5620	1780	Practically non-toxic
Chronic toxicity					
Northern Bobwhite Quail – reproduction (Rodgers, 1995a)	–	–		300	
Mallard – reproduction (Rodgers, 1995b)	–	–		300	

[a] ppm is equal to mg PBO per kg diet.

of up to 300 mg PBO per kg diet had no adverse effects on reproductive performance or on their offspring. Statistically significant effects were observed at 1200 mg PBO per kg diet for adult body weight, adult food consumption, number of eggs laid, number of eggs cracked and eggshell thickness (Rodgers, 1995a, b). The limit doses (exposure levels) for the acute studies and the chronic dietary NOEC values shown in Table 8.3 were used in characterizing risk.

3. EXPOSURE ASSESSMENT

The exposure to a chemical that an organism in the environment receives is dependent upon the route of exposure, the magnitude of residues in the exposure matrix, and the rate and duration of ingestion, or the length of exposure, to the contaminated matrix. This may include ingestion of contaminated diet, water or air, or direct contact with a contaminated matrix, e.g. surface water or sediment. Exposure is quantified by measuring or estimating the expected environmental concentration (EEC), defined as the magnitude of residues in various environmental matrices (e.g. surface water, pore water, sediment or wildlife food items).

The EEC in a given matrix is a function of the estimated use rate of the chemical in the field and the fate and transport of the chemical in the environment. The distribution and degradation of a compound is a function of the physicochemical properties (e.g. solubility, octanol/water partition coefficient, Henry's Law constant), the environmental fate processes and the use pattern (i.e. method of application, rate, frequency of application) (Table 8.4). Aquatic EECs are

Table 8.4. Physicochemical and environmental fate properties of PBO used in EXAMS modelling

Parameter	Value	Reference
Water solubility	14.34 mg L^{-1}	(Hazleton Laboratories, 1989)
Aqueous hydrolysis	Stable	(Williams, 1991)
Vapor pressure	1.59×10^{-7} mm Hg	(Bowman, 1989)
Henry's Law constant	2.35×10^{-4} atm m^3 mole^{-1}	(Brookman *et al.*, 1995)
Octanol/water partition coefficient (K_{OW})	45 100 (log K_{OW} = 4.65)	(Brookman *et al.*, 1995)
K_{OC}	399–830 µg absorbed	(Brookman *et al.*, 1995)
Aqueous photolysis ($t_{1/2}$)	8.5 h	(Brookman *et al.*, 1995)
Soil photolysis ($t_{1/2}$)	≈1.0 day	(Brookman *et al.*, 1995)
Aerobic soil metabolism ($t_{1/2}$)	14 days	(Brookman *et al.*, 1995)
Aerobic aquatic metabolism ($t_{1/2}$)	213 days	(Brookman *et al.*, 1995)

expressed as a probability and often include historical weather data, soil types and typical timing of application. Dietary exposures are a function of the residues in feed items.

3.1. Aquatic Exposure

Because there are no direct applications of PBO to water, the exposure of aquatic organisms to PBO is limited to spray drift at the time of application to adjacent lands and runoff associated with rainfall. Regarding drift the EPA uses a default assumption of 5% drift, which is defined as 5% of the actual application rate applied to the entire surface of the water body (pond) on a per acre basis. The compound applied is assumed to be instantaneously at equilibrium within the water column (Urban and Cook, 1986). In modelling aquatic exposure, the amount of chemical in runoff is also added to the water body (pond) based on historical rainfall data, the properties of the chemical, and other factors.

Runoff is generally estimated using the pesticide root zone model (PRZM) (Mullins *et al.*, 1993). Laboratory data on adsorption and desorption, as octanol/water partitioning, photolysis (see Table 8.4) indicate that PBO has a high affinity for organic-rich surfaces and will rapidly sorb to suspended particles and sediment. Therefore, PBO is expected to have a low potential for leaching or runoff under recommended use conditions. The runoff scenario used here assumes that 5% of the applied material runs off from 10 acres of treated field into a 2-m deep 1 hectare pond (i.e. a 10 : 1 ratio of drainage system area to receiving water area) (Urban and Cook, 1986).

The exposure analysis modelling system (EXAMS) is used to predict the distribution of residues in the various compartments of the aquatic system, including the water column and sediment pore water. Model calculations using PBO-specific data, specifically the physicochemical properties presented in Table 8.4, indicate that 75% of the PBO from runoff is associated with particles, and deposits immediately to the sediment compartment of the pond. The remaining 25% of the PBO in this runoff enters the water column (Burns and Cline, 1985; Brookman *et al.*, 1995). The assumptions used to estimate the EEC in water for PBO, including the information about chemical and biological degradation processes, are reported in Table 8.4 and discussed in more detail in Chapter 7.

For the EXAMS calculation, the proposed maximum use rate of PBO was 0.37 lb (168 g). PBO per acre. The rate of drift can be calculated as follows.

5% of the rate applied $\times$ maximum application rate = rate of drift

$$0.05 \times 0.37 \text{ lb PBO acre}^{-1} = 0.019 \text{ lb PBO acre}^{-1}$$

or in SI units,

$$0.05 \times 0.168 \text{ kg PBO ha}^{-1} = 0.0084 \text{ kg ha}^{-1}$$

Based on the assumptions above and the equation for drift, the maximum rate of entry of PBO into aquatic systems via drift is 0.018 lb PBO per acre (or 0.0084 kg ha^{-1}) water. The results of this calculation are used as the input rate per acre (surface area) of water and then adjusted for depth, degradation rates, and other factors (Mullins *et al.*, 1993).

The maximum rate of entry into surface water (2 m average depth) via runoff is 0.019 lb PBO per acre $\times$ 10 acres of drainage = 0.19 lb PBO per acre (or 0.084 kg ha^{-1}) water. For the model simulation, PBO loads were introduced in the environment consistent with 10 applications, 1 week apart, beginning in June. EXAMS then simulated the fate and distribution of PBO following such a treatment regimen.

Runoff

5% of the rate applied $\times$ maximum application rate $\times$ acres treated = total mass PBO in runoff

$$0.05 \times 0.37 \text{ lb PBO acre}^{-1} \times 10 \text{ acres} = 0.185 \text{ lb PBO}$$

or in SI units,

$$0.05 \times 0.168 \text{ kg PBO ha}^{-1} \times 10 \text{ ha} = 0.084 \text{ kg PBO}$$

Results of the EXAMS simulations indicate that approximately 92% of the total mass of PBO is predicted to bind to sediment, where it is largely unavailable for uptake by aquatic organisms (i.e., not bioavailable) or to leach into ground water (Brookman *et al.*, 1995). Pore-water concentrations are predicted by EXAMS and presented as predicted peak, 96-h and 21-day EEC values of 0.0235, 0.0230 and 0.0217 mg L^{-1}. The resultant peak (highest yearly instantaneous) concentration of PBO in the water column (dissolved and complexed) is 0.0071 mg L^{-1}. (The highest yearly instantaneous peak concentration is the highest value from the 90th percentile year using 30 years of weather data and ranking the yearly peak values (Society of Environmental Toxicology and Chemistry, 1994).) The 96-h and 21-day maximum integrated average EECs for PBO in the water column are 0.00577 mg L^{-1} and 0.0051 mg L^{-1}, respectively (Brookman *et al.*, 1995). During the risk assessment, these 96-h and 21-day average values will be compared with the appropriate acute or chronic aquatic endpoints from aquatic toxicology studies to determine margins of safety. The margin of safety (MOS, effects/exposure) is defined here as the appropriate effects value (e.g. LC_{50} or chronic NOEC for the most sensitive species tested), divided by the appropriate exposure value (EEC). This is also sometimes called the margin of exposure (MOE) (Klaassen, 1996). Toxicity values for water column species will be compared to predicted pore-water EECs.

3.2. Avian Exposure

Estimations of the EEC of PBO in avian food crops were derived in one of two ways: actual measurement of PBO residues after spraying on crops, or by using empirical relationships between the application rate of a pesticide and the resulting peak concentration on various types of food crops (Kenaga, 1973) (Table 8.5).

Residue data were obtained for a wide variety of crops as a part of the PBO data development programme. Applications of PBO were made weekly at the maximum allowable rate for 10 weeks. Crops were harvested immediately after the final treatment and PBO levels determined. The highest PBO residues were observed in the leafy green vegetable crop group. Individual spinach samples showed levels as high as 38.7 ppm (Hattermann, 1996). This value was used in subsequent risk assessments as a representative worst-case scenario.

Predictions of PBO residues were also made for crops by using the nomogram developed by Kenaga (Kenaga, 1973). A portion of this work is incorporated into the US Environmental Protection Agency, Office of Pesticide Programs, Standard Evaluation Procedure for Ecological Risk Assessment (Urban and Cook, 1986). Concentrations on the plant are dependent upon the application rate and the crop of interest. The Kenaga nomogram relates application rates (lb per acre) to measured residues, taken shortly after application, in different crop types. Leafy crops with large biomass per unit of soil surface will exhibit lower concentrations of a pesticide than will plants such as short rangegrass that have

Table 8.5. Expected environmental concentration – terrestrial PBO in avian diet (mg PBO per kg food)

Food category	Residues after use at × the application rate	
	1.0 ×	10 ×
Leaves and leafy crops[a]	46	460
Short rangegrass[a]	88	880
Spinach[b]	–	38.7

[a] Calculated using Kenaga nomograms (Kenaga, 1973)
[b] Calculated using measured spinach residues (Hatterman, 1996)

a relatively small amount of biomass per unit area of ground surface. As shown in Table 8.5, the measured residues of PBO in spinach (a leafy crop) (Hattermann, 1996) was compared with predicted values in leafy crops and short range-grass. The latter is a standard value used by the USEPA because it represents the worst-case scenario predicted by Kenaga (Kenaga, 1973).

This assessment included the estimate of the residues of PBO following application at the recommended application rate and at 10 times the single application rate. This incorporates two highly conservative assumptions. First, the 10× rate simulates the maximum number of applications allowed on the label and incorporates the unrealistic assumption that no degradation or dissipation occurs in the field (or all 10 applications are made at the same time). This is a worst-case scenario considering the degradation predicted by the environmental fate studies and the actual levels of PBO observed in the field, as well as the low likelihood that 10 applications of PBO would ever be made prior to harvest. The second extremely conservative assumption used in this exposure assessment is that 100% of the avian diet is derived from treated crop.

In summary, the terrestrial EECs for three potential avian food items are calculated for 1.0× and 10.0× the maximum labelled application rate (Table 8.5). Although these EECs for dietary exposure represent redundant worst-case scenarios, the worst-case scenario for actual PBO data indicates exposure level 10 to 20 times lower than the levels predicted by Kenaga (Kenaga, 1973), indicating that the estimate of exposure using the latter approach is conservative for PBO.

4. RISK CHARACTERIZATION

Risk is generally characterized by a comparison of exposure and effects. This is most frequently reported as a ratio of effect to exposure level and described here as the MOS. For example the EEC for an environmental matrix, e.g. water, would be divided by the NOEC for the corresponding species of interest, e.g. *Daphnia magna*. The species with the lowest acute or chronic toxicity value is used in the risk characterization. For terrestrial species, the dietary exposure from acute or chronic studies is compared with the expected exposure an animal would receive eating food items containing PBO at residue levels either predicted by the Kenaga nomograms (Kenega, 1973) or measured in the field (Hattermann, 1996).

The risk criteria currently used by the Environmental Fate and Effects Division of the Office of Pesticide Programs, USEPA, are summarized in Table 8.6. For aquatic species the acute risk criteria are 1/10 the LC_{50} and 1/2 the LC_{50} for the most sensitive species tested. Below 1/10 the LC_{50} the risk is assumed acceptable. Between 1/10 LC_{50} and 1/2 LC_{50} the risk is assumed to be mitigated by restricted use status. An EEC above 1/2 LC_{50} is assumed to be unacceptable. For the purposes of calculating MOS values for PBO the more conservative 1/10 LC_{50} values from column 2 of Table 8.6 will be used. Although these criteria are currently used by the EPA, they may undergo some reform in the near future (see Society of Environmental Toxicology and Chemistry, 1994, for a discussion of alternative approaches).

For avian species the key risk criteria are 1/5 of the LC_{50} and the LC_{50}. If the EEC is below 1/5 of the LC_{50}, then the risk is acceptable. If the EEC is between 1/5 of the LC_{50} and the LC_{50} itself, then risk is presumed to be mitigated by restricted use status. An EEC above the LC_{50} results in a determination of unacceptable risk. For the purposes of calculating MOS values for avian species, the risk criteria of 1/5 the LC_{50} (column 2, Table 8.6) is used.

Finally, chronic risk criteria are based on the NOEC from avian reproduction studies or aquatic full or partial life cycle studies. As shown in column 2 of Table 8.6, an EEC below the NOEC is considered to be an acceptable risk and the EEC above or equivalent to the chronic NOEC is an unacceptable risk (Urban and Cook, 1986).

The MOS values reported in this chapter are the multiples of the EEC values necessary to reach the respective risk criteria (Table 8.6, column 2), based on the appropriate effects values. The MOS values provide a measure of how far the exposure level is from the respective risk criteria. MOS values are discussed below to highlight the gap between the very conservative assessment of risk and the regulatory risk criteria defined by the USEPA.

4.1. Aquatic Risk

Acute MOS values for aquatic and estuarine/marine species range from 4.0 to 106 (Table 8.7). The lowest MOS value (4.0) is based on the EC_{50} of 0.23 mg PBO L^{-1} for the most sensitive species (oyster) tested and the modelled 96-h EEC of 0.00577 mg PBO L^{-1} in the water column. Thus, even the comparison of toxicity to the most sensitive organism tested to the conservatively modelled PBO levels yields an acceptable MOS of 4.0 (Table 8.7). The key LOC for acute

Table 8.6. Regulatory risk criteria[a]

Toxicity	Presumption of no risk	Presumption of risk that may be mitigated by restricted use[b]	Presumption of unacceptable risk
Acute			
Birds	EEC < 1/5 LC_{50}	1/5 LC_{50} ≤ EEC, LC_{50}	EEC ≥ LC_{50}
Aquatic organisms	EEC < 1/10 LC_{50}	1/10 LC_{50} ≤ EEC < ½ LC_{50}	EEC ≥ ½ LC_{50}
Chronic	EEC < chronic NOEC	N/A	EEC ≥ chronic effect levels including reproductive effects

[a] Adapted from Urban and Cook (1986).
[b] Restricted use is a classification of a pesticide whereby its use is limited to applicators who have been certified by the USEPA through EPA-approved training programmes.

Table 8.7.   Acute aquatic toxicity and MOS

Study	1/10 LC$_{50}$ or 1/10 EC$_{50}$	MOS[a]
Freshwater		
Rainbow trout	0.61	106
Bluegill	0.54	93
Daphnia magna	0.05	8.8
Daphnia magna (sediment)[b]	0.05	4.4
Estuarine/marine		
Sheepshead minnow	39.4	68.2
Mysid shrimp	0.05	8.4
Eastern oyster	0.02[c]	4.0
Eastern oyster (sediment)[b]	0.02	1.0

[a] 1/10 LC/EC$_{50}$ compared with modelled 96-h PBO level (0.00577 mg L^{-1}) (Brookman *et al.*, 1995).
[b] 1/10 EC$_{50}$ compared with modelled 96-h sediment pore-water value (0.0230 mg L^{-1}) (Brookman *et al.*, 1995).
[c] Most sensitive species.

aquatic toxicity (as shown in Table 8.6) is EEC < 1/10 LC$_{50}$ and all MOS values for aquatic species are 4.0 times the EEC. No acute risk is presumed.

The chronic MOS values for PBO are listed in Table 8.8 for *Daphnia magna* and two fish species. The NOECs are reported in Table 8.8 and show MOS values ranging from 5.9 to 35.3. The lowest MOS reported here shows that the NOEC is 5.9 times higher than the EEC which surpasses the LOC of EEC ≤ NOEC.

Toxicity for sediment-dwelling organisms is not a standard data requirement under US Federal Insecticide Fungicide and Rodenticide Act (FIFRA) in the USA. Here EECs (pore-water) from EXAMS modelling are compared to aquatic toxicity values for water column species. Acute MOS values were calculated for *Daphnia magna* and Eastern oyster are 4.4× and 1.0×, respectively. The EEC is equal to the 96-h EC$_{50}$ from an oyster shell deposition study. Based on the conservative nature of the exposure estimates and the risk characterization, and the fact that the toxicity value is an effect and not a lethal concentration (with growth and not mortality as the endpoint), a presumption of no risk is concluded for sediment-dwelling organisms.

There is built-in conservatism in the current method of aquatic risk assessment. Laboratory studies are designed to provide constant aqueous exposures to determine the inherent toxicity of a compound and for comparison with toxicity of other compounds under the same conditions. However, these flow-through studies represent worst-case exposures, since PBO dissipates rapidly in the environment. The acute and chronic studies were conducted using flow-through test systems with continuous generation of new exposure solutions. These systems are designed to maintain exposures at consistent levels which are usually artificially high. Researchers were only able to maintain 55% to 91% of the nominal exposure levels in the acute studies and 50% to 80% in the chronic *D. magna* life cycle studies (Holmes and Smith, 1992a; Putt, 1994), despite continuous

Table 8.8.   Chronic aquatic toxicity and MOS

Study	NOEC (mg L^{-1})	MOS[a]
Daphnia magna – 21-day life cycle I	0.030	5.9
Daphnia magna – 21-day life cycle II	0.066	12.9
Fathead minnow – 35-day early life stage	0.18	35.3

[a] MOS is the ratio of the NOEC compared with modelled 21-day average PBO level (0.0051 mg L^{-1}) (Brookman *et al.*, 1995).

addition of PBO to the test system. This observed loss of test substance is consistent with the rapid photodegradation of PBO measured in the aqueous photolysis study ($t_\frac{1}{2}$= 8.4 h; see Table 8.6). When concentrations of a chemical cannot be maintained in a flow-through system (at levels below the limit of solubility), then very rapid dissipation would be expected in ambient waters.

PBO partitions to the sediment and suspended particulates, reducing bio-availability, and is subject to photochemical, biological and other degradative processes. It is inappropriate and overly conservative to compare peak concentrations predicted by models with sustained artificially high concentrations used in laboratory testing. At the same initial level of PBO, the total exposure in a laboratory system will be much greater because it is sustained. The magnitude and duration of exposure in ambient aquatic systems is much lower because of degradation and dissipation processes. Therefore, the MOS values calculated by this conservative methodology are substantially lower than would be expected in real life situations.

The terrestrial use of PBO is not expected to present any appreciable risk to aquatic organisms. PBO residues that would be predicted to occur in water from spray drift and runoff following ground application would be well below the lowest NOECs determined in laboratory studies. Furthermore, the assessment of risk and derivation of MOS values are based on comparisons of laboratory studies with highly conservative, modelled peak EECs. These levels would immediately begin to decrease in ambient aquatic systems. In addition, PBO binds tightly to sediment, which limits the concentration of PBO that is available in the water column and in pore-water, thus further limiting the actual exposure of aquatic organisms. However, even under these conservative scenarios, the MOS values for the most sensitive acute (96-h oyster EC$_{50}$; Table 8.6) and chronic (reproduction in *D. magna*; Table 8.8) endpoints are 1064 and 8.7, respectively, indicating acceptable acute and chronic MOS values for sediment dwelling organisms.

4.2. Avian Risk

The avian dietary risk assessment is based on a comparison of toxicity data from laboratory studies and measured or estimated residues in different categories of avian food items. The acute dietary (LC$_{50}$) values and chronic NOECs from avian reproduction studies are compared with the residues measured or calculated for specific categories of avian food. This includes a highly conservative assumption that the given, contaminated food item represents 100% of food intake.

The acute dietary LC_{50}s are reported as limit tests (Table 8.3) (i.e. 5620 mg per kg diet). For both mallard ducks and Northern bobwhite the LC_{50} is much higher. In both of these acute dietary studies there were only two mortalities (mallard), although lower weight gains and transient behavioural effects were observed in both species. The NOECs for these acute studies were 1000 mg PBO per kg feed for mallard duck and 1780 mg PBO per kg feed for Northern Bobwhite Quail. Acute MOS values for each species and food category and two application rates (1.0× and 10.0× of the maximum single application rate) are presented in Table 8.9. As shown in Table 8.9, the acute MOS values range from 1.3 to 29 (base on 1/5 LC_{50}) even when considering the conservative assumptions utilized in this assessment.

Similar calculations were performed for the chronic endpoints (see Table 8.10). These MOS values were based on an NOEC of 300 mg PBO per kg feed for both the Northern bobwhite and the mallard duck. For the highest labelled application rate (1×) of 0.37 lb PBO acre^{-1} chronic MOS values ranged from 3.4× to 6.5×, using the Kenaga nomogram (Urban and Cook, 1986; Kenaga, 1973). The lowest MOS values of 0.3 and 0.7 were calculated for the 10× PBO application rate of 3.7 lb PBO acre^{-1}. Use of the measured residue value for spinach results in a chronic MOS value of approximately 8.0.

Comparison of results of dietary toxicity studies and estimated or calculated residues in avian food categories provide more than adequate margins of safety. As shown above, when a series of redundant worst-case assumptions are made and the highest residue food item (short rangegrass) is used, the acute and chronic MOS values are 12.8 and 3.4, respectively (Tables 8.9 and 8.10). When actual residues (spinach) are used, the acute and chronic MOS values are 29 and 8.0, respectively. Further, when an additional conservative 10× factor is applied to the exposure (i.e. 10 applications), the acute MOS values are 1.3 or greater.

Table 8.9.   Acute avian MOS – terrestrial exposure

Species	Food category	Acute MOS[a]	
		0.37 lb acre^{-1} [b]	3.7 lb acre^{-1} [c]
Northern Bobwhite Quail	Leaves and leafy crops[d]	24.4	2.4
	Short rangegrass[d]	12.8	1.3
	Spinach[e]	–	29
Mallard	Leaves and leafy crops[d]	24.4	2.4
	Short rangegrass[d]	12.8	1.3
	Spinach[e]	–	29

[a] 1/5 LD/LC_{50} compared with EEC for each crop using residue values found in Table 8.1.
[b] Maximum labelled application rate.
[c] 10 times the maximum application rate.
[d] Estimated according to Kenaga (Kenaga, 1973).
[e] Estimated using measured spinach residue values (Hatterman, 1996).

Table 8.10. Chronic avian MOS – terrestrial exposure

Species	Food category	Chronic MOS[a]	
		0.37 lb acre^{-1} [b]	3.7 lb acre^{-1} [c]
Northern Bobwhite	Leaves and leafy crops[d]	6.5	0.7
Quail	Short rangegrass[d]	3.4	0.3
	Spinach[e]	–	8.0
Mallard	Leaves and leafy crops[d]	6.5	0.7
	Short rangegrass[d]	3.4	0.3
	Spinach[e]	–	8.0

[a] NOEC (300 ppm) compared with EEC for each crop.
[b] Maximum labelled application rate.
[c] 10 times the maximum application rate.
[d] Estimated according to Kenaga (Kenaga, 1973).
[e] Estimated using measured spinach residue values (Hatterman, 1996).

These MOS values are more than adequate based on EPA policy that states if the EEC < chronic NOEC, then the pesticide poses negligible risk. Even the chronic MOS values of 0.7 for leafy crops and 0.3 for short rangegrass for the 10× application scenario are adequate to support the position that PBO is safe when used according to the current label, given the low likelihood of 10 applications being made and the conservative assumptions discussed above (i.e. 100% of bird diet from crop and no degradation of residues). The MOS values resulting from this analysis indicate that the use of PBO does not pose any unacceptable risk of chronic or acute toxicity to avian species exposed via the diet.

Another possible route of exposure for birds is from the ingestion of PBO-contaminated water. The risk from this type of exposure is assessed by comparing the peak aquatic EEC values with dietary intake rates. A typical Northern bobwhite with a body weight of 0.2 kg ingests about 0.05 kg water per day. If we use a worst-case scenario, all of the water comes from an aquatic system containing PBO at the instantaneous peak EEC of 0.0071 mg L^{-1} water (1 L is 1 kg). Thus the bird ingests 0.05 kg × 0.0071 mg/kg = 0.00036 mg per 0.2 kg body weight = 0.0018 mg per kg body weight. This is then compared with the NOEC from the acute oral LD$_{50}$ of 486 mg per kg body weight. The MOS is therefore calculated: 486 mg/kg/kg b.wt. = 0.0018 mg/kg/bodyweight = 27 × 10^4. Comparing this same instantaneous peak EEC of 0.0071 mg/kg to the chronic NOEC of 300 mg PBO per kg feed results in an MOS of 4.2 × 10^4. The terrestrial use of PBO will present essentially no risk to Northern bobwhite, mallard ducks, or other avian species due to drinking water from areas which have PBO residues from runoff or spray drift.

5. SUMMARY AND CONCLUSION

A comprehensive assessment of risk to aquatic and terrestrial organisms according to PBO label directions has been conducted based on assessments of hazard and exposure followed by the US Environmental Agency, Office of Pesticide Programs. The exposure of aquatic organisms was estimated using standard models and results were compared with those of standard testing with aquatic and estuarine species. MOS values indicate that the exposure is well below the LOCs used to determine risk (Table 8.6). Acute MOS values range from 4.4 to 106 for aquatic organisms and 1 to 68.2 for estuarine species including sediment-dwelling organisms (Table 8.7). Chronic aquatic MOS values ranged from 5.9× to 35.3× based on fish and invertebrate life cycle or life stage studies (Table 8.8). This risk assessment indicates that use of PBO presents no significant ecological risk to aquatic organisms.

The risk to avian wildlife species was based on a comparison of standard toxicity tests with estimates of dietary exposure to Northern bobwhite and mallard ducks. Worst-case exposure estimates were made based on Kenaga (Kenaga, 1973) and empirical data (Hattermann, 1996). Acute MOS values ranged from 1.3× to 29× for mallard and Northern bobwhite using 10× the highest labelled application rate (Tables 8.9 and 8.10). Chronic avian risk was estimated from two avian reproduction studies and estimated dietary exposure. The MOS values, based on the 10× application rate, ranged from 0.7× to 8.0× for both species, indicating that there is a presumption of no risk (Tables 8.5, 8.9 and 8.10).

This risk assessment shows that highly conservative exposure assessments based on high use rates and maximum frequency, combined with results from effect studies which are also conservative (i.e. exaggerated rates and sustained exposure), indicate a presumption of no risk. The risk assessment for avian species is also conservative in the magnitude and duration of exposure.

Residues from a leafy crop immediately after treatment were compared with acute and chronic studies where exposure is maintained for time frames that are unreasonable for a compound like PBO, which dissipates rapidly in the environment. Furthermore, birds would not feed solely on the crop and field treated. Thus there is a presumption of no risk to avian wildlife species from use of PBO.

REFERENCES

Bowman, B. (1989). Determination of the Vapor Pressure of Piperonyl Butoxide. Unpublished report no. 38007 from Analytical Bio-Chemistry Laboratories, Inc. Undertaken for the PBO Task Force, Washington DC, USA.

Brookman, D.J., Curry, K.K, Hobson, J.F., Kent, D.J. and Smith, C.A. (1995). Data Waiver Request for the Requirement of a Study on Aquatic Field Dissipation of Piperonyl Butoxide. Unpublished report from Technology Sciences Group Inc. Undertaken for the PBO Task Force, Washington DC, USA.

Burns, L.A. and Cline, D.M. (1985). Exposure Analysis Modeling System: Reference Manual for EXAMS II. pp. 443. EPA/600/3-86/034.

Campbell, S., Lynn, S.P. and Smith, G.J. (1991). An Acute Oral Toxicity Study in the Northern Bobwhite Quail (*Colinus virginianus*). Unpublished report no. 306-103 from Wildlife International Ltd. Undertaken for the PBO Task Force, Washington DC, USA.

Grimes, J., Lynn, S.P. and Smith, G.J. (1991a). A Dietary LC_{50} Study with Piperonyl Butoxide in the Northern Bobwhite Quail. Unpublished report no. 306-101 from Wildlife International Ltd. Undertaken for the PBO Task Force, Washington DC, USA.

Grimes, J., Lynn, S.P. and Smith, G.J. (1991b). A Dietary LC_{50} Study with Piperonyl Butoxide in the Mallard. Unpublished report no. 306-102 from Wildlife International Ltd. Undertaken for the PBO Task Force, Washington DC, USA.

Hattermann, D.R. (1996). Amended Piperonyl Butoxide Analytical Phase on the Raw Agricultural Commodity Residue Evaluation of Piperonyl Butoxide & Pyrethrins Applied As Pyrenone Crop Spray to Leafy Vegetables. Biological Test Center & Pharmaco LSR International Inc., vol. II. (Protocol No. 18009A003). Undertaken for the PBO Task Force, Washington DC, USA.

Hazleton Laboratories, Inc. (1989). Determination of Water Solubility of Piperonyl Butoxide – Final Report. HLA 6001-385. Undertaken for the PBO Task Force, Washington DC, USA.

Holmes, C.M. and Smith, G.J. (1992a). A Flow-Through Life-Cycle Toxicity Test with Piperonyl Butoxide in Daphnids (*Daphnia magna*). Unpublished report no. 306A-104 from Wildlife International Ltd. Undertaken for the PBO Task Force, Washington DC, USA.

Holmes, C.M. and Smith, G.J. (1992b). A 48-Hour Flow-Through Acute Toxicity Test with Piperonyl Butoxide in Daphnids (*Daphnia magna*). Unpublished report no. 306A-105A from Wildlife International Ltd. Undertaken for the PBO Task Force, Washington DC, USA.

Holmes, C.M. and Smith, G.J. (1992c). A 96-Hour Flow-Through Acute Toxicity Test with Piperonyl Butoxide in the Bluegill (*Lepomis macrochirus*). Unpublished report no. 306A-101 from Wildlife International Ltd. Undertaken for the PBO Task Force, Washington DC, USA.

Holmes, C.M. and Smith, G.J. (1992d). A 96-Hour Flow-Through Acute Toxicity Test with Piperonyl Butoxide in the Rainbow Trout (*Oncorhynchus mykiss*). Unpublished report no. 306A-106 from Wildlife International Ltd. Undertaken for the PBO Task Force, Washington DC, USA.

Holmes, C.M. and Smith, G.J. (1992e). A 96-Hour Flow-Through Acute Toxicity Test with Piperonyl Butoxide in Sheepshead Minnow (*Cyprinodon variegatus*). Unpublished report no. 306A-108A from Wildlife International Ltd. Undertaken for the PBO Task Force, Washington DC, USA.

Holmes, C.M. and Smith, G.J. (1992f). A 96-Hour Flow-Through Acute Toxicity Test with Piperonyl Butoxide in Mysid Shrimp (*Mysidopsis bahia*). Unpublished report no. 306A-109 from Wildlife International Ltd. Undertaken for the PBO Task Force, Washington DC, USA.

Holmes, C.M. and Smith, G.J. (1992g). A 96-Hour Shell Deposition Test with Piperonyl Butoxide in the Eastern Oyster (*Crassostrea virginica*). Unpublished report no. 306A-110A from Wildlife International Ltd. Undertaken for the PBO Task Force, Washington DC, USA.

Kenaga, E.E. (1973). Factors to be considered in the evaluation of the toxicity of pesticides to birds in their environment. In: *Environmental Quality and Safety,* vol. II, pp. 166–181. Academic Press, New York.

Klaassen, C.D. (1996). (Fifth Edition). *Cassarett and Doull's Toxicology: The Basic Science of Poisons*, 5th edn. McGraw Hill, New York.

Machado, M.W. (1994). PBO Technical Task Force Blend PB200 – The Toxicity to Fathead Minnow (*Pimephales promelas*) During an Early Life-Stage Exposure. Unpublished report no. 94-5-5264 from Springborn Laboratories, Inc. Undertaken for the PBO Task Force, Washington DC, USA.

Mullins, J.A., Carsel, R.F., Scarbrough, J.E. and Iver, A.M. (1993). PRZM-2 A Model for Predicting Pesticide Fate in the Crop Root and Unsaturated Soil Zones: Users Manual for Release 2.0: EPA/600/R-93/046. Environmental Research Laboratory, Office of Research and Development, US Environmental Protection Agency, Athens, Georgia.

Putt, A.E. (1994). PBO Technical Task Force Blend PB200 – The Chronic Toxicity to *Daphnia magna* Under Flow-Through Conditions. Unpublished report no. 94-5-5270 from Springborn Laboratories, Inc. Undertaken for the PBO Task Force, Washington DC, USA.

Rand, G.M. (1995). *Effects, Environmental Fate, and Risk Assessment: Fundamental of Aquatic Toxicology* 2nd edn. Washington.

Rodgers, M.H. (1995a). Piperonyl Butoxide: Effects on Reproduction in Bobwhite Quail After Dietary Administration. Unpublished report no. PBT 2/950167 from the Huntingdon Research Centre Ltd. Undertaken for the PBO Task Force, Washington DC, USA.

Rodgers, M.H. (1995b). Piperonyl Butoxide: Effects on Reproduction in the Mallard Duck After Dietary Administration. Unpublished report no. PBT 2/950168 from the Huntingdon Research Centre Ltd. Undertaken for the PBO Task Force, Washington DC, USA.

Society of Environmental Toxicology and Chemistry (SETAC). (November 1994). Aquatic Dialogue Group: Pesticide Risk Assessment & Mitigation – Final Report. Society of Environmental Toxicology and Chemistry (SETAC).

Sved, D.W., Holmes, C.M. and Smith, G.J. (1992). A Bioconcentration Study with Piperonyl Butoxide in the Bluegill (*Lepomis macrochirus*). Unpublished report no. 306A-102 from Wildlife International Ltd. Undertaken for the PBO Task Force, Washington DC, USA.

Swigert, J. Personal communication. James Swigert is a consulting Aquatic Toxicologist and former manager of aquatic toxicology laboratories.

US Environmental Protection Agency (1985). Acute Toxicity Test for Freshwater Fish – Hazard Evaluation Division Standard Evaluation Procedure. Office of Pesticide Programs. EPA-540/9-85-06.

Urban, D., and Cook, N. (1986). Hazard Evaluation Division Standard Evaluation Procedure Ecological Risk Assessment. EPA Report no. 540/9-85-001.

Williams, M.D. (1991). Hydrolysis of Piperonyl Butoxide as a Function of pH at 25°C. ABC Laboratories, Inc. Report no. 39164. Undertaken for the PBO Task Force, Washington DC, USA.

9

The Absorption, Distribution, Metabolism and Excretion of Piperonyl Butoxide in Mammals

ANDREW COCKBURN and DAVID NEEDHAM

1. INTRODUCTION

A primary goal of risk assessment is to predict the potential toxicity of chemicals and establish exposure levels for humans that ensure an adequate margin of safety. For environmental health products and agrochemicals we have to consider the factory worker, the pest control officer or farmer, the householder or bystander, and individuals who may become exposed to residues sometimes found in food.

Risk assessment can utilize two approaches. The first compares information on the health of a large group of persons who have probably been exposed to the substance with that of a similar group who probably have not (epidemiology). This has the disadvantage that such data may only be gained retrospectively, i.e. after marketing, and hence after exposure has already taken place. Moreover, epidemiological models are imprecise tools with which to establish degrees of risk, partly because of the large size of the sample required to get an accurate picture, and partly because of the difficulty in obtaining adequate control populations and accurate reporting of the data. The alternative uses animals as surrogates in order to obtain the necessary data, under controlled conditions, to allow extrapolation to humans. Here two types of information are required. First we need to learn what the substance does to the animal, the type of toxicity caused by either short-term high exposure, or long-term low exposure (acute and chronic toxicity), and second, we need to learn what the animal does to the substance once absorbed (is it metabolized, and how is it excreted?). Such data on the fate of the molecule are gained by studying toxicokinetics. Without this information the toxicologist is 'flying blind' and cannot make a proper risk assessment.

The purpose of this chapter is to describe the fate of piperonyl butoxide (PBO) in both laboratory and food-producing animals and to show how understanding the fate of the compound in these species contributes to an overall safety assessment of the product.

PBO (Fig. 9.1) is an economically important, widely used synergist of natural pyrethrins and synthetic pyrethroid insecticides with an excellent record of human health and safety over the 40 years or so since its introduction. Most pyrethroids have a good record of environmental and human safety; in the absence of PBO, a much greater tonnage of more potent, synthetic insecticides would be required to achieve the same levels of insect control, with potential adverse consequences for the natural environment.

As mentioned earlier, humans are likely to come into contact with PBO through a variety of different scenarios. Fortunately, the human body has the ability to deal with resultant exposure in a number of ways. Firstly, the body surface and portals of entry are well equipped to restrict the entry of foreign substances. Any xenobiotic absorbed by inhalation, ingestion or dermally is then subjected to biotransformation by a wide variety of metabolizing enzyme systems found in most tissues of the body, particularly in the liver, into compounds more readily excreted in urine and faeces. For many compounds, the 'drug' metabolizing enzymes, particularly cytochromes P450, can be induced, leading to an increase in the overall metabolic capacity. This change can lead to faster metabolism of the compound, resulting in a possible alteration of the rate of clearance from the body or changes in the distribution within tissues or metabolic profile. Toxicokinetic studies are used to answer four basic questions.

1. Is PBO absorbed?
2. Is PBO metabolized?
3. What is its disposition in tissues and does it persist?
4. Is the animal toxicology relevant for risk assessment for humans?

2. TOXICOKINETICS OF PBO

Toxicokinetic studies in rodents are conducted to meet governmental regulatory requirements and as mechanistic studies to aid in the interpretation of toxicological findings. The major agrochemical and environmental health markets are in Europe, the USA and Japan/Asia, which are regulated by the European Union (EU), the United States Environmental Protection Agency (USEPA) and the Japanese Ministry of Agriculture, Fisheries and Food (JMAFF)/the Ministry of Health and Welfare, respectively. These authorities each require a slightly

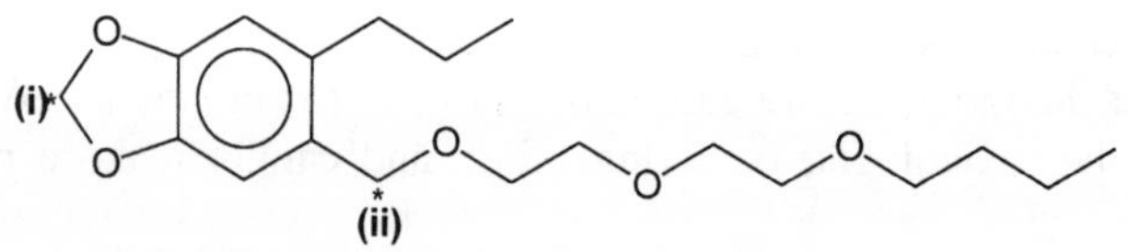

Figure 9.1. PBO. (i) ^{14}C radiolabel position on methylenedioxy ring. (ii) ^{14}C radiolabel position on the α-carbon.

different set of toxicokinetic studies to be carried out (EC Commission Directive 87/302/EEC (1987); US Environmental Protection Agency Pesticide Assessment Guidelines, Subdivision F (1982) and Japanese MAFF Guideline (1985)), a situation further complicated by the growth of specific regulatory requirements in individual member countries of Europe and individual states in the USA.

Agrochemicals and environmental health products are applied to crops to prevent disease, reduce weed burden or to kill pests such as insects and rodents, either in the field or for the protection of stored commodities. Some residues of the applied substances may thus end up in human food in grains or fruit, as prepared meals, or on the parts of plants not normally consumed by humans (straw, cotton seed, etc.) which are used for animal feed. By-products of food processing can also end up as animal feed (citrus pulp, oil seed rape cake, etc.), exposing food-producing animals to agrochemicals or their residues. The animals may also be exposed directly to the compound through the use of animal health products such as insecticides. For these reasons there is an additional requirement to conduct studies on major groups of food-producing animals (ruminants and poultry primarily) to determine the potential for exposure of humans to residues if they consume milk, eggs or meat. Here it is necessary to determine not only the magnitude of the total residue in such products, but also the nature of the residues, in order to assess the potential toxicological hazard to the consumer.

The fate of PBO has been examined over a number of years in the rat and mouse, and to a more limited extent in the goat and hen. The rate and extent of absorption, the magnitude and extent of disposition in the tissues, and the rate and route of excretion are relatively well documented in rodents using a variety of dose routes and vehicles, as is the nature and magnitude of the residues in the eggs, milk and meat of food-producing animals. However the metabolic fate is less well understood.

A number of specialized mechanistic investigations have also been conducted on PBO to clarify the major toxicological findings seen at high dose levels in rodents, which include liver enlargement, liver nodules/liver tumours and thyroid effects. This work has played a crucial part in enabling a full understanding of the biological action of PBO, thus facilitating the necessary risk assessment of the compound for crop and environmental health use.

The findings obtained for PBO from both regulatory toxicokinetic studies, as well as mechanistic studies and their complementary impact, are described below.

2.1. Toxicokinetic Studies

2.1.1. Absorption and Elimination

Casida *et al.* (1966) and Kamienski and Casida (1970) showed that PBO was well absorbed following oral, gavage administration in a solution of dimethylsulfoxide to both rats and mice. Following a single oral dose of either 1.7 mg kg^{-1} (mice) or 3.4 mg per kg body weight (rats), 65.4% of the dose was found in the urine of mice and 73.3% in the urine of rats when the compound was radiolabelled in the α-carbon position of the 2-(2-butoxyethoxy)ethoxymethyl side chain. The importance of the methylenedioxy ring in the metabolism of PBO was clearly seen when the compound was radiolabelled with ^{14}C in the methylenedioxy ring; only 6.1–6.3% of the dosed radioactivity was excreted in the urine, the majority being found in expired [^{14}C]carbon dioxide (65.4 and 73.3% for the mouse and rat, respectively). This clearly showed that the major route of metabolism of PBO involved the opening of the methylenedioxy ring followed by loss of the methylene group into the endogenous metabolic pool. This is also believed to be the basis of the initial inhibition of the cytochrome P450 enzyme system, which is essential for the compound's efficacy as a synergist. Current evidence suggests that it is the carbene intermediate formed from the methylene group that complexes with the Fe^{2+} ion of cytochrome P450 (Wilkinson *et al.*, 1984; Ortiz de Montellano and Reich, 1986) to lead to this effect.

At a higher oral dose level, the amount of the α-carbon-labelled dose excreted in the urine of rats fell and faecal excretion increased. Only 38% of a single oral dose of 500 mg kg^{-1} was excreted in the urine (with 62% in faeces) over a 7-day period. This may not necessarily signify a reduction in the amount of PBO absorbed, since biliary excretion of PBO metabolites has been shown to be the major route of elimination following intravenous administration (Fishbein *et al.*, 1969).

The toxicokinetics of PBO has also been studied in the goat and hen following daily oral administration for 5 days at a dose level of 10 or 100 mg kg^{-1} diet. In the goat the dose was excreted mainly in the urine (79.1 and 72.6%, respectively) with the remainder being found in the faeces. Milk was a minor route of elimination, accounting for less than 1% of the total dose (Selim, 1995a).

2.1.2. Distribution

Following a single oral dose of PBO radiolabelled in the α-carbon position, the concentration of radioactive residues was determined in the tissues of rats 168 hours after dosing. Apart from the gastrointestinal tract, the highest concentration of radioactivity was found in the liver (1.5–1.6 ppm and 8.4–9 ppm for male and female rats dosed with 50 or 500 mg PBO per kg body weight, respectively). The residues in the remaining tissues were generally an order of magnitude lower (Selim, 1991). Apart from a slight lowering of the radioactive residue concentrations, there was little effect on the disposition or magnitude of the residues following oral pretreatment of the rats with unlabelled PBO at a dose rate of 50 mg per kg body weight for 13 days before administering a single radiolabelled dose. The nature of the residue in the tissues was not determined in this study. However in an earlier study it was reported that following an intravenous dose of PBO, unchanged compound was only detectable in the fat and lungs, where it accounted for less that a quarter of the total radioactive residue (Fishbein *et al.*, 1969).

The concentration and nature of the radioactive residues in the milk and edible tissues of a goat were examined following the daily oral administration of α-carbon-radiolabelled PBO at either 0.34, 3.62 or 32.8 mg per kg body weight daily for 5 days, the animals being killed approximately 22 hours after the final dose. As with the findings from the rats, the highest concentration of radioactive residues was found in the liver (0.36, 2.01 and 56.2 ppm, respectively, for the three dose levels). Extraction and examination of the tissue residues showed that at the intermediate dose level unchanged PBO was a major component in the milk, liver and fat, but was only a very minor component of the residue present in the kidney (Selim, 1995a).

2.1.3 Metabolism

PBO contains four potential sites for metabolism:

1. the methylenedioxy ring;
2. the phenyl ring;
3. the 2-(2-butoxyethoxy) ethoxymethyl side chain; and
4. the propyl side chain.

Though any or all of these sites could be involved in the overall metabolism of PBO, the results of Kamienski and Casida tend to suggest that at low doses, the major site of metabolism is the methylenedioxy ring, which results in the methylene group being excreted as carbon dioxide (Kamienski and Casida, 1970). However, the subsequent metabolism of PBO is very complex, with over 18 different metabolites being reported in the urine of mice following dosing with the α-carbon-labelled compound and 12 following dosing with the methylenedioxy-labelled compound. Two metabolites were tentatively identified in the urine: 6-propylpiperonylic acid (MDB acid) and the corresponding glycine conjugate. These metabolites represented less that 0.5% of the applied dose. A further two metabolites were tentatively identified from incubations with mouse liver microsomes (Fig. 9.2).

Fishbein *et al.* (1969) reported that, following intravenous administration of PBO to rats, there were 13 metabolites in the bile and 11 in the urine when the compound was labelled in the methylenedioxy ring, and 24 metabolites in the bile and 26 in urine when the compound was labelled in the α-carbon position. These workers did not identify any of the metabolism products.

In a more recent study the metabolism of PBO in the goat has been examined as part of the identification of PBO-derived tissue residues. Four metabolites were isolated and identified in the urine by mass spectrometry (Fig. 9.3); all were metabolic products produced by successive oxidation of the 2-(2-butoxyethyl)ethoxymethyl side chain and contained the methylenedioxy ring as an intact moiety (Selim, 1995a).

Analysis of the residues present in the edible tissues of the goat showed the presence of over 20 different metabolic products, though unchanged PBO was the major component in the fat and one of the major components in the milk.

In the case of the hen unchanged PBO was the only component found in egg whites and was a major part of the residue in the yolks (Selim, 1995b). The

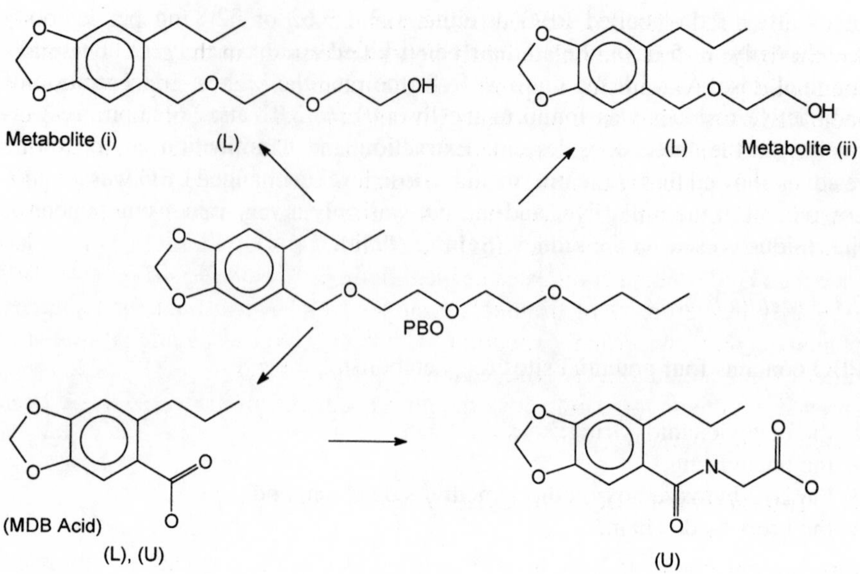

Figure 9.2.   Tentative metabolism of PBO in the mouse. L, liver metabolite; U, urine metabolite.

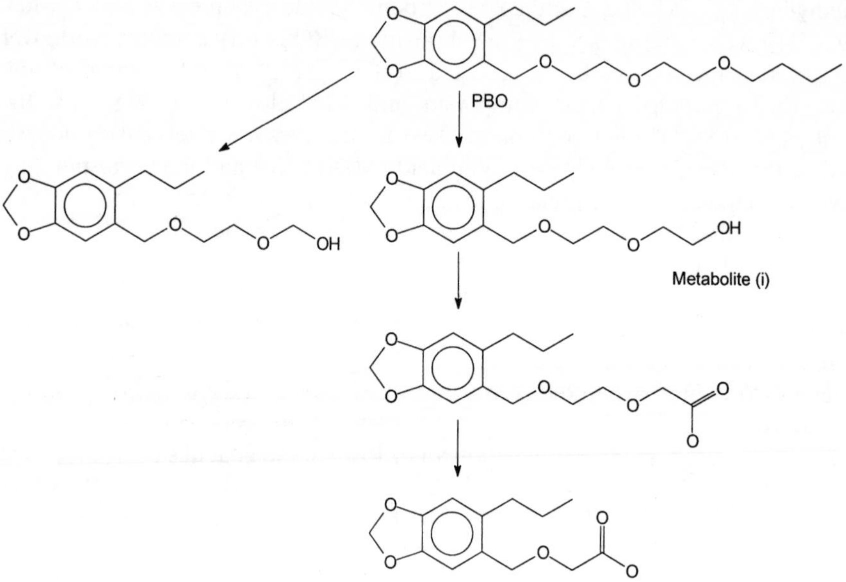

Figure 9.3.   Metabolic pathway of [^{14}C]PBO in the goat.

remaining residues in the yolk and other edible tissues appeared to be similar to those found in the goat.

Overall it would appear that whilst the site of metabolism and interactions with the hepatic mixed-function oxidase system are well understood, the actual metabolism of PBO, as described by the metabolic products formed, is poorly known in the rodent and only slightly better understood in the goat. The major metabolic pathway involves opening of the methylenedioxy ring yet the only identified metabolites are compounds with metabolism taking place exclusively on the 2-(2-butoxyethyl)ethoxymethyl side chain. The profile of metabolites seen in the goat tissues also shows oxidation taking place along the 2-(2-butoxyethyl)ethoxymethyl side chain. A similar metabolic profile was seen in leaf lettuce following spraying with PBO (Selim, 1994), with several of the lettuce metabolites being conjugates of identified goat metabolites (Fig. 9.4). This similarity in the nature of the likely residues from two distinct food sources, plants and animals, strongly supports the use of the rat as an animal model in order to validate the effects of ingestion of not only PBO *per se*, but also crops sprayed with the compound, since the crop residues will have effectively been tested in the toxicology programme.

2.2. Interaction of PBO with Cytochrome P-450

Enzyme induction can have a significant impact on the resultant toxicological profile of animals dosed at high levels over the prolonged periods required for

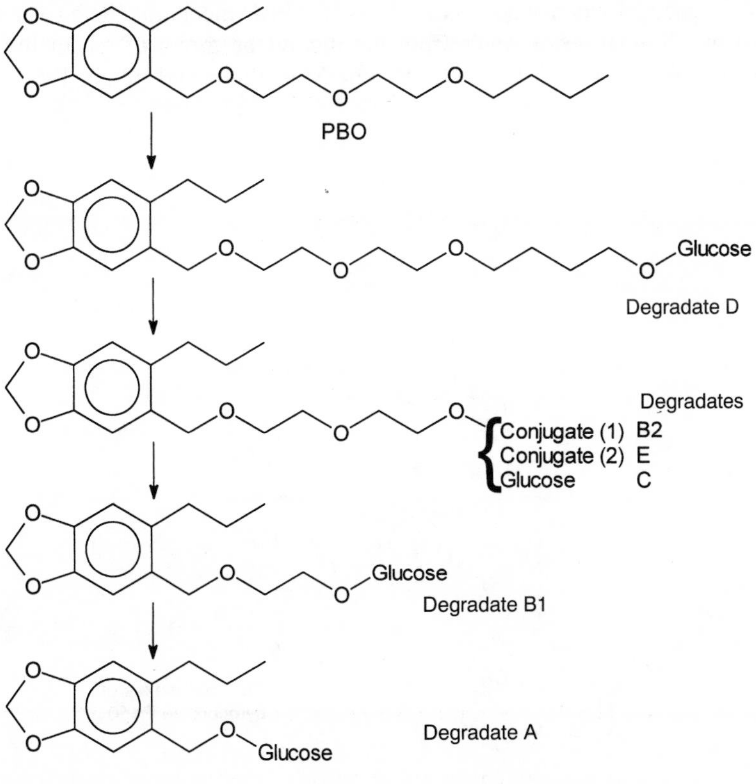

Figure 9.4.   Postulated metabolic pathway of [^{14}C]PBO in leaf lettuce.

toxicity testing. If this can be clearly demonstrated qualitatively and adequately quantitated, this information can be factored into weight of evidence considerations of the relevance of the toxicological findings for humans. Such mechanistic understanding is often critical in order to support optimum classification and regulation by the governmental authorities.

The effect of PBO on the hepatic microsomal mixed function oxidase (MFO) system has been the focus of extensive study over many years. Following the initial administration of PBO there is an immediate inhibition of the cytochromes P450, which in mice only recovers to predose levels at about 24 hours, followed by an overall induction of cytochrome P450 which reaches a maximum value 36–48 hours after treatment (Skrinjaric-Spoljar *et al.*, 1971). In the house fly, however, the extent of the inhibition is much greater, and the duration much longer, with predose levels of cytochrome P450 still not being reached 64 hours after application. This is the basis for the biological activity of PBO.

In the early phase of the interaction of PBO with cytochrome P450 it is believed that the inhibition is caused by the formation of an activated PBO metabolite that binds to the iron atom present in the cytochrome P450 (Wilkinson *et al.*, 1984; Ortiz de Montellano and Reich, 1986), producing a stable complex that is not readily receptive to further metabolism (see Fig. 9.5). The available spectral evidence points to the reactive PBO metabolite being a carbene formed by elimination of water from the methylene group of the methylenedioxy ring (Wilkinson *et al.*, 1984; Ortiz de Montellano and Reich, 1986).

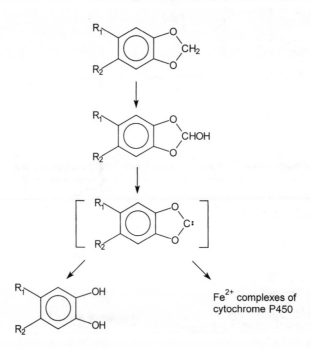

Figure 9.5.   Proposed route of bioactivation of PBO by cytochrome P450.

In further studies to elucidate the enzyme induction potential of PBO, groups of male CD-1 mice were fed diets containing PBO at 10, 30, 100 or 300 mg kg^{-1} day^{-1} for 42 days. Three further groups of mice were also included: control, positive control (0.05% w/w phenobarbitone in the diet) and a group implanted with mini-osmotic pumps containing 5-bromo-2'-deoxy-uridine (BrDU) to label replicating DNA. Statistically significant increases were seen in liver weight at 7 and 42 days in animals fed PBO at 300 mg kg^{-1} day^{-1} (23% and 28%, respectively), in total cytochrome P450 expressed per gram of liver (40%, 43% and 73% in the 30, 100 and 300 mg kg^{-1} day^{-1} groups, respectively), and in some MFO enzyme activities in the two highest dose groups compared with controls. The specific isoenzymes (forms of the same enzyme) of CYP450 examined were those of the CYP1A, CYP2B and CYP3A subfamilies. A transient increase in replicative DNA synthesis was observed, at 7 days only, in mice fed the highest dose of PBO, indicating a threshold of c. 100 mg kg^{-1} day^{-1}. Histological examination of the livers revealed that only the 300 mg kg^{-1} day^{-1} dose group had midzonal hypertrophy of the hepatocytes after 7 days, whereas both the 100 and 300 mg kg^{-1} day^{-1} groups had this effect after 42 days of treatment.

In the phenobarbitone positive control mice there was also a marked but transient increase in replicative DNA synthesis at 7 days and increased liver weights (30% and 29% at 7 and 42 days, respectively, with centrilobular hypertrophy at both time points). The degree of induction of total hepatic cytochrome P450 and the various isoenzymes examined was greater than that seen for PBO on a molar basis.

Similar studies were performed in groups of male rats of the Fisher 344 strain fed at nominal dose levels of 100, 550, 1050 and 1850 mg kg^{-1} day^{-1} for either 7 or 42 days. A group of eight rats from each of the four PBO-treated and two control groups (control diet or phenobarbitone 0.05% w/w) were implanted with mini osmotic pumps containing BrDU to label replicating DNA on day 0, and killed on day 7. The remaining eight rats from each of the treatment groups were implanted with an osmotic pump containing BrDU on day 35 of treatment and killed on day 42.

PBO caused statistically significant dose-related increases in absolute and relative liver weights at 550 mg kg^{-1} day^{-1} and above and relative liver weights were increased at 100 mg kg^{-1} day^{-1}. Body weights were significantly reduced in all animals fed 550 mg kg^{-1} day^{-1} PBO and above at both time points. There were significant dose-related increases in liver microsomal protein content and total cytochrome P450 content at 7 and 42 days, although the latter was only statistically significant in the three highest dose groups and the degree of induction was less than that seen at day 7. Replicative DNA synthesis was increased at the two intermediate dose levels at 7 days and remained significantly elevated in the 1050 mg kg^{-1} day^{-1} dose group at 42 days. Clinical chemistry analysis of serum indicated liver damage at both time points in a dose-dependent manner. Histopathological examination of the livers showed periportal/midzonal hypertrophy in all animals treated for 7 or 42 days at dose levels above 550 mg kg^{-1} day^{-1} and changes in the bile duct/oval cell epithelium at 7 days with bile duct/oval cell proliferation in the majority of animals at 42 days. These findings were not seen at $100 \text{ mg kg}^{-1} \text{ day}^{-1}$ at either time point. Scattered individual cell necrosis was reported in PBO-fed animals at the two highest dose levels. Phenobarbitone again produced a similar pattern of liver effects to those seen with PBO, namely potent enzyme induction, significant increase in relative liver weight and centrilobular hypertrophy.

In conclusion, these studies demonstrate that PBO produces induction of xenobiotic MFO metabolizing enzymes which is associated with liver enlargement, hypertrophy and hyperplasia. The findings are closely similar to those seen with phenobarbitone and have a clear threshold below which no effects are seen. The absolute no effect levels were $100 \text{ mg kg}^{-1} \text{ day}^{-1}$ in the rat and $10 \text{ mg kg}^{-1} \text{ day}^{-1}$ in the mouse.

3. RELEVANCE OF ADME STUDIES WITH PBO FOR ASSESSING HUMAN SAFETY

An integrated approach to the use of absorption, distribution, metabolism and excretion (ADME) studies in conjunction with toxicity testing protocols is essential to confirm: (1) the relevance of the animal models used to extrapolate any potential hazard to humans; (2) the physiological reason for certain types of toxicological findings; (3) the safety of milk, meat, eggs and crops that may have been exposed to PBO or its residues; and (4) an appropriate residue to permit the establishment of suitable methodologies for Public Health Laboratory analysis of the residues present in such animal and crop products to determine average daily intake and hence to ensure consumer safety.

3.1. Relevance of Animal Species Studied to Humans

Despite the lack of a clearly defined metabolic pathway for PBO, the studies so far conducted tend to support the premise that the compound is absorbed and excreted similarly by all species examined. The compound appears to be well absorbed in rats and mice, and unchanged PBO can be identified in the fat of rats and goats (Fishbein *et al.*, 1969; Selim, 1995a). The profile of the metabolites in mice and in goats shows that, apart from opening of the methylenedioxy ring, the metabolism of PBO proceeds via successive oxidations on the 2-(2-butoxyethoxy)ethoxymethyl side chain. These data, suggest that there is no reason to assume that humans will handle PBO differently from the other species examined.

3.2. Explanation of Toxicological Findings

A comprehensive package of toxicology studies has been conducted with PBO. The findings from these indicate a very safe toxicity profile. Until recently there had been no evidence of any oncogenic potential in several rodent chronic and oncogenicity studies. However, a recent re-registration study for EPA employing dose levels of up to 300 mg kg^{-1} day^{-1} in the CD-1 mouse showed that there was a dose-related increase in liver weight and benign proliferative lesions (hyperplasia and adenoma) in the liver. In rats, the most consistent subchronic finding was increased liver weights. At the end of the 2-year combined chronic and oncogenicity study, apart from very marked increases in liver weight, hypertrophy, hyperplasia, focal cells and eosinophilic cells, there was a low incidence of thyroid hyperplasia and tumours.

In the presence of a negative package of genotoxicity studies, what was the toxicological significance of the findings – could PBO be an oncogen? Clearly this was a critical issue and it became necessary to elucidate the nature of these findings in order to gain the optimum risk assessment and regulatory acceptance of PBO for humans. It is well known that liver hyperplasia and adenoma occur in mice as a result of chronic exposure to liver enzyme inducers (Tennekes *et al.*, 1979; Schulte-Hermann, 1974; IARC, 1979, 1987). Phenobarbitone is an archetypal inducer of microsomal enzymes of the cytochrome P450 2B subfamily, a distinct and different spectrum of isoenzymes from that induced by 3-methylcholanthrene, which induces the CYP1A subfamily. The CYP1A subfamily has been correlated with the metabolic activation of carcinogens to their carcinogenic form (Ioannides *et al.*, 1984). Phenobarbitone treatment has been shown to induce hepatocellular adenomas and carcinomas in male and female mice and adenomas in rats. The response in mice appears to parallel the spontaneous incidence of hepatic tumours in the various strains examined; mice also appear to be more sensitive than rats. However, phenobarbitone is not a human liver carcinogen (Clemmesen and Hjalgrim-Jensen, 1978, 1980; Olsen *et al.*, 1993) and has not been shown to bind to or damage DNA. Moreover relatively extensive human epidemiological studies at enzyme-inducing doses have indicated no increased cancer risk: thus the relevance of the hepatic tumours in mice for humans is doubtful (McClain, 1990). The same conclusion may be drawn for PBO, which demonstrated a phenobarbitone-like induction profile in mice (Section 2.2). The spectrum of isoenzymes examined included those most commonly reported to be induced by the nongenotoxic mouse liver oncogen phenobarbitone and both compounds produced a transient increase in replicative DNA synthesis and hypertrophy of the liver of CD-1 mice at high dose levels. A no observed effect level (NOEL) for induction of hepatic enzymes by PBO was 10 mg kg^{-1} day^{-1}, this being consistent with the clear NOEL for liver changes seen in the Bushy Run mouse oncogenicity study (Hermanski and Wagner, 1993). Thus the results of the mode of action study confirm that chronic dietary administration of PBO could have been predicted to lead to the observed effects reported in the CD-1 mouse oncogenicity study. The mode by which hepatoadenomas in the mouse are produced by PBO is likely to be similar to that of other MFO inducers such as phenobarbitone, and clearly only occurs above a defined physiological threshold of increased MFO activity.

The finding of a low incidence of thyroid tumours in rats could theoretically have been due to either a genotoxic or nongenotoxic mechanism. The former would be of serious regulatory concern but seemed unlikely in the presence of an uneventful package of mutagenicity studies against chromosome, gene and

DNA endpoints. It is also well known that thyroid tumours arise in rats as a consequence of altered pituitary–thyroid homeostasis. This commonly results from induction of hepatic enzymes which increase the catabolism of circulating thyroid hormones (T_3 and T_4), in turn increasing thyroid stimulating hormone (TSH) secretion and hence thyroid activity in an attempt to maintain homeostasis. The net effect of long-term stimulation of the thyroid by increased levels of TSH is a greater risk of developing tumours in thyroid follicular cells. This promoting effect is greater in rats than in mice, with males more often developing a higher tumour incidence than females. The marked sensitivity of the rat thyroid gland to the tumour-promoting effects of increased circulating levels of TSH is in marked contrast to findings in humans, where there is no convincing evidence that treatment with drugs or exposure to environmental chemicals, such as PBO, that induce hepatic microsomal enzymes, lead to an increased risk for the development of thyroid cancer (Curran and de Groot, 1991).

The hepatic mechanistic studies in the rat were thus of great value to show that the probable mechanism exists for PBO to cause such thyroid effects in the rat as a consequence of its potent microsomal enzyme-inducing activity. The fact that PBO showed a closely comparable profile to the archetypal inducer phenobarbitone, which also induces increased thyroid activity, serves only to reinforce the likelihood that this is a classic secondary mechanism. The primary effect of both xenobiotics is inextricably linked to their biological property of liver enzyme induction. This perturbation in turn leads to remote secondary effects on the thyroid. By recognizing and understanding this mechanism it may be appreciated that the levels of PBO to which humans might be exposed are more than 1000 times below the threshold for any effect in the liver. In the absence of any primary hepatic effect there can be no risk to humans of any secondary (thyroidal) effect, to which they are in any case significantly less sensitive to than the rat.

3.3. Safety of PBO or Its Residues to Exposed Humans

Interpretation of findings from the metabolic profiling studies in both plants and food-producing animals show that the basic metabolic pathway is similar in the two matrices. The main route of metabolism was oxidation on the 2-(2-butoxyethoxy)ethoxymethyl chain, which has also been shown to be a metabolic pathway in the rodent. Since these metabolites are formed in the rat, they have therefore been tested for up to 2 years in long-term reproduction, teratology and mutagenicity rodent studies. The findings from such animal studies can be used to extrapolate the hazard to humans. By employing appropriate safety factors it is then possible to set acceptable daily intake/exposure levels to assure human safety from all possible combined sources of exposure. This final part of the process, which requires estimates of exposure to be established, is known as risk assessment and may also involve risk management controls, such as protective clothing to ensure safety. For PBO the acceptable daily intake (ADI) is 0.2 mg per kg body weight (WHO JMPR, 1996). The product's wide use, together with an extensive database including detailed mechanistic studies, built over the last three decades, confirms that, as would be predicted, the product has an excellent record of safety to humans.

3.4. Target Residue for Public Health Analysis

Until a more detailed metabolic profile of PBO has been established, the analytical method for residue assessment is based on the amount of the parent compound (PBO) that is present. This will give a reasonable approximation of the level in milk. In the case of crops, acid hydrolysis to methylenedioxy benzyl alcohol can be used to quantitate the majority of the residues present.

4. DISCUSSION

An integrated approach to the use of ADME studies in conjunction with extensive toxicity testing of PBO has been crucial for the development of an optimal regulatory position for the compound. The toxicokinetic studies explain how PBO is metabolized by the body, and how this process in turn leads to its efficacy as synergist for insecticides and also to its activity as a mixed function oxidase inducer in mammals. The latter property accounts for the principal toxicological findings seen with the compound, namely liver tumours in mice and weak thyroid effects in rats.

By understanding this it may be readily appreciated that the toxicological findings in rats and mice are indeed predictable and, because of a differing response in insects, humans are protected by a wide margin of safety. This has of course been borne out in practice by a very long history of safe use, encompassing the last 40 years or so.

Studies have also shown a consistent metabolic pattern for PBO in plants and animals resulting from β-oxidation of the 2-(2-butoxyethyl)ethoxymethyl side chain. In consequence the residues to which humans and animals may be exposed have essentially been evaluated in the full spectrum of toxicological tests covering the potential for adverse effects on skin, eyes, allergy, organs and tissues, pregnancy and reproduction, heredity and oncogenicity. With the benefit of mechanistic studies it has been established that PBO is devoid of intrinsic toxicity. The liver and thyroid effects which did occur were secondary to the potent enzyme-inducing potential of the molecule. They were also threshold mediated, and hence not relevant to humans when PBO is used as directed.

The compound has a good overall toxicokinetic profile, being well absorbed and distributed throughout the body. While some PBO enters the fat the majority is metabolized at four main sites of attack (the methylenedioxy ring, the phenyl ring, the propyl side chain and the 2-(2-butoxyethoxy)ethoxymethyl side chain) and rapidly excreted. Cleavage of the methylenedioxy ring results in the excretion of $^{14}CO_2$, while β-oxidation of the 2-(2-butoxyethoxy)ethoxymethyl side chain results in some metabolites still containing the methylenedioxy ring. The overall metabolic picture is extremely complex with metabolites being excreted via the urine, bile and faeces. By radiolabelling in different parts of the molecule some 20–40 metabolites have been reported to occur. Though many of these metabolites are considered minor in quantitative terms, a number of studies over the years indicate that full toxicology in rodents and nonrodents will have assessed their impact. The fact that there has been no unexplained toxicity with PBO in mammals indicates that PBO is degraded to residues of little or no toxicological significance. Moreover the finding that food-producing animals and lettuce produce a similar spectrum of metabolites to the toxicological species means that any residues consumed by humans are likely to have been taken into account toxicologically and are thus safely covered within the government approved ADI.

In conclusion, the results from in-depth studies of the metabolic fate of PBO amassed over a number of years have provided essential evidence as to the safety to humans of this potent and valuable insecticide synergist.

REFERENCES

Casida, J.E., Engel, J.L., Essac, E.G., Kamienski, F.X. and Kuwatsuka, S. (1966). Methylene-^{14}C-dioxyphenyl compounds: Metabolism in relation to their synergistic action. *Science NY* **153**, 1130–1133.
Clemmesen, J. and Hjalgrim-Jensen, S. (1978). Is phenobarbital carcinogenic? A follow-up of 8076 epileptics. *Ecotoxicol. Environ. Safety.* **1**, 457–470.
Clemmesen, J. and Hjalgrim-Jensen, S. (1980). Epidemiological studies of medically used drugs. *Arch. Toxicol. Suppl.* **3**, 19–25.
Curran, G. and de Groot, L.J. (1991). The effect of hepatic enzyme-inducing drugs on thyroid hormones and the thyroid gland. *Endocrin. Rev.* **12**, 135–150.
Fishbein, L., Falk, H.L., Fawkes, J., Jordan, S. and Corbett, B. (1969). The metabolism of piperonyl butoxide in the rat with 14-C in the methylendioxy or α-methylene group. *J. Chromatogr.* **41**, 61–79.
Hermansky, S.J. and Wagner, C.L. (1993). Chronic Dietary Oncogenicity Study with Piperonyl Butoxide in CD-1 Mice. Unpublished report no. 91 NO134 from the Bushy Run Research Centre. Undertaken for the PBO Task Force, Washington DC, USA.
IARC (1979). *Some Halogenated Hydrocarbons.* IARC Monographs on the Evaluation of the Carcinogenic Risk of Chemicals to Humans, vol. 20 IARC, Lyon.
IARC (1987). IARC Monographs on the Evaluation of Carcinogenic Risks to Humans, updating of IARC monographs vols **1** to **42**. Suppl. **7**. IARC, Lyon.
Ioannides, C., Lum, P. and Parke, D.V. (1984). Cytochrome P448 and the activation of toxic chemicals and carcinogens. *Xenobiotica* **14**, 119–137.
Kamienski, F.X. and Casida, J.E. (1970). Importance of demethylation in the metabolism in vivo and in vitro of methylenedioxyphenyl synergists and related compounds in mammals. *Biochem. Pharmacol.* **19**, 91–112.
McClain, R.M. (1990). Mouse liver tumours and microsomal enzyme inducing drugs. Experimental and clinical perspectives with phenobarbital. In: *Mouse Liver Carcinogenesis. Mechanisms and Species Comparison*, pp. 345–365.
Olsen, J.H., Wallin, H., Boici, J.D., Rask, K., Schulgen, G. and Fraumani, J.F. (1993). Phenobarbital, drug metabolism and human cancer. *Cancer Epidemiol. Prevent.* **2**, 449–452.
Ortiz de Montellano, P.R. and Reich, N.O. (1986). Inhibition of cytochrome P450 enzymes. In: *Cytochrome P450: Structure, Mechanism and Biochemistry* (Ortiz de Montellano, P.R., ed.), pp. 273–314. Plenum Press, New York.
Schulte-Hermann, R. (1974). Induction of liver growth by xenobiotic compounds and other stimuli. *Crit. Rev. Toxicol.* **3**, 97–158

Selim, S. (1991). Absorption, Distribution, Metabolism and Excretion (ADME) Studies of Piperonyl Butoxide in the Rat. Unpublished report no. PO1825 from Biological Test Center, Irvine, CA 92714, USA. Undertaken for the PBO Task Force, Washington DC, USA.

Selim, S. (1994). Nature of ^{14}C-Piperonyl Butoxide Residues in Leaf Lettuce. Unpublished report no. PO792010 from Biological Test Center, Irvine, CA 92714, USA. Undertaken for the PBO Task Force, Washington DC, USA.

Selim, S. (1995a). Nature of the Residue in Meat and Milk of ^{14}C-Piperonyl Butoxide in the Goat Following Oral and Dermal Administrations. Unpublished report no. P0993007 from Biological Test Center, Irvine, CA 92714, USA. Undertaken for the PBO Task Force, Washington DC, USA.

Selim, S. (1995b). Nature of the Residue in Meat and Eggs of ^{14}C-Piperonyl Butoxide in Laying Hens Following Oral and Dermal Application. Unpublished report no. P0993008 from Biological Test Center, Irvine, CA 92714, USA. Undertaken for the PBO Task Force, Washington DC, USA.

Skrinjaric-Spoljar, M., Matthews, H.B., Engel, J.L. and Casida, J.E. (1971). Response of hepatic microsomal mixed-function oxidases to various types of insecticide chemical synergists administered to mice. *Biochem. Pharmacol.* **20**, 1607–1618.

Tennekes, M.A., Wright, A.S. and Dix, K.M. (1979). The effects of Dieldrin, diet and other environmental components on enzyme function and tumour incidence in livers of CF-1 mice. *Arch. Toxicol. Suppl.* **2**, 197–212.

WHO JMPR (1996) Report of the FAO Panel of Experts on Pesticide Residues in Food and the Environment and the WHO Expert Group on Pesticide Residues. FAO Plant Production and Protection Paper 133, pp. 168–171. Rome.

Wilkinson, C.F., Murray, M. and Marcus, C.B. (1984). In: *Reviews of Biochemical Toxicology* 6 (Hodgson, E., Bend, J.R. and Philpot, R.M., eds.), pp. 26–64. Elsevier Science, New York.

10

Plant Metabolism and Crop Residue Studies of Piperonyl Butoxide and its Metabolites

SAMI SELIM and ROBERT TESTMAN

1. INTRODUCTION

Piperonyl butoxide (PBO) in combination with pyrethrum and pyrethroids is used to control insect pests on a wide variety of food crops. Defining the nature and magnitude of the residues in crops resulting from such uses is an important part of the evaluation of the safety of the product to humans and to the environment.

The metabolism of PBO has been studied in three different crops: lettuce, cotton and potatoes. Once the metabolites had been identified, crop residues of PBO and its closely related metabolites were determined in approximately 70 commodities from field studies on eight crop groups.

2. ANALYSIS OF PBO AND ITS METABOLITES

To provide adequate accountability in metabolism studies of the distribution of parent compound and metabolites in each crop, ^{14}C-radiolabelled PBO was used. The use of radiolabelled PBO allows the distinction of parent compound and metabolites from naturally occurring plant compounds.

PBO was ^{14}C-radiolabelled (*) at the benzylic carbon of the butylcarbityl side chain, as shown below.

A range of analytical methods was developed to characterize the nature of metabolites formed in crops both from the ^{14}C metabolism studies and field crop residue studies.

PIPERONYL BUTOXIDE
ISBN 0-12-286975-3

2.1. Fluorescence

PBO fluoresces with an emission wavelength of ~345 nm when excited at ~288 nm (Schreiber-Deturmeny *et al.*, 1993). To determine the structural characteristics responsible for the fluorescence, the fluorescence of the compounds shown in Table 10.1 was determined. This shows that the fluorescent characteristic is a result of an intact methylenedioxybenzene double ring structure.

Table 10.1.   Fluorescence of PBO and related compounds (excitation at 288 nm, emission at 345 nm)

Compound	Structure	Fluorescent?
PBO		Yes
Safrole		Yes
Dihydrosafrole		Yes
Piperonyl alcohol		Yes
1,3-Benzodioxole		Yes
Guaiacol		No
Catechol		No
3-Hydroxy-4-methoxy-benzaldehyde		No
3,4-Dihydroxy-benzaldehyde		No

2.2. Hydrolysis

PBO was found to be unstable when subjected to acid hydrolysis. When [^{14}C]-PBO was refluxed with 1 mol L^{-1}. HCl for 1 hour, a single fluorescent radioactive compound was formed. This compound was isolated by preparative high-performance liquid chromatography (HPLC) and identified by gas chromatography–mass spectroscopy (GC-MS), high-resolution MS, nuclear magnetic resonance (NMR), and comparison of HPLC retention time with a synthetic standard.

The acid hydrolysis product was identified as 2-propanyl-4,5-methylenedioxybenzyl alcohol (hydroxymethyldihydrosafrole or HMDS), with the structure shown below:

2.3. Mass Spectral Analysis

The mass spectrum of PBO (Fig. 10.1) presents characteristic fragments which provide certain structural information. In particular, positive ion electron ionization (EI) and thermospray techniques produce a base peak at *m/z* 176–177. The proposed structure of this base peak is shown below:

m/z 177

The *m/z* 176–177 fragment is significant in that its presence in a metabolite spectrum would indicate that the modifications to the PBO molecule must be limited to the butylcarbityl side chain. Negative ion thermospray produces a large M-1 ion for PBO and some metabolites, giving molecular weight information.

3. METABOLISM STUDIES

Radiolabelled PBO was incorporated into a typical end use insecticide formulation and sprayed 4–6 times per crop at approximately weekly intervals at an application rate of 0.5 lb active ingredient (a.i) per acre. This represented the maximum expected exposure regime for field crops.

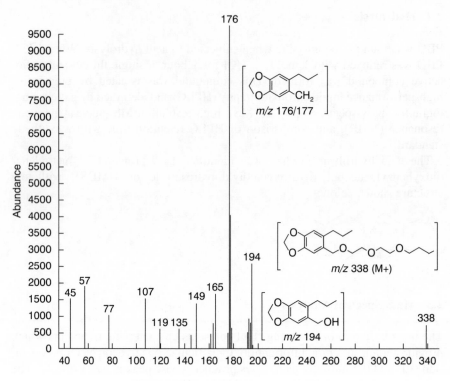

Figure 10.1.   Mass spectra of PBO.

3.1. Metabolism in Leaf Lettuce

3.1.1. Harvest and Total Radioactive Residues

Half of the treated lettuce plants were harvested on the day of the fifth and last application (day 0). The remaining plants were harvested 10 days later (day 10). The plants were homogenized and the total radioactive residue (TRR) of each was determined. The mean TRR (PBO equivalents) for the treated day 0 and treated day 10 plants were 36.6 ppm and 23.0 ppm, respectively (Table 10.2).

3.1.2. Extraction and Analysis

Samples of the day 0 and day 10 homogenates were extracted and partitioned as shown schematically in Fig. 10.2. The petroleum ether and Sep-Pak aceto-nitrile fractions were analysed by HPLC (Fig. 10.3). The HPLC analyses showed a similar metabolite profile, with a lower percentage of PBO in the day 10 sample (24.4% of TRR) than in the day 0 sample (49.3% of TRR). The HPLC profile of the petroleum ether extracts showed that the predominant radioactivity extracted into the petroleum ether was parent compound (6.3 ppm for day 10). Numerous metabolites were observed in the HPLC profile of the Sep-Pak

Table 10.2. Total radioactive residue (TRR) and PBO in metabolism study matrices

Crop	Matrix	TRR (ppm)	PBO (ppm)
Cotton	Leaves	142.0	23.9
	Hulls	7.0	1.2
	Seeds	0.4	<0.1
	Lint	0.5	<0.1
Lettuce	Leaves (day 0 Harvest)	36.6	18.0
	Leaves (day 10 Harvest)	23.0	6.3
Potato	Leaves	616.9	240.9
	Tubers	0.5	ND

ND, none detected.

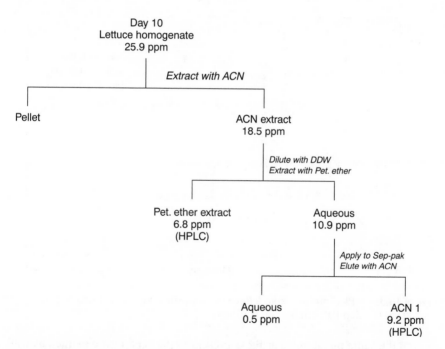

Figure 10.2. Extraction of lettuce leaf homogenate. (ACN, acetonitrile; DDW, distilled deionized water).

acetonitrile extracts with five major metabolite peaks ranging in concentration from 0.2 to 2.0 ppm (day 10).

The residual pellet following acetonitrile extraction was extracted with $1 \, \text{mol} \, L^{-1}$ NaOH or refluxed with $1 \, \text{mol} \, L^{-1}$ HCl and the extracts partitioned with ethyl acetate or applied to C_{18} Sep-Paks. The HPLC profiles of the extracts

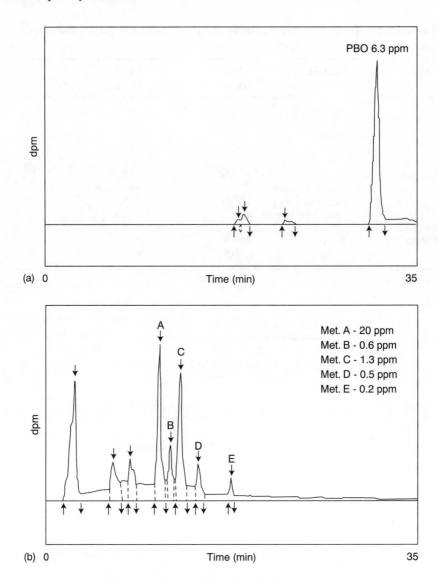

Figure 10.3. HPLC profiles from lettuce metabolism study. (a) Petroleum ether extract; (b) C$_{18}$ Sep-Pak acetonitrile partition.

showed the same metabolites as those present in the acetonitrile extract, as well as several metabolites which were highly polar or nonextractable from the aqueous phase of the acid-hydrolysed extract, indicating a significant degradation to highly polar metabolites.

3.1.3. Metabolite Characterization

The five major metabolite peaks were extensively purified by repeated prepara-tive HPLC. Metabolite B was determined to consist of two metabolites during the HPLC purification. These metabolites were designated as Metabolite B1 and Metabolite B2. All the metabolites were found to fluoresce at the wavelengths characteristic of PBO. The metabolites were individually acid-hydrolysed, with all producing HMDS. The fluorescence and acid hydrolysis data indicate that the methylenedioxybenzene ring structure and propanyl side chain must be intact, and that the differences between the metabolites must be limited to the butylcarbityl side chain.

3.1.4. Metabolite Identification

The purified metabolites were analysed by liquid chromatography–mass spectrometry (LC-MS) using positive ion thermospray, negative ion thermospray and particle beam EI interfaces. Sample spectra for metabolite A are shown in Fig. 10.4. The positive ion thermospray produced a m/z 177 fragment, characteristic of the MS of PBO, confirming that the analysed peak is a metabolite of PBO that contains the HMDS structural skeleton with modification limited to the butylcarbityl side chain. This result is in agreement with the earlier fluorescence and acid hydrolysis results. The negative ion thermospray showed an M-1 fragment at m/z 355, giving a molecular weight of 356. The particle beam EI showed a characteristic PBO fragment at m/z 177 and a peak at m/z 194. The difference between the molecular weight of 356 and the 194 is 162, which corresponds to a C_6 sugar conjugated to the alcohol resulting from the cleavage of the ether bond closest to the benzene ring.

The analysis of the six metabolites by LC-MS using positive ion thermospray produced similar mass spectra with a strong peak at m/z 177 in all samples. The negative ion thermospray gave M-1 fragments for the six metabolites. Particle beam EI showed the characteristic m/z 177 peak, as well as an aglycone peak corresponding to the structural representation below:

From the data, a metabolic pathway was postulated as shown in Fig. 10.5.

3.2. Metabolism in Potato

Potato tubers were harvested 8 days after the fourth and last application of the $[^{14}C]$-PBO formulation. The potato tubers were composited and a sample was homogenized and analysed for TRR. The concentrations of TRR in potato tubers and leaves are shown in Table 10.2 and indicate that limited translocation from the leaves to the tubers had occurred.

A tuber homogenate sample was extracted twice with acetone and once with

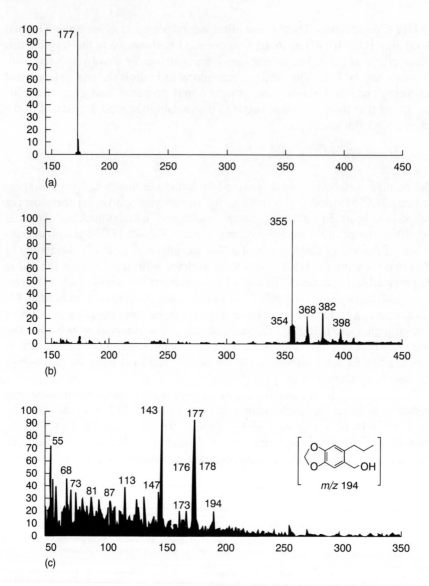

Figure 10.4.   Mass spectra of lettuce metabolite A. (a) Positive ion thermospray; (b) negative ion thermospray; (c) particle beam EI.

methanol, and the extracts partitioned as shown schematically in Fig. 10.6. Metabolites in the acidified and nonacidified ethyl acetate extracts were characterized by HPLC (Fig. 10.7). At least fifteen metabolite peaks with concentrations ranging from 0.006 to 0.018 ppm were observed. No PBO was found. The majority of the residues from the tuber pellet were polar and insoluble in organic solvents. Phenylhydrazine derivatization failed to precipitate the acid-hydrolysed ^{14}C residues as osazones, indicating that the radioactivity was not incorporated into sugars or starch. These results characterized the residues of

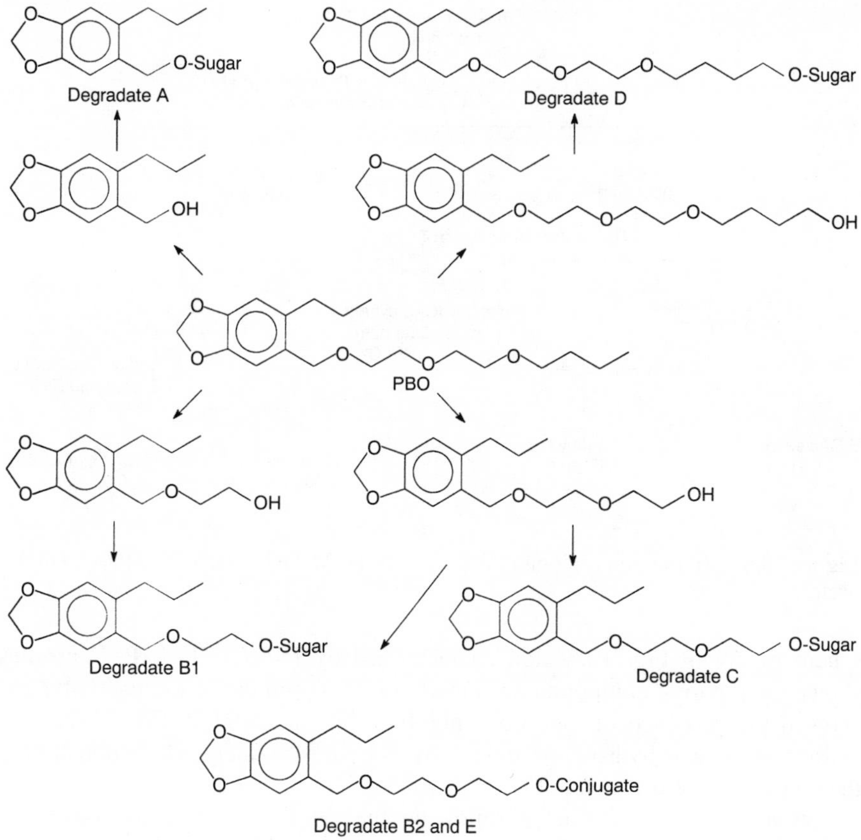

Figure 10.5. Proposed metabolic pathway in lettuce.

PBO in potato tubers as highly polar in nature with concentrations of individual metabolites observed in the tubers less than 0.020 ppm.

Analysis of the potato leaf extracts by HPLC showed that parent compound accounted for over 50% of the radioactivity and that multiple metabolite peaks were present. The metabolites did not match with those observed in the potato tuber, indicating that further metabolism has occurred to the translocated residues.

3.3. Metabolism in Cotton

Cotton bolls and leaves were harvested 15 days after the sixth and last application of the ^{14}C-PBO formulation. The cotton bolls were manually separated into hulls, seeds and lint. Samples were composited by commodity type and analysed for TRR. The concentrations of TRR are shown in Table 10.2, indicating limited translocation had occurred from the leaves and hulls to the seeds and lint.

A composited leaf homogenate sample was initially extracted as shown

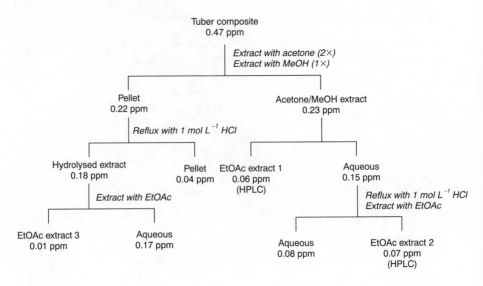

Figure 10.6. Extraction of potato tuber composite. MeOH, methanol; EtOAc, ethyl acetate.

schematically in Fig. 10.8, and characterized by HPLC. The HPLC profiles showed that parent compound accounted for 23.9 ppm of the radioactivity, and showed the presence of numerous highly polar metabolites (Fig. 10.9). The extracts were acid-hydrolysed and the hydrolysates re-analysed, demonstrating the conversion of the metabolites to HMDS.

Cotton hulls were extracted following the same procedure as above and analysed, showing that PBO accounted for 1.23 ppm of the TRR.

HPLC analysis of cotton seed homogenate following the extract scheme shown in Fig. 10.10 showed the presence of PBO (0.086 ppm) and two polar metabolites (ca. 0.04 ppm).

Following extraction (Fig. 10.11), analysis of an initial MeOH extract of the cotton lint composite sample by HPLC (Fig. 10.12) revealed the presence of a major polar metabolite peak, accounting for 0.191 ppm of the radioactivity in the extract, and one nonpolar peak which appeared to be PBO, accounting for 0.047 ppm of the radioactivity in the extract. Metabolites in other extracts were found to be too polar for extraction and HPLC analysis.

3.4. Discussion of Plant Metabolism

The distribution of radioactivity observed in the metabolism studies shows that very little translocation of PBO-related residues from sites of application of the parent compound occurs in plants. The metabolism of the applied PBO on leaf surfaces occurs at a moderate rate with significant amounts of PBO present 8–15 days after the last application. Metabolism is predominantly via cleavage of the ether linkages of the butylcarbityl side chain of PBO, followed by conjugation with sugars or other polar molecules.

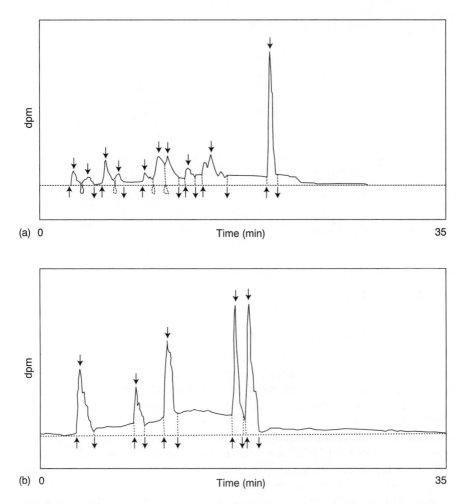

Figure 10.7.   HPLC profiles from potato metabolism study. (a) Acidified ethyl acetate extract; (b) nonacidified ethyl acetate extract.

PBO in water is also degraded to HMDS by the action of sunlight (Selim, 1995). It is therefore likely that PBO applied to plants will be susceptible to photodecomposition as well as to metabolism and that the isolated metabolites could result from a combination of the two mechanisms. The residues which were isolated from potato tubers and cotton seeds were significantly altered to highly polar entities incapable of producing HMDS when acid-hydrolysed.

4. CROP RESIDUE STUDIES

In order to determine the distribution of PBO and its principal metabolites (as HMDS) in crops *in situ*, a series of field studies was carried out between 1992 and 1996 in different geographical regions of the USA on cotton plus eight crop

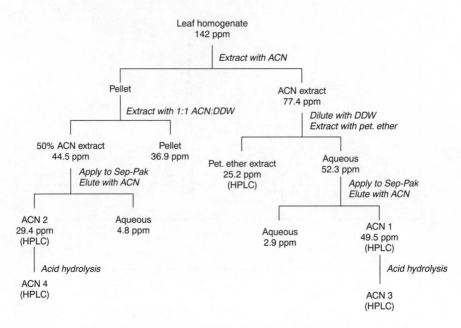

Figure 10.8.   Extraction of cotton leaf homogenate.

groups: brassica, citrus, cucurbits, fruiting vegetables, leafy vegetables, legume vegetables, root and tuber and small fruit. Appropriate crops were selected to represent each crop group in accordance with Environmental Protection Agency (EPA) guidelines.

The representative formulation selected for the crop field trial was Pyrenone® Crop Spray (Roussel-Uclaf, now AgrEvo Environmental Health) containing 6% pyrethrins, 60% PBO, and 34% emulsifiers and solvents. Ten applications, greatly in excess of normal commercial usage (each 0.5 lb acre^{-1} (56 mg/m^2) actual PBO per application), were made by commercial applicator to each crop before harvest using standard agricultural equipment. Each crop was harvested and stored frozen until residue analysis was conducted. Between 3 and 15 representative commodities per crop group were analysed.

4.1. Analytical Method

The analytical method used for this residue determination required both sensitivity and selectivity for the compound(s) of interest. The scientific literature revealed a variety of analytical methods for PBO which were of potential interest, although none were deemed appropriate for analysis of low concentration of PBO residues.

Several published methods showed acceptable recoveries from specific agricultural commodities using GC with flame ionization or electron capture detection (Kawana *et al.*, 1976; IUPAC, 1979; Simonaitis, 1983), GC-MS (Cave, 1981) and HPLC with UV detection (Krause and August, 1983; Nijhuis *et al.*,

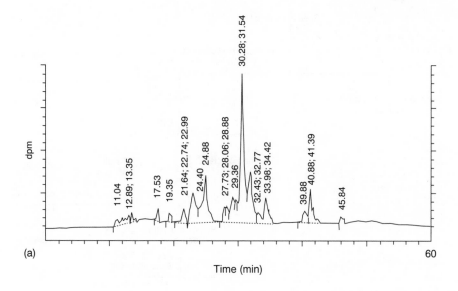

(a)

Time (min)

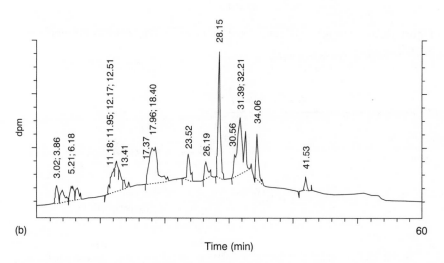

(b)

Time (min)

Figure 10.9. HPLC profiles from cotton leaf metabolism study. (a) Degradate fraction 1; (b) degradate fraction 2.

1985). However, these methods required extensive sample preparation and column clean-ups and did not analyse metabolites.

Liquid chromatography with fluorescence detection had been used for the analysis of PBO in grain crops with minimal sample preparation (Isshiki *et al.*, 1977) and was further developed with the use of HPLC (Krause, 1983).

For the current investigation, HPLC with fluorescent detection was chosen to analyse for PBO and major metabolites. The analysis of the metabolites used

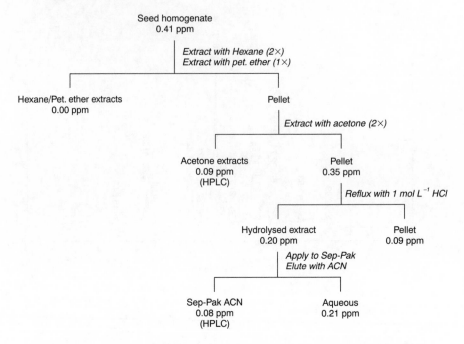

Figure 10.10. Extraction of cotton seed homogenate. (Pet. ether, petroleum ether)

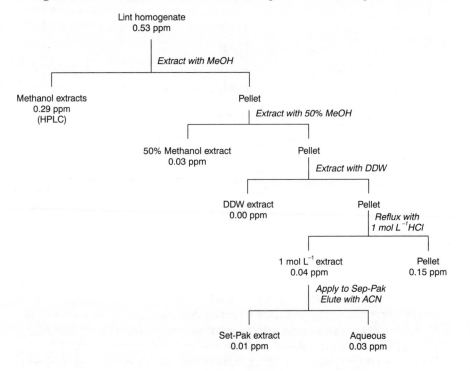

Figure 10.11. Extraction of cotton lint homogenate.

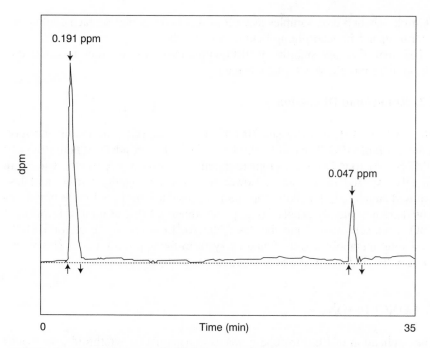

Figure 10.12. HPLC profile from cotton lint metabolism study. Methanol extract.

acid hydrolysis to convert the metabolites to a single compound, HMDS, which was then quantitated by HPLC using a fluorescence detector. A schematic representation of the extraction and hydrolysis method is shown in Fig. 10.13. The HPLC analysis was conducted using a C_{18} column and an isocratic acetonitrile/water mobile phase with either 70% acetonitrile (PBO) or 50% acetonitrile (HMDS). The analytical method was subjected to thorough validation against

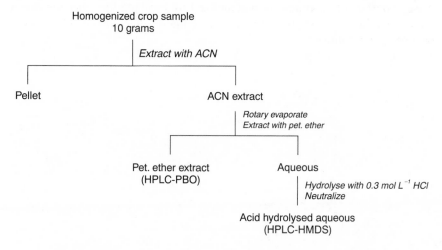

Figure 10.13. PBO and degradate residue method in field crops.

[14]C-radiolabelled crop samples as well as fortified samples of each commodity prior to its use in determining field residue levels.

The freezer storage stability of PBO within each raw agricultural commodity for up to 12 months was also confirmed.

4.2. Results and Discussion

The residue levels of PBO and HMDS for the assayed commodities are presented in Table 10.3. The results show that the highest residues of both PBO and HMDS were found on leafy crops (e.g. lettuce, mustard greens, etc.) which are directly exposed to the spray. Lower residues are found on fruits and less exposed stems (e.g. broccoli) and minimal residues are found in commodities which are not directly exposed (e.g. potato tubers). Little or no translocation of PBO or closely related metabolites occurred. Low residue levels of HMDS-producing metabolites were found away from the application site. The results confirm the findings of the radiolabelled metabolism studies.

5. CONCLUSIONS

The application of PBO to food crops as a synergist to pyrethroid insecticides results in total residues ranging from very low residues in root crops to significant (>20 ppm) residues in leafy vegetables. This broad range of residue values ties together with the distribution pattern observed in the metabolism studies to show that very little translocation of PBO-related residues occurs in plants. In addition, those residues which appear to have been translocated are significantly degraded to highly polar entities incapable of producing HMDS when acid-hydrolysed. The vast majority of the PBO-derived residues present are found in areas of direct application, especially the leaves. PBO residues in these exposed commodities are metabolized by breaking the ether linkages of the butylcarbityl side chain and conjugating the resultant alcohols with sugars or other polar entities.

Table 10.3 PBO and degradate residue level in treated field crops

Crop group	Crop	Commodity type[a]	Location(s) (US state)	PBO levels found (ppm)	Degradate levels found (ppm PBO equiv.)
Brassica	Broccoli	RAC	CA, OR, AK	0.6–2.3	<0.1–0.4
	Cabbage	RAC	CA, FL, NY	<0.1–6.4	0.5–2.6
	Mustard green	RAC	GA, TX	25.1–38.2	NA
	Mustard seed	RAC	GA	<0.1–2.1	<0.1–0.4
Citrus	Grapefruit	RAC	FL, TX	0.3–1.4	<0.1–0.1
	Lemon	RAC	AZ, CA	1.1–3.1	NA
	Orange	RAC	CA, FL, TX	0.5–1.0	<0.1–0.2
	Orange dry pulp	PC	FL	47.7–59.7	NA
	Orange juice	PC	FL	<0.1	NA
	Orange molasses	PC	FL	4.9–5.0	NA
	Orange oil	PC	FL	138.4–146.4	NA
Cotton	Forage	RAC	AZ, CA, LA, MS, TX	10.7–36.8	2.9–10.2
	Seed	RAC	AZ, CA, LA, MS, TX	<0.1–0.2	<0.1
	Hulls	PC	TX	0.1	NA
	Meal	PC	TX	<0.1	NA
	Crude oil	PC	TX	0.5–0.7	NA
	Refined oil	PC	TX	1.3–2.7	NA
	Soapstock	PC	TX	0.2–0.5	NA
Cucurbits	Cantaloupe	RAC	AZ, CA	0.4–0.8	<0.1–0.1
	Cucumber	RAC	MI, NC	<0.1–0.7	0.1–0.2
	Summer squash	RAC	FL, GA, NJ, TX	0.1–0.3	<0.1–0.1
Fruiting vegetables	Pepper	RAC	CA, NC, TX	0.2–1.4	<0.1–0.2
	Tomato	RAC	FL, MI, NJ	0.2–1.0	<0.1–0.1
	Tomato juice	PC	CA	0.9–1.5	NA
	Tomato puree	PC	CA	2.3–3.4	NA
	Dry pomace	PC	CA	249.6–346.6	NA
	Wet pomace	PC	CA	42.8–62.0	NA

Table 10.3 *Continued*

Crop group	Crop	Commodity type[a]	Location(s) (US state)	PBO levels found (ppm)	Degrade levels found (ppm PBO equiv.)
Leafy vegetables	Celery with leaves	RAC	CA, MI	6.1–23.4	<0.1–1.8
	Celery w/o leaves	RAC	CA, MI	1.0–3.7	<0.1–0.4
	Head lettuce with wrap	RAC	CA, FL	3.2–5.0	0.7–1.3
	Head lettuce w/o wrap	RAC	CA, FL	<0.1–0.5	<0.1
	Leaf lettuce	RAC	AZ, FL	16.1–23.1	1.8–2.9
	Spinach	RAC	CO, TX	27.9–38.7	4.8–8.4
Legume vegetables	Dry bean forage	RAC	CO, ND	9.4–24.6	9.4–20.0
	Dry bean hay	RAC	CO, ND	10.7–21.4	5.3–20.8
	Dry bean seed	RAC	CO, ND	<0.1–0.1	<0.1
	Dry bean vine	RAC	CO, ND	10.6–25.5	5.7–42.7
	Dry pea forage	RAC	TX, WA	<0.1–41.6	<0.1–16.2
	Dry pea hay	RAC	TX, WA	1.2–48.0	6.7–60.1
	Dry pea seeds	RAC	TX, WA	0.1–0.6	<0.1
	Dry pea vine	RAC	TX, WA	27.2–96.5	4.0–78.6
	Succulent bean hay	RAC	FL, WI	6.2–41.6	30.1–48.2
	Succulent bean pods	RAC	FL, WI	0.3–2.2	0.5–6.1
	Succulent pea vine	RAC	FL, WI	14.0–28.2	NA
	Succulent bean waste	PC	NY	39.7–56.2	NA
	Succulent pea hay	RAC	CA, MN	25.2–152.8	14.4–20.6
	Succulent pea pods	RAC	CA, MN	1.0–5.1	0.4–1.7
	Succulent pea vine	RAC	CA, MN	35.0–47.4	6.3–8.3
Root and tuber	Carrot	RAC	TX	0.6–1.1	<0.1
	Potato	RAC	CO, ID, ME	<0.1–0.1	<0.1
	Radish leaves	RAC	FL	34.7–37.4	10.3–11.2
	Radish roots	RAC	FL	0.2–0.3	<0.1
	Sugar beet leaves	RAC	MN, ND	8.4–14.5	3.1–4.6
	Sugar beet roots	RAC	MN, ND	<0.1	<0.1
	Sugar beet pulp	PC	CA	0.2–0.3	NA
	Sugar beet molasses	PC	CA	<0.1	NA
	Refined sugar	PC	CA	<0.1	NA

Table 10.3 *Continued*

Crop group	Crop	Commodity type[a]	Location(s) (US state)	PBO levels found (ppm)	Degradate levels found (ppm PBO equiv.)
Small fruit	Blackberry	RAC	OR	2.7–2.9	1.6–2.1
	Blueberry	RAC	MI, NC	4.2–5.5	1.2–4.4
	Cranberry	RAC	MA	2.8–4.2	NA
	Grape	RAC	NY	7.7–9.6	NA
	Grape dry pomace	PC	CA	54.5–80.6	NA
	Grape wet pomace	PC	CA	22.3–30.7	NA
	Grape juice	PC	CA	0.2	NA
	Raisin	PC	CA	13.9–14.9	NA
	Raisin waste	PC	CA	23.4–33.8	NA
	Strawberry	RAC	FL, OR	1.3–3.1	0.2–0.5

NA, not analysed.
[a] RAC, Raw Agricultural Commodity
PC, Processed Commodity

REFERENCES

Cave, S. (1981). Simultaneous estimation of bioresmethrin and piperonyl butoxide by gas–liquid chromatography with chemical-ionization mass spectrometry. *Pestic. Sci.* **12**, 156–160.

Isshiki, K., Tsumura, S. and Watanabe, T. (1977). Analytical method for piperonyl butoxide in agricultural products. II. Determination by high-speed liquid chromatography. *Shokuhin Eiseigaku Zasshi* **18**, 159–163.

IUPAC (1979). IUPAC reports on pesticides. 9. Recommended methods for the determination of residues of pyrethrins and piperonyl butoxide. *Pure Appl. Chem.* **51**, 1615–1623.

Kawana, K., Nakaoka, T. and Fukui, S. (1976). Detection of piperonyl butoxide in marketed rice grains. *Kanagawa-ken Eisei Kenkyusho Kenkyu Hokoku* **6**, 33–36.

Krause, R. (1983). Determination of fluorescent pesticides and metabolites by reversed-phase high-performance liquid chromatography. *Chromatogr.* **255**, 497–510.

Krause, R. and August, E.M. (1983). Applicability of a carbamate insecticide multiresidue method for determining additional types of pesticides in fruits and vegetables. *Assoc. Off. Anal. Chem.* **66**, 234–240.

Nijhuis, H., Heeschen, W. and Hahne, K. (1985). Determination of pyrethrum and piperonyl butoxide in milk by high-performance liquid chromatography (HPLC). *Pyreth. Post* **16**, 14–17.

Schreiber-Deturmeny, E., Pauli, A. and Pastor, J. (1993). Determination of safrole, dihydrosafrole and chloromethyldihydrosafrole in piperonyl butoxide by high-performance liquid chromatography (HPLC). *J. Pharmaceutical Sci.* **82**, 813–816.

Selim, S. (1995). Aqueous Photolysis of Piperonyl Butoxide. Unpublished report no. P0594010 from Biological Test Center, Irvine, CA 92714, USA. Undertaken for the PBO Task Force, Washington DC, USA.

Simonaitis, R. (1983). Recovery of piperonyl butoxide residues from bread made from cornmeal and wheat flour. *Pyreth. Post* **15**, 66–70.

11
The Evaluation of Synergistic Action in the Laboratory and Field

DUNCAN STEWART

1. INTRODUCTION

The evaluation of synergism in the laboratory must always be related to its practical use in formulated products. This chapter aims to present some basic principles using products in commercial use and reference is made to both published and unpublished experimental data.

2. DEFINITION OF SYNERGISM

Yamamoto (1973) defined synergism as 'greater than additive toxicity'. He also stated that a 'pesticide synergist is a term usually used for compounds that are non-toxic or negligibly toxic at the dosage employed but which serve to enhance the toxicity of a pesticide chemical when combined with it in a formulation'. Yamamoto also noted that 'the effectiveness of an insecticide synergist is commonly expressed by the ratio of the LD_{50} of the insecticide alone to the LD_{50} of the insecticide with the synergist.

Hewlett (1968) distinguished between synergism and potentiation, saying that 'if one compound is toxic on its own, and that another on its own is not', then 'if a combination of the two is more toxic than the insecticidal component alone, synergism will be said to occur. Whereas if two compounds are each separately toxic to the insect, and if the toxicity of the combination is greater than expected from the sum of the toxicities of the separate components, potentiation will be said to have taken place.'

Brown et al. (1967) noted that the administration of two compounds to the insect sometimes gives an effect greater than the sum of the individual activities; this was termed potentiation. When one of the compounds had little or no effect when used alone, it was deemed to be a synergist.

PIPERONYL BUTOXIDE
ISBN 0-12-286975-3

2.1. A Brief Review

Review papers by Metcalf (1967) and Casida (1970) drew attention to the vast amount of research aimed at identifying new synergists and evaluating them against compounds across the whole spectrum of the insecticidal groups that were known.

Metcalf listed attempts to relate synergism 'to such diverse factors as stabilisation of droplet size, reduction [sic] of rate of knockdown, stimulation of flight activity, prevention of deterioration of the toxicant, increased penetration into the insect or formation of molecular complexes between synergist and insecticide'. He added that 'recent investigations have made it very clear that virtually all examples of pronounced synergism are related to interference by the synergist of the *in vivo* metabolic detoxification of the insecticide'.

Casida reviewed synergism by mixed function oxidase compounds, stressing that

> practical use of synergists is closely associated with pyrethrum because they greatly enhance the effectiveness of this expensive natural product. Only a relatively few compounds are used as synergists even though three decades of screening have revealed hundreds of candidate compounds of this type, not only for pyrethrum or other pyrethroids, but also for certain other insecticide chemicals (especially methyl carbamates).

He also observed that 'there are only four methylene dioxyphenyl (MDP) synergists used commercially in the United States: PBO, Sulphoxide, *N*-Propyl Isome and Tropital. PBO, introduced in 1947, dominates'.

3. EVALUATION METHODOLOGY

There is only one method that gives sufficient accuracy of dose application for a basic understanding of the relationship between an insecticide and a synergist. The measured droplet technique of applying insecticide in solvent to a selected site on an insect is the generally accepted procedure. The use of apparatus such as that described by Glynne Jones and Lowe (1956) gives the best results. This was commercially produced by Burkhard Manufacturing and an updated microprocessor-controlled version is now available (Burkhard Manufacturing Co. Ltd, Woodcock Hill Industrial Estate, Rickmansworth, Herts, WD3 1PJ, UK).

The work of Nash (1954) and later Glynne Jones and Chadwick (1960) illustrate the principles. Both teams used carbon dioxide to facilitate the handling of the flies. Nash applied her dose to the dorsum of the thorax, while the others treated the ventral surface of the abdomen. The difference in susceptibility of the sexes was noted in both cases. Glynne Jones and Chadwick fully assayed the synergist (S421), which had having noted toxicity, but stated that none of the others showed an appreciable effect at the highest concentrations used.

The LD_{50} values obtained for S421 and unsynergized pyrethrins against female flies provide enough data to allow critical examination of the degree of true synergism achieved, as opposed to the purely insecticidal joint activity of the two compounds. The formulae of Sun and Johnson (1960) can be simplified to the formula below, which calculates the expected LD_{50} of a mixture of two insecticides if purely additive action is assumed. The result is the harmonic mean of the two LD_{50}s within which the proportions of the compounds are those of the mixture.

$$\frac{\text{Proportion of A} + \text{Proportion of B}}{\left(\dfrac{\text{Proportion of A}}{LD_{50} \text{ of A}}\right) + \left(\dfrac{\text{Proportion of B}}{LD_{50} \text{ of B}}\right)} = \begin{array}{l}\text{Expected } LD_{50} \text{ of mixture}\\ \text{(total of insecticides)}\end{array}$$

Using the above formula, Table 11.1 can be calculated. It can be seen that the true factors of synergism (FOS) are considerably smaller than those calculated conventionally.

There are a number of points in addition to those quoted by Busvine (1971) to be observed in a topical application exercise to minimize variation in results.

• Test insects should be of the same sex, at the same stage in the life cycle and same state of nutrition. Instars likely to mature should be tested soon after moulting to avoid further changes before the assessment of mortality, the timing of which should be chosen to be at the point of maximum effect.
• The use of carbon dioxide or cooling for the handling of insects is to be avoided. To minimize loss of dosage, insects should be able to stand after treatment. Gentle suction is the most appropriate holding method. Holding a smaller insect by the head can be effective.

Table 11.1. The relative activity of pyrethrins and S421 alone and in different mixtures

	Observed LD_{50} (µg per insect)	Conventional FOS[a]	Expected LD_{50} (µg per insect)	True FOS
Pyrethrins	0.86[b]			
S421	4.50			
Pyrethrins + S421 (1:1)	0.59 + 0.59	1.45	0.72 + 0.72	1.22
Pyrethrins + S421 (1:5)	0.42 + 2.10	2.05	0.44 + 2.20	1.05
Pyrethrins + S421 (1:10)	0.20 + 2.00	4.30	0.30 + 3.00	1.50

[a] Factor of synergism.
[b] Mean of values from Glynne Jones and Chadwick (1960).

- The site of application should be chosen so that the droplet spreads to the same degree on each insect. The dorsum of the thorax or pronotum are preferable.
- A volatile solvent such as acetone is unsuitable as it may result in variable dosing. Solvents with high boiling points can affect results (Webb and Green, 1945). Carter *et al.* (1975) showed that the use of mineral oil in a direct spray could result in greater synergism than when kerosene was the carrier.
- The experiment should be designed to produce data suitable for probit analysis and full statistical evaluation as outlined by Finney (1971). For individual regression lines of each material or mixture, five dose levels are desirable; these should be selected to give between 10% and 90% mortality.
- The compounds and mixtures for comparison should be tested to a balanced design, using three or more replicate tests on each mixture. This allows analysis to remove day-to-day variation. A larger exercise inevitably takes a longer time to complete. Fewer replicates can give rise to imbalances in the analysis owing to exaggerated weighting coefficients. A regression line with bad dose–effect correlation will result in a very low weighting coefficient and in the reverse case a very high one. Unless there are further replicates, severe distortions in the relative efficacies will be calculated.

4. SYNERGISTIC ISOBOLES

The use of topical application to individual house flies in evaluating the intrinsic synergism between a given insecticide and a given synergist is practicable and can provide a satisfactory measure of the success of a subsequent product in giving this degree of synergism.

Hewlett (1969) and Tammes (1964) referred to isoboles. They illustrated the concept with figures in which the axes were in terms of the doses of compounds A and B, respectively. Isoboles were drawn showing the full range of combinations of the two materials that would give equitoxic effects. These could show additive, synergistic or antagonistic action. Most workers who have published data from a range of mixtures have restricted themselves to comparatively low ratios. Isoboles drawn on an arithmetic scale give curved isoboles for synergism and antagonism.

Wickham *et al.* (1974) changed the configuration of the axes to the logarithmic scale and showed that this allows a better presentation and understanding of the regular rate of increase in factor of synergism achieved by increasing the ratio of synergist to insecticide. They also examined a wide enough spectrum of ratios to show the features discussed below. They quoted data for mixtures of bioallethrin and PBO.

These and subsequent data were produced using topical application to a uniform susceptible laboratory strain of house flies. The same solvent was used throughout: 2-ethoxyethanol. The data were obtained over a relatively short period but with few exceptions compounds were not tested concurrently, and the data are therefore not comparable between compounds. However, for each set of materials, the ranges of synergist ratios were produced as an individual set of data which could be processed as suggested by Finney (1971).

The data summarized in Fig. 11.1 show logarithmic isoboles for the synergism of pyrethrins by PBO at the LD_{50} and LD_{95} levels and the calculated additive isoboles at the same levels. The data were collated and analysed according to the procedures outlined above. The ratios of the respective LD_{95}s and LD_{50}s are representative of the slopes of the probit/log dose regressions of the toxicants. The smaller the ratio, the steeper is the regression. In this case, the variation obscures any significant relationship between the LD_{95}/LD_{50} ratio and the logarithmic ratio of pyrethrins to PBO (Fig. 11.2).

It can be seen that at very low levels of PBO there is no significant synergism, only variation in the activity of pyrethrins alone. The synergistic isoboles then diverge from the additive isoboles until very high ratios are reached, when they turn horizontal and merge with those of their respective additive isoboles at the level of PBO activity. Tables 11.2 and 11.3 list the conventional factors of synergism over the range of ratios by comparison with the level of pyrethrins alone. Also listed are the slightly lower factors taking into account the additive effects of PBO at those same absolute levels.

Both these methods of presentation give rise to unrealistically high factors of synergism. A more revealing approach is to calculate the factor of synergism in respect of the additive effect at that ratio. This can be seen to show a maximum factor of synergism (Fig. 11.3).

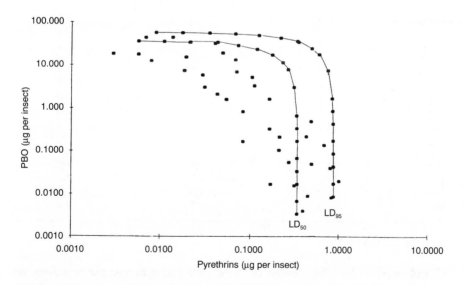

Figure 11.1. Curves of equitoxic effect (isoboles), calculated using simple additive action, and observed data of synergism at LD_{50}s and LD_{95}s achieved by topical application to house flies using pyrethrins/PBO.

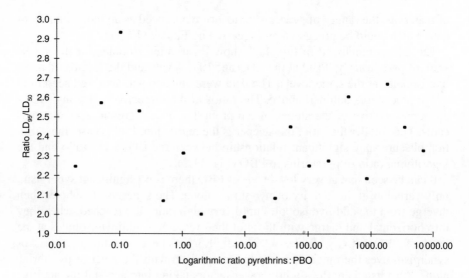

Figure 11.2.   Ratios of LD_{95} to LD_{50} indicating slopes of the regression lines over a range of pyrethrins : PBO ratios using topical application to house flies.

Table 11.2.   Factors of synergism (FOS) for pyrethrins/PBO at LC_{50} by topical application to house flies

PBO: pyrethrins (*n*:1)	Conventional FOS	Additive FOS at absolute level of PBO	Additive FOS by ratio
0.01	0.9	0.9	0.9
0.02	0.8	0.8	0.8
0.05	1.1	1.1	1.1
0.10	2.0	2.0	2.0
0.20	1.2	1.2	1.2
0.50	1.6	1.6	1.6
1	1.6	1.6	1.6
2	4.1	4.1	4.0
10	4.1	4.0	3.8
30	6.4	6.1	5.0
50	8.0	7.5	5.4
100	11.0	10.1	5.7
200	11.6	9.7	4.1
400	18.3	14.6	3.9
800	17.5	10.1	2.1

5. CALCULATIONS

Calculations of the effectiveness of insecticide and synergist are based on the formulae of Sun and Johnson (1960). Table 11.4 combines data from Tables 11.2 and 11.3 and should be consulted to aid the appreciation of the formulae quoted below.

Table 11.3. Calculations of factors of synergism (FOS) for topical application to house flies

Ratio	Pyr (LD$_{50}$) (µg per insect)	PBO (LD$_{50}$) (µg per insect)	Additive PYR (LD$_{50}$) by ratio	FOS by convention	Additive PYR (LD$_{50}$) at absolute PBO (µg per insect)	Real FOS at absolute PBO	Real FOS by ratio
A	B	C	D	E	F	G	H
Alone	0.338	36.398					
0.01	0.395	0.004	0.338	0.9	0.338	0.9	0.9
0.02	0.450	0.009	0.338	0.8	0.338	0.8	0.8
0.05	0.313	0.016	0.338	1.1	0.338	1.1	1.1
0.10	0.169	0.017	0.338	2.0	0.338	2.0	2.0
0.20	0.271	0.054	0.337	1.2	0.337	1.2	1.2
0.50	0.209	0.105	0.336	1.6	0.337	1.6	1.6
1	0.212	0.212	0.335	1.6	0.336	1.6	1.6
2	0.082	0.164	0.332	4.1	0.336	4.1	4.0
10	0.082	0.821	0.309	4.1	0.330	4.0	3.8
30	0.053	1.593	0.264	6.4	0.323	6.1	5.0
50	0.042	2.120	0.231	8.0	0.318	7.5	5.4
100	0.031	3.070	0.175	11.0	0.309	10.1	5.7
200	0.029	5.840	0.118	11.6	0.284	9.7	4.1
400	0.019	7.400	0.072	18.3	0.269	14.6	3.9
800	0.019	15.440	0.040	17.5	0.195	10.1	2.1

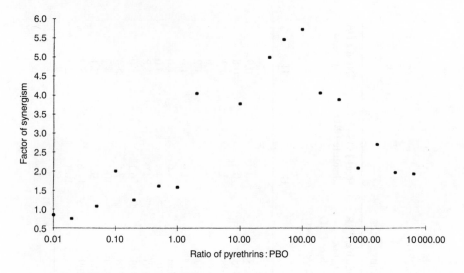

Figure 11.3.   Real factors of synergism at a range of ratios of pyrethrins to PBO using topical application to house flies.

Table 11.4.   Data for bioresmethrin/PBO alone and at a range of ratios by topical application to house flies

Ratio	BRM (LD$_{50}$) (μg per insect	PBO (LD$_{50}$) (μg per insect)	BRM (LD$_{95}$) (μg per insect)	PBO (LD$_{95}$) (μg per insect
Alone	0.017	42.0530	0.039	85.1200
0.01	0.016	0.0002	0.031	0.0003
0.05	0.025	0.0013	0.057	0.0029
0.50	0.021	0.0105	0.044	0.0220
5	0.020	0.1000	0.037	0.1850
30	0.015	0.4500	0.029	0.8700
200	0.011	2.2000	0.023	4.6000
500	0.008	4.0000	0.016	8.0000
1500	0.006	9.0000	0.014	21.0000

BRM, bioresmethrin.

$$\frac{\text{Proportion of insecticide} + \text{Proportion of synergist}}{\left(\dfrac{\text{Proportion of insecticide}}{\text{LD}_{50} \text{ of insecticide}}\right) + \left(\dfrac{\text{Proportion of synergist}}{\text{LD}_{50} \text{ of synergist}}\right)} = \begin{array}{l}\text{Total insecticide} \\ \text{mixture (LD}_{50})\end{array}$$

This can be reduced (at the LD$_{50}$) to:

$$\left(\frac{\text{Dose of insecticide}}{\text{LD}_{50} \text{ of insecticide}}\right) + \left(\frac{\text{Dose of synergist}}{\text{LD}_{50} \text{ of synergist}}\right) = 1$$

This can be transposed under the same terms to:

Dose of insecticide = (1 − Dose of synergist/LD_{50} of synergist)
$\times$ LD_{50} of insecticide

Referring to the data in Table 11.3:

D = 1/(1/0.338 + A/36.398), E = 0.338/B, F = (1 − C/36.398) × 0.388
G = F/B, and H = D/B

6. COMPARATIVE SYNERGISM

The data represented in Fig. 11.4 and in Tables 11.4 and 11.5 were obtained with bioresmethrin, a very active pyrethroid. These data demonstrate that there is a clear relationship between the slope of the regression line and the logarithmic ratio of bioresmethrin to PBO. Also there are much smaller factors of additive synergism with bioresemethrin.

A corollary of this is that, for a more active compound like bioresmethrin, a conventional ratio of insecticide to synergist such as 1:5 will not give any synergism at the LD_{50} level, and the higher levels of insecticide necessary to achieve LD_{95}s also means that, for a fixed ratio, a higher level of synergist is present and a greater factor of synergism can result. This reinforces the fact that it is the absolute level of synergist that is important, not just the ratio.

The data in Tables 11.2 and 11.5 show that the more active bioresmethrin is

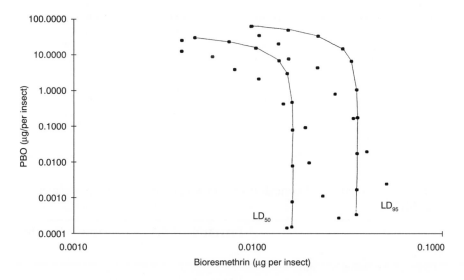

Figure 11.4. Curves of equitoxic effect (isoboles), calculated using simple additive action, and observed data of synergism at LD_{50}s and LD_{95}s achieved by topical application to house flies using bioresmethrin/PBO.

Table 11.5. Factors of synergism (FOS) for bioresmethrin/PBO at LC_{50} by topical application to house flies

PBO : bioresmethrin ratio	Conventional FOS	Additive FOS at absolute level of PBO	Additive FOS by ratio
0.01	1.1	1.1	1.1
0.05	0.7	0.7	0.7
0.50	0.8	0.8	0.8
5	0.9	0.8	0.8
30	1.1	1.1	1.1
200	1.5	1.5	1.4
500	2.1	1.9	1.8
1500	2.8	2.2	1.8

Table 11.6. Data for pyrethrins/PBO alone and at a range of ratios by topical application to house flies

Ratio	Pyrethrins (LD_{50}) (μg per insect)	PBO (LD_{50}) (μg per insect)	Pyrethrins (LD_{95}) (μg per insect)	PBO (LD_{95}) (μg per insect)
Alone	0.338	36.398	0.873	57.626
0.01	0.395	0.004	0.830	0.008
0.02	0.450	0.009	1.009	0.020
0.05	0.313	0.016	0.806	0.040
0.10	0.169	0.017	0.495	0.050
0.20	0.271	0.054	0.687	0.137
0.50	0.209	0.105	0.433	0.217
1	0.212	0.212	0.491	0.491
2	0.082	0.164	0.165	0.329
10	0.082	0.821	0.164	1.636
30	0.053	1.593	0.111	3.330
50	0.042	2.120	0.104	5.210
100	0.031	3.070	0.069	6.930
200	0.029	5.840	0.067	13.340
400	0.019	7.400	0.048	19.360
800	0.019	15.440	0.042	33.920

Table 11.7. Summary of comparative data with pyrethrum and bioresmethrin with PBO

	Pyrethrins : PBO	Bioresmethrin : PBO
Maximum recorded additive factor of synergism by ratio	5.7	1.9
Recorded ratio	100:1	3200:1
Ratio of LD_{50} synergist alone : LD_{50} pyrethroid alone	108	2768

less synergized than pyrethrins irrespective of ratio or absolute level of PBO. The ratios at which the maximum factors of synergism are recorded for pyrethrins and bioresmethrin are of the same order as the ratios of the respective efficacies of the pyrethroids and PBO when tested alone (see Tables 11.4, 11.6 and 11.7).

Examination of other data obtained across the full range of pyrethroids tested independently reveals that, in general, these findings are held in common. The data in Table 11.8 and Fig. 11.5 show that there is an interestingly close relationship between the pyrethroid : PBO ratios at which the maximum real synergism occurred and the relative potencies of the appropriate pyrethroid and PBO alone tested alongside. These data indicate that, in a research programme, it is inadvisable to use a fixed ratio of synergist to insecticide.

Although the individual pyrethroids in the study were tested independently, the data in Tables 11.8 and 11.9 demonstrate basic principles of synergistic action. Table 11.8 shows the pyrethroids in order of their maximum additive factor of synergism by ratio. The conventional factors of synergism at these ratios are larger. It will be noted that the levels of PBO at these maximum factors of synergism are basically in reverse order and range from approximately 1 to 13 µg per insect.

Table 11.9 compares the conventional factors of synergism achieved at a number of constant levels of synergist with the pyrethroids in this study. There is a wider range of factors of synergism at 1.0 µg than at either of the two higher levels. The different ranges and orders of magnitude of the factors of synergism reflect the different rates of increase of synergism between the insecticides.

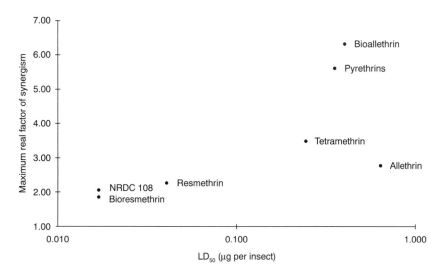

Figure 11.5.   Relationship of maximum real factors of synergism of a range of pyrethroids to their activity alone at LD_{50} levels.

Table 11.8. Maximum real factors of synergism for a range of pyrethroids and the ratios of PBO at which they were achieved

Insecticide	LD_{50} pyrethroid alone (µg per insect)	LD_{50} PBO alone (µg per insect)	Max. real FOS by ratio	Ratio at recorded max. FOS	LD_{50} of pyrethroid at max. FOS (µg per insect)	PBO level at max. ratio
Bioresmethrin	0.017	42.1	1.85	3200	0.004	12.80
NRDC 108	0.017	37.0	2.00	800	0.006	4.80
Resmethrin	0.040	42.8	2.30	1600	0.007	11.20
Allethrin	0.620	32.5	2.90	50	0.110	5.50
Tetramethrin	0.238	33.1	3.57	50	0.049	2.45
Pyrethrins	0.338	36.4	5.70	100	0.031	3.10
Bioallethrin	0.380	32.5	6.40	20	0.048	0.96

Table 11.9. Comparison of conventional FOS for a range of pyrethroids at fixed levels of PBO

Insecticide	LD_{50} pyrethroid alone	LD_{50} PBO alone	Max. real FOS by ratio	Conventional FOS[a] at 1.0µg PBO	Conventional FOS[a] at 3.0µg PBO	Conventional FOS[a] at 10.0µg PBO
Bioresmethrin	0.017	42.1	1.85	1.2	2.7	6.9
NRDC 108	0.017	37.0	2.00	2.7	3.8	4.8
Resmethrin	0.040	42.8	2.30	2.3	3.7	5.0
Allethrin	0.620	32.5	2.90	4.7	6.4	8.3
Tetramethrin	0.238	33.1	3.57	4.1	6.7	9.0
Pyrethrins	0.338	36.4	5.70	7.0	10.2	14.1
Bioallethrin	0.380	32.5	6.40	8.8	10.9	12.8

[a] By manual interpolation from synergized LD_{50} values.

7. COST-EFFICIENCY

The work reviewed here does not consider the actual cost of the insecticides and synergist. Instead, comparisons are made using a range of relative costs of insecticide to synergist compared with synergistic activity for given ratios of insecticide–synergist mixtures.

The data in Table 11.10 and Fig. 11.6 are constructed from the topical application data in Table 11.2, and show the costs over a range of ratios of synergized mixtures of an insecticide relative to the cost of the insecticide used alone. A number of relative prices of insecticide to synergist have been chosen which encompass a range wider than that likely to be found in practice. Table 11.10 refers to mixtures of pyrethrins and PBO; the data for bioresmethrin are similarly treated in Table 11.11.

These data come from tests made under controlled laboratory conditions and may seem to be irrelevant to practical pest control. However, they serve to show the principle that the cost-efficiency of using a synergist in a product is dependent on both the degree of synergism that is conferred and the relative prices of the insecticide and synergist. If factors of synergism like those found by topical application to house flies of pyrethrins/PBO could be achieved under field conditions with a product, there would be a cost benefit.

The degree of synergism to be found in practice is dictated by the formulation of the product, the manner of presentation to the pest and the susceptibility of the pest strain.

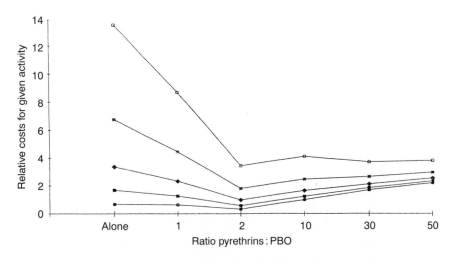

Figure 11.6. Cost-efficiency relationships for pyrethrins/PBO mixtures using topical application data to house flies at the LD_{50} level. Price ratios: ‐, 2 : 1; ‐, 5 : 1; ‐, 10 : 1; ‐, 20 : 1; ‐, 40 : 1.

Table 11.10. Cost-efficiency relationships for pyrethrins/PBO mixtures using topical application data to house flies at the LD_{50} level

Ratio Pyr : PBO	Pyr (LD_{50}) (μg per insect)	PBO (LD_{50}) (μg per insect)	Cost for Pyr/PBO at price ratio 2 to 1	Cost for Pyr/PBO at price ratio 5 to 1	Cost for Pyr/PBO at price ratio 10 to 1	Cost for Pyr/PBO at price ratio 20 to 1	Cost for Pyr/PBO at price ratio 40 to 1	Cost for Pyr/PBO at price ratio 100 to 1
Alone	0.338	36.40	0.68	1.69	3.38	6.76	13.52	33.790
1	0.212	0.21	0.64	1.27	2.33	4.46	8.70	21.432
2	0.082	0.16	0.33	0.57	0.99	1.81	3.45	8.374
10	0.082	0.82	0.99	1.23	1.64	2.46	4.11	9.031
30	0.053	1.59	1.70	1.86	2.12	2.66	3.72	6.903
50	0.042	2.12	2.20	2.33	2.54	2.97	3.82	6.360
100	0.031	3.07	3.13	3.22	3.38	3.68	4.30	6.140
200	0.029	5.84	5.90	5.99	6.13	6.42	7.01	8.760
400	0.019	7.40	7.44	7.49	7.59	7.77	8.14	9.250
800	0.019	15.44	15.48	15.54	15.63	15.83	16.21	17.370
1600	0.008	12.64	12.66	12.68	12.72	12.80	12.96	13.430
3200	0.006	17.92	17.93	17.95	17.98	18.03	18.14	18.480
6400	0.003	18.56	18.57	18.57	18.59	18.62	18.68	18.850

Table 11.11. Cost-efficiency relationships for bioresmethrin/PBO mixtures using topical application data to house flies at the LD_{50} level

Ratio BRM : PBO	BRM (LD_{50}) (μg per insect)	PBO (LD_{50}) (μg per insect)	Cost for BRM/PBO at price ratio 2 to 1	Cost for BRM/PBO at price ratio 5 to 1	Cost for BRM/PBO at price ratio 10 to 1	Cost for BRM/PBO at price ratio 20 to 1	Cost for BRM/PBO at price ratio 40 to 1
Alone	0.017	42.05	0.03	0.09	0.17	0.34	0.68
0.5	0.021	0.01	0.05	0.12	0.22	0.43	0.85
5.0	0.020	0.10	0.14	0.20	0.30	0.50	0.90
30.0	0.015	0.45	0.48	0.53	0.60	0.75	1.05
200.0	0.011	2.20	2.22	2.26	2.31	2.42	2.64

8. THE SYNERGISTIC SPECTRUM

The practicalities of the use of synergists were reviewed by Brown *et al.* (1967). These authors drew attention to the fact that very large factors of synergism can be demonstrated in the laboratory by direct application of the toxicant and synergist to individual insects, defined as 'intrinsic synergism'.

Glynne Jones and Green (1959) had shown that *Calandra granaria* (now *Sitophilus*) and *C. oryzae* differed in the degree to which PBO could synergize pyrethrins applied topically. An approximately three-fold difference in factor of synergism could be seen at a 20:1 ratio. Mount and Pierce (1973) compared pyrethrins and two synthetic pyrethroids against pairs of mosquito species of three genera in a wind tunnel, producing an unpredictable range of relative responses. Consistently less synergism was shown against *Aedes aegypti* than against *A. taeniorhyncus*. A similar feature was shown against the two *Culex* species, but to a lesser degree. Pyrethrins were more synergized against *Anopheles quadrimaculatus* than were the synthetics, while the converse was observed with *A. albimanus*.

Chadwick (1971) showed that when comparing application of pyrethroids alone and synergized to *Blattella germanica* by an overall dusting method and a direct spray method, the former showed a consistently higher rate of synergism. Davies *et al.* (1970) recorded a lesser synergism against house flies in spray tests than by topical application.

Farnham (1977) worked with an isomer of tetramethrin and two synergists, applying the compounds independently over a range of time intervals. He demonstrated that a lower factor of synergism was obtained when the synergists were applied after the insecticide. The best results were shown to be with application either concurrently or in the hours immediately preceding insecticide application.

9. SYNERGISTS AFFECT THE RESISTANCE TO INSECTICIDES

Synergism in a susceptible strain can be shown to overcome naturally occurring breakdown mechanisms. Some forms of resistance result from an increase in such breakdown mechanisms. Resistance may also be due to a restriction on the speed of entry of an insecticide, thereby exaggerating the effect of normal levels of breakdown enzymes, e.g. *pen* (Farnham, 1973). The scope of the use of synergists in cases of resistance and the possibility that a wider range of insecticides could become available for use was reviewed by Oppenoorth (1971).

A comprehensive study by Forrester *et al.* (1993) compared the degree of synergism conferred by a wide range of synergists. These authors used larvae of a susceptible and a resistant strain of *Helicoverpa armigera* (Hübner). It was shown in the laboratory, using pretreatment with a very high dose of synergist, that even the highest resistance factor could be reduced to one or less. Although this work was performed with high doses of PBO, they had been shown to be inactive on the pest. The ratios of synergist to insecticide were therefore of a high order – 50 µg of synergist to as little as 0.002 µg of insecticide. Fieldwork involving the same insecticide has used much lower dose relationships of 200–600 g PBO per hectare to 0.5 g per hectare of the insecticide with good suppression of resistance.

It might be expected that the slopes of the regression lines of the synergized resistant strain would be steeper. There are instances of this, but the feature is not consistent. Raffa and Priester (1985) showed a greatly increased slope with synergized fenvalerate against larvae of a pyrethroid-resistant strain of *Spodoptera exigua*. They also worked with a bimodal population of methomyl-resistant *Heliothis virescens*. This heterogeneous strain displayed the typical two separate dose–response curves of an emergent resistant strain. The use of a synergist steepened the overall slope to that of the susceptible strain.

Further studies following the same approach were carried out with *Blattella germanica*, the German cockroach. Umeda *et al.* (1988) compared the use of PBO, an oxidative inhibitor, with that of NIA 16388, both an oxidase and esterase inhibitor. The latter synergized permethrin in this case, while the former had little effect.

Cochran (1987) tested bendiocarb and pyrethrins with synergists against resistant *Blattella germanica,* successfully rendering eight strains susceptible.

Raffa and Priester also referred to two further benefits of the use of synergists. Routine use can be advantageous in the prevention of selection for resistance, and the use of a synergist which is not as active on a predator as on the pest can be employed in integrated pest management (IPM).

10. GUIDELINES ON THE PRACTICAL ASSESSMENT OF SYNERGISTS

When testing a potential product to check on the achievable practical synergism, the tests carried out must be realistic. The target species should be allowed to behave in the presence of the product much as it would in a real scenario. If the insecticide will be picked up by the insect in flight, the product must be presented as a cloud and not as a contamination of the cage in which the test insects are held. A space spray must have had the opportunity to mature by evaporation or fallout. Test insects confined on residual deposits will individually gather a wide range of different doses from a film and subsequently lose different proportions of this amount depending on their activity and behaviour (Adams *et al.*, 1992).

Under field conditions, only a proportion of the insect population will be exposed to the treatment. The practical synergistic study answers the question: 'Will the synergist give an increase in effect on that section of the infestation that it reaches?' Effort must be made to minimize the variation that will occur in the field. Target pests need to be allowed equal access to the treatments. The 'field' ideally should be brought into the laboratory. The 'laboratory' may have to be a large ventilated chamber.

If the study is on a residual product, it should be noted that the character of the deposit will change with time. Some insects are flushed out onto the wet freshly sprayed surface, while others will only cross it after it has dried. In the first instance a synergist might be advantageous. In the second situation, there is the possibility that an added nonvolatile liquid may act as a solvent, reacting with the substrate to remove the insecticide by absorption.

Modern chemical microanalytical techniques introduce the possibility of quantifying the doses picked up by individual insects. Specific aspects of testing space sprays and residual products are discussed below.

10.1. Space Sprays

The majority of products which contain a synergist are used in the formulation of aerosols or are employed as space sprays. In practice, insects in flight gather droplets of the sprayed product as they impact on their bodies and wings. Variation in test methods can emphasize one aspect of a product performance in relation to another. A close appreciation of the mechanics of published test methods will reveal the reasons for what might otherwise appear to be contradictions in the results.

10.1.1. Aerosols

White *et al.* (1992) have published their findings on indoor aerosol space sprays which make distinctions between those tests which are allied to the British Standard Method (1967) and those which follow the American CSMA pattern (1971). BS 4172:1967 has been revised in the light of the above studies as BS 4172:1993.

The principal difference between the BS 4172 and the CSMA method lies in the time of release of the test insects and their consequent exposure to the spray. The small size of the Peet–Grady chamber also contributes to a lack of the full expression of formulation effects. It was originally designed (Peet and Grady, 1928) for the evaluation of coarse hand sprays produced by venturi action.

Formulation changes in respect of solvents or valves/actuators in aerosols can give rise to significant differences in the droplet size spectrum and the distribution of the aerosol in space and time. White *et al.* (1992) showed that aerosols composed of larger droplets will lose a high proportion soon after discharge. A low-boiling solvent in the formulation will promote more efficient collection by a flying insect in the short term but will at the same time cause greater loss of airborne droplets due to fallout. A rapidly evaporating solvent will give rise to small droplets, which largely remain in the air, but also have distinctly reduced impaction efficiency.

BS 4172 involves the release of the test insects from floor level in the centre of the test chamber, which is the size of an average room, 'immediately' after spraying. The actively flying insects sample the spray cloud distribution on their way to their first resting site. A much reduced flight activity pattern follows. This approach favours the knockdown aspect of the product type, while also achieving a full appreciation of the killing power, but only examines the aerosol cloud in its early stages.

In the CSMA method, which uses a chamber of $8\,m^3$, the insects are released first and have largely settled down before the spray is discharged. Most larger droplets will have settled out before the majority of the test insects can sample them. While gross differences in knockdown may be distinguished at the resulting slower reaction times of the CSMA method, finer distinctions, which are normally observed by discriminating purchasers who prefer fast action, can be missed. However, the delayed response to the spray allows a better analysis of the length of life of the aerosol cloud. Users who are less interested in knockdown but who wish merely to see an area free of flying insects on their return after a treatment are adequately served by this test method.

A further feature of the CSMA approach is only to collect for mortality assessment those insects knocked down at the end of the exposure period. This discounts the possibility of a 100% kill in some cases where a slower-acting insecticide has been used in the formulation. BS 4172 involves collecting and observing all insects. This labour-intensive aspect gives a better appreciation of the killing power of a product.

A full appreciation of the influence of the inclusion of a synergist as part of the involatile component in a fly spray will be gained by a study in which insects are released over a range of times. This should include prerelease as in the CSMA specification to some 5 minutes after spraying. This will establish the knockdown and kill profiles suggested by White *et al.* (1993).

Wickham *et al.* (1974) described a study of aerosol formulations on the knockdown and kill of house flies. Knockdown and kill pyrethroids were tested alone and in mixtures with PBO under strict laboratory conditions following the methodology given in BS 4172. Formulations of mixtures of knockdown pyrethroids, kill pyrethroids and a synergist were also tested. Unlike their topical application studies, the work on the different knockdown and kill pyrethroids was integrated and expressed as though they had been tested against an insect population of consistent susceptibility. The data were therefore comparable.

Table 11.12 lists factors of synergism at LC_{95} levels and KC_{50} (4 min) for a range of pyrethroids in aerosols against house flies. An appreciation of the degrees of synergism achieved at the LC_{95} level for pyrethrins and bioresmethrin can be obtained by reference to earlier tables on topical application. The pyrethrins aerosols gave a factor of synergism of 3.0 at a low ratio of synergist

Table 11.12. Aerosol performance for house flies (*Musca domestica*) using the BS 4172:1967 method

Insecticide	Factor of synergism for kill at LC_{95} using 0.5% PBO	Factor of synergism for knockdown at KC_{50} (4.0 min) using 0.5% PBO
Pyrethrins	3.0	1.4
Tetramethrin	1.6	1.1
Bioallethrin	1.8	1.2
Bioresmethrin	1.9	–

to insecticide (2.5:1), while bioresmethrin (at 7:1) showed a factor of synergism of 1.9. Table 11.6 showed similar factors of synergism for pyrethrins around that ratio. In contrast, bioresmethrin did not show a factor of synergism of 1.9 until ratios in excess of 1:200 (Table 11.4). Synergism for knockdown at the KC_{50} (4 min) level is of a lower order.

Wodageneh and Matthews (1981) alluded to the benefits to be gained by the inclusion of an involatile component in a spray. Tsuda and Okuno (1985) demonstrated it but White *et al.* (1993) measured the true potential. They showed that odourless petroleum distillate, the solvent employed in the aerosols tested by Wickham *et al.* (1974), evaporates quickly except for a small proportion of higher-boiling fractions. The presence of PBO (vapour pressure 0.117 mPa at 20°C) confers a larger droplet size even at 0.5% w/w in the aerosol (2.5% w/w in the filling solution which gives the droplets). White *et al.* (1993) showed that the impaction efficiency of droplets under these conditions increased with size. The droplet size increase because of the additional PBO would have been of greater importance in the case of bioresmethrin than for the pyrethrins, where there was further low-boiling solvent in the 25% w/w pyrethrins extract used in the formulations.

Table 11.13 shows factors of synergism in aerosols from tests similar to those of Wickham *et al.* (1974) but with mosquitoes (R. Slatter and M.D.V Moss, unpublished results). Comparison with the effects seen against house flies reveals higher factors of synergism for kill with tetramethrin, pyrethrins and bioresmethrin, even though the last two were the most effective alone. Bioallethrin was synergized to a much lesser degree. Knockdown synergism is again less than that seen with kill for pyrethrins and tetramethrin, but, in contrast, antagonism is observed for bioallethrin. It is interesting to observe that synergism and antagonism can occur at the same time, although on different parameters.

10.1.2. Large-scale Space Sprays

Any test method employed to investigate the potential value of the inclusion of a synergist must involve a procedure which allows the unique characteristics of the formulation to be demonstrated. Physical measurements of the spectrum of any new formulation should be taken and adjustments made to the sprayer to

Table 11.13.   Aerosol performance for mosquitoes (*Culex quinquefasciatus*) adopting the BS 4172:1967 method

Insecticide	Factor of synergism for kill at LC_{95} using 0.5% PBO	Factor of synergism for knockdown at KC_{50} (4.0 min) using 0.5% PBO
Pyrethrins	8.3	1.4
Tetramethrin	3.5	1.7
Bioallethrin	1.6	0.5
Bioresmethrin	5.0	–

achieve a droplet size as similar as possible to that of the reference formulation. Unless the two spectra are similar at inception, any beneficial effects could be clouded by physical factors such as early loss due to fallout.

The vagaries of field conditions make significant distinctions between formulations questionable. An appreciation of these and of how formulation can also play a part can be gained from David (1946), Taha *et al.* (1979), Martin *et al.* (1977) and Groome *et al.* (1989). Deposits of spray on the mesh of cages is dependent on wind speed and give a distorted value to spray efficacy owing to the extra tarsal contact with the deposits. Meshes were used in the field as a means of assessing sprays from both ground and aerial applications by Groome *et al.* (1989) and Hursey and Allsopp (1984).

Davies (1974) used caged house flies in a large barn in which there was virtually no air movement other than that caused by the sprayer, which was not discharging directly at the cages. The relative efficacies of the killing pyrethroids and the synergized knockdown pyrethroids under these conditions can be seen to agree well with those quoted by Wickham (1976) and Davies *et al.* (1970), who used the same strain of house flies. Although no direct comparison was made in the above study of unsynergized and synergized compounds, bioresmethrin was added to synergized pyrethrins and checks using the formulae from Wickham *et al.* (1974) allow a comparison of the degree of observed activity relative to that of the expected additive action. Davies showed that little insecticide was deposited on the cage mesh under her conditions. Synergism of the bioresmethrin in this exercise can be detected.

10.1.3. *Space Sprays for Control of Crawling Insects*

Space spraying against infestations of crawling insects in enclosed areas is a widely used control measure. By virtue of their habits the target pests are often not exposed to the majority of the spray, and in many cases the efficacy of such treatments relies heavily on the expellency or 'flushout' properties of the product.

Smaller droplets will diffuse into cracks, crevices and other areas where the target insects may be hidden at the time of spraying. Following 'flushout', the insects will be exposed to fallout of the remainder of the suspended droplets and to a residual deposit from the larger droplets which have already settled out.

A flushout (FO) test was described by Chadwick and Evans (1973). The apparatus involved an artificial harbourage from which cockroaches would be expelled into an arena. The distance to which the insects were dispersed and the speed with which this occurred allowed the calculation of FO_{50}s in terms of time. Fuchs (1988) commented that the above technique did not necessarily give equal dosing to individual insects. Using his different, though much less realistic, presentation of treated paper discs, he tested a wide range of pyrethroids. Only pyrethrins were examined for dose-effect and checked for the interaction with PBO which was shown to depress flushout.

Products used to control crawling insects must involve investigation at points other than just at floor level. Carter and Dodd (1979) placed glass containers with test insects at floor level and at 1.5 and 4.5 m above floor level in a 150 m^3 experimental chamber and a 6000 m^3 food processing plant. These were left exposed for 45 minutes before collection. A dose–effect relationship was shown in the first case where four treatment levels were sprayed. Only two dose levels were used in the plant area. PBO was added to the permethrin at a 1:1 ratio in the second test, but the quoted data do not allow for interpretation of synergism. The assessment of synergistic effects under these circumstances can prove very difficult.

Slatter *et al.* (1981) made a similar approach when evaluating a new spray dispersal apparatus. Physical measurements were made of the spray distribution, which was applied at four rates, with exposure of the insects for 2 hours.

10.1.4. Aerosols Giving Residual Status

Crawling insect killer (CIK) products serve two approaches to the control of pests: direct attack on individuals or visible groups of insects and the placing of a residual deposit. Chadwick and Dixon (1976) described a test cabinet designed for the direct spraying of cockroaches to assess knockdown and mortality following treatment. This approach allows the investigation of CIK formulations in which the solvents as well as the insecticide can play a part. No investigation of synergism was attempted.

The residual action can be studied by spraying a range of surfaces and storing some samples for future exposures. The application should be made in a manner which is considered practical.

10.2. Residual Products

Residual products will be presented to the target species over a range of time. The initially laid down deposit will be 'wet'. This is followed by a 'curing' period during which the product may become gradually more firmly attached to the substrate and therefore less easily picked up. Poor experimental formulations can result in extremely low control.

Volatile components of a deposit will be lost relatively quickly and others may be degraded or absorbed by the surface to which they have been applied and disappear or become ineffective after a short interval. Stable compounds with a long residual life may be evaluated over a limited period using a set regime which can produce data allowing estimation of the length of effective control.

Carter and Chadwick (1978) described a method of direct spraying of crawling insects using the Potter tower (Potter, 1952). This procedure is not particularly relevant as the targets are rarely exposed at the time of application. A more practical approach was outlined by Lucas (1991). Insects will normally only meet a freshly sprayed deposit following flushout from nearby harbourages. A 'wet walkover' test has been developed involving pipetting onto glass plates or tiles a fixed volume of the diluted product. The volume should conform to the recommended application rate.

The insects can be exposed to this wet deposit, using a 'walk on/walk off' technique. The test specimens are slowly and carefully transferred, using the enclosing cylinder, from a clean tile to a freshly treated tile so that contact is as natural as possible. After a 30-second period they are gently moved back again. Following observation for knockdown, the insects are held in suitable containers with food and water for mortality assessment. Analysis of insects collected during a field trial confirmed the pickup of active ingredient is in close agreement with the laboratory tests.

The exposure period used by most workers is 30 minutes. This may be unrealistic from a practical point of view as contact under field conditions, where treatments are restricted to limited areas, is often transitory. Adams *at al.* (1992) showed that excitation by pyrethroids could quickly result in insects leaving a treated area and frequently knockdown would occur at some distance from the point of pickup. Extremely short exposure times as advocated by Lucas (1991) are more appropriate. He described the increased pickup due to the activation of permethrin in contrast to that of bendiocarb. This activity was measured by an actograph, a device sensitive to movement, generating an electrical signal whose intensity is directly proportional to speed and amount of movement. Analyses of *Blattella germanica* exposed for 30 seconds or 2 minutes to deposits of these materials showed that the longer period resulted in greater pickup of permethrin owing to its activating effect. A comparison between the efficacy of two products over a range of exposures can result in different relative potencies. Standardization at a 30-second 'walk on/walk off' exposure is to be recommended.

Insects are usually held in groups of ten. Adams showed that cross-contamination occurs between individuals. A female cockroach could transfer a dose sufficient to kill a male. A case could be made for holding insects separately unless the material has such a slow action that return of the insects to their harbourage would occur in the field before immobilization.

Carter and Chadwick (1978) had earlier referred to a testing regime to establish data for a graph of 'apparent' LC_{50} against time. As the deposit degrades, plates treated initially at increased levels would be required to give a regression line. The slope of the line is a measure of the residuality of the product. Laboratory and field trials were carried out with several permethrin formulations, both with and without synergist. The deposit levels achieved in the field were monitored by the analysis of strategically placed filter papers. The population was assessed using a logarithmic system of scoring of flushed out insects. This highlighted well the problems encountered in working under real conditions and how they can confuse direct comparisons.

Perhaps the ultimate laboratory approach to the realistic comparison of products was suggested by Lucas. He used 'arenas', 3 m × 3 m × 2 m high, in temperature and humidity controlled rooms. The arenas were built of corrugated cardboard. Harbourages were provided where sheets overlapped. Plastic sheets formed into 'bowls' contained the structures and were treated to prevent escapes. An infestation was introduced composed of all stages of the life cycle. Twenty-four hour video monitoring showed natural diurnal habits and revealed the behaviour and reactions to the treated areas. Population levels were assessed by 'Roatel' traps, which allow re-release after counts have been made.

REFERENCES

Adams, A.J., Bowyer, R.J. and Cleverly, A. (1992). Influence of post-contact cockroach behaviour upon efficacy of residual control treatments. *Proceedings of the XIX International Congress of Entomology,* Beijing, China.
Brown, N.C., Chadwick, P.R. and Wickham, J.C. (1967). The role of synergists in the formulation of insecticides. *Int. Pest Control* 10–13.
BS 4172: 1967. *Specification for the Insecticidal Efficacy of Aerosols Against Flies.* British Standards Institution, London.
BS 4172: 1993. *Hand-held Pressurised Aerosol Dispensers Against Houseflies.* Part I. Specification for insecticidal efficiency. Part II. Method for determination of insecticidal efficiency. British Standards Institution, London.
Busvine, J.R. (1971). *A Critical Review of the Techniques for Testing Insecticides.* Commonwealth Agricultural Bureaus, 2nd edn, Farnham Royal, UK.
Carter, S.W. and Chadwick, P.R. (1978). Permethrin as a residual insecticide against cockroaches. *Pestic. Sci.* **9**, 555–565.
Carter, S.W. and Dodd, G.D. (1979). The use of Ultra Low Volume Pyrethroid Sprays for Insect Control in Stored Products. *5th British Pest Control Conference,* Session 4, Paper 11.
Carter, S.W., Chadwick, P.R. and Wickham, J.C. (1975). Comparative observations on the activity of pyrethroids against some susceptible and resistant stored products beetles. *J. Stored Prod. Res.* **11**, 135–142.
Casida, J.E. (1970). Mixed-function oxidase involvement in the biochemistry of insecticide synergists. *J. Agric. Food Chem.* **18**, 753–772.
Chadwick, P.R. (1971). Activity of some new pyrethroids against *Blattella germanica* L. *Pestic. Sci.* **2**, 16–19.
Chadwick, P.R. and Dixon, K. (1976). Crawling insect killer aerosols: a method of testing knockdown and some effects of formulation. *Aerosol Rep.* **16**, 368.
Chadwick, P.R. and Evans, M.E. (1973). Laboratory and field tests with some pyrethroids against cockroaches. *Int. Pest Control* Jan–Feb, 11–16.
Cochran, D.G. (1987). Effects of synergists on bendiocarb and pyrethrins resistance in the German cockroach (Dictyoptera: *Blattidae*). *J. Econ. Entomol.* **80**, 728–732.
Chemical Specialities Manufacturers Association (CSMA) (1971). Aerosol and pressurised space spray insecticide test method for flying insects. *Soap Chem. Spec. Blue Book,* p. 158.
David, W.A.L. (1946). Factors influencing the interaction of insecticidal mists and flying insects. *Bull. Entomol. Res.* **37**, 1–27.
Davies, M.S. (1974) Evaluations of synthetic pyrethroids through a thermal fogger for control of houseflies. *Int. Pest Control* May/June **16**, 4–8.
Davies, M.S., Chadwick, P.R., Holborn, J.M., Stewart, D.C. and Wickham, J.C. (1970). Effectiveness of the (+)-trans-chrysanthemic acid ester of (+)-allethrolone (bioallethrin) against four insect species. *Pestic. Sci.* **1**, 225–227.
Farnham, A.W. (1973). Genetics of resistance of pyrethroid-selected houseflies, *Musca domestica* L. *Pestic. Sci.* **4**, 513–520.
Farnham, A.W. (1977). The effect of various intervals between treatments on the synergism of a pyrethroid insecticide, (IR, trans)-tetramethrin, in houseflies. *Proceedings of British Crop Protection Conference Pests and Diseases* Vol. 1, p. 149–153.
Finney, D.J. (1971). *Probit Analysis,* 3rd edn. Cambridge University Press, Cambridge.
Forrester, N.W., Bird, L.J. and Layland, J.K. (1993). Management of pyrethroid and endosulfan resistance in *Helicoverpa armigera* (Lepidoptera: *Noctuidae*) in Australia. Section 10. Pyrethroid resistance: resistance breaking pyrethroids. *Bull. Entomol. Res.* (supplement 1, September), 132 pp.

Fuchs, M.E.A. (1988). Flushing effects of pyrethrum and pyrethroid insecticides against the German cockroach (*Blattella germanica* L.). *Pyreth. Post* **17**, 3–7.

Glynne Jones, G.D. and Chadwick, P.R. (1960). A comparison of four pyrethrin synergists. *Pyreth. Post* **5**, 22–30.

Glynne Jones, G.D. and Green, E.H. (1959). A comparison of toxicities of pyrethrin synergised pyrethrins to *Calandra oryzae* L. and *Calandra granaria* L. *Pyreth. Post* **5**, 3.

Glynne Jones, G.D. and Lowe, H.J. (1956). An inexpensive addition to a micrometer syringe for the semi-automatic production of small measured drops. *Laboratory Practice* **5**, 69.

Groome, J.M., Martin, R. and Slatter, R. (1989). Advances in the control of public health insects by the application of water-based Ultra Low Volume space sprays. *Int. Pest Control* Nov/Dec, 137–140.

Hewlett, P.S. (1968). Synergism and potentiation in insecticides. *Chem. & Ind.* no. **22**, 701–706.

Hewlett, P.S. (1969). The toxicity to *Tribolium castaneum* (Herbst) (Coleoptera: Tenebrionidae) of mixtures of pyrethrins and piperonyl butoxide: fitting a mathematical model. *J. Stored Prod. Res.* **5**, 1–9.

Hursey, B.S. and Allsopp, R. (1984). The eradication of tsetse flies (*Glossina* spp.) from Western Zimbabwe by integrated aerial and ground spraying. Tsetse and Trypanosomiasis Control Branch, Department of Veterinary Services, Zimbabwee.

Lucas, J.R. (1991). Modern developments in the efficacy testing of cockroach control products – bringing the field into the laboratory. *CSMA Mid Year Meeting,* Chicago, May 1991 (unpublished).

Martin, S.J.S., Stewart, D.C. and Invest, J.F. (1977). Control of Mansonia species in Gabon using ULV pyrethroids. *Mosquito News* **37**, 395–403.

Metcalf, R.L. (1967). Mode of action of insecticide synergists. *Ann. Rev. Entomol.* **12**, 229–256.

Mount, G.A. and Pierce, N.W. (1973). Toxicity of selected adulticides to six species of mosquitoes. *Mosquito News* **33**, 368–370.

Nash, R. (1954). Studies on the synergistic effect of piperonyl butoxide and isobutyl-undecylene-amide on pyrethrins and allethrin. *Ann. Appl. Biol.* **41**, 652–663.

Oppenoorth, F.J. (1971). Resistance in insects: the role of metabolism and the possible use of synergists. *Bull. World Health Org.* **44**, 195–199.

Peet, C.H. and Grady, A.G. (1928). Studies in insecticidal activity. I. Testing insecticides against flies. *J. Econ. Entomol.* **21**, 612–617.

Potter, C. (1952). An improved laboratory spraying apparatus for applying direct sprays and surface films with data on the electrostatic charge on atomised spray fluids. *Ann. Appl. Biol.* **39**, 1–7.

Raffa, K.F. and Priester, T.M. (1985). Synergists as research tools and control agents in agriculture. *J. Agric. Entomol.* **2**, 27–45.

Slatter, R., Stewart, D.C., Martin, R. and White, A.W.C. (1981). An evaluation of Pestigas BB – a new system for applying synthetic pyrethroids as space sprays using pressurised carbon dioxide. *Int. Pest Control* Nov/Dec, 162–164.

Sun, Y.P. and Johnson, E.R. (1960). Analysis of joint action of insecticides against house flies. *J. Econ. Entomol.* **53**, 261–266.

Taha, A.M., Banoub, W.F., Stewart, D.C. and Martin, S.J.S. (1979). Fly control with ULV permethrin in Cairo. *PANS* **25**, 371–377.

Tammes, P.M.L. (1964). Isoboles, a graphic representation of synergism in pesticides. *Neth. J. Plant Pathol.* **70**, 73–80.

Tsuda, S. and Okuno, Y. (1985). Solvents and insecticidal efficacy of the aerosol containing tetramethrin and d-phenothrin. *J. Pestic. Sci.* **10**, 621–625.

Umeda, K., Yano, T. and Hirano, M. (1988). Pyrethroid-resistance mechanism in German cockroach, *Blattella germanica* (Orthoptera: *Blattidae*). *Appl. Entomol. Zool.* **23**, 373–380.

Webb, J.E. and Green, R.A. (1945). On penetration of insecticides through the insect cuticle. *J. Exp. Biol.* **22**, 8–13.

White, A.W.C., Martin, R., Stewart, D.C. and Wickham, J.C. (1992). Assessing the performance against houseflies of indoor aerosol space sprays. Part I. Factors examined during tests with houseflies in free flight. *Pestic. Sci.* **34**, 153–162.

White, A.W.C., Martin, R. and Lowe, J.A. (1993). The importance of particle size analysis in spray systems. *Spray Technol.* **3**, 30–36.

Wickham, J.C. (1976). Independent and joint biological activity of isomeric forms of synthetic pyrethroids. *Pestic. Sci.* **7**, 273–277.

Wickham, J.C., Bone, A.J. and Stewart, D.C. (1974). The application of computer-based techniques in the evaluation of pesticide products. *Pestic. Sci.* **5**, 353–362.

Wodageneh, A. and Matthews, G.A. (1981). The addition of oil to pesticide sprays. Effect on droplet size. *Trop. Pest Manag.* **27**, 121–124.

Yamamoto, I. (1973). Mode of action of synergists in enhancing the insecticidal activity of pyrethrum and pyrethroids. In: *Pyrethrum – Natural Insecticide International Symposium Proceedings. Mode of Action of Synergists,* p.195–210. Academic Press, NY, USA.

12

The Mode of Action of Piperonyl Butoxide with Reference to Studying Pesticide Resistance

ANDREW W. FARNHAM

1. INTRODUCTION

The methylenedioxyphenyl compound piperonyl butoxide (PBO) has been widely used in the field as an additive to some insecticides in the control of arthropod pests. Following the discovery of its synergistic properties it was commercially employed as an adjunct for natural pyrethrins, the combination having a much greater insecticidal activity than the natural product alone. As will be discussed later in this chapter, the degree of potentiation of insecticidal activity is related to the ratio of components in the mixture. As the proportion of PBO increases, so the amount of natural pyrethrins required to evoke the same kill decreases. The insecticidal activity of other pyrethroids, particularly of knockdown agents, can also be enhanced by addition of PBO and it is included in aerosol products for household use in flying insect control. A commercially cost-effective ratio is used, often in the range 5 : 1 to 10 : 1 PBO : pyrethroid. It is unlikely that PBO improves the initial rapid knockdown activity of such pyrethroids, but its inclusion does improve the lethal effects. The enhancement of the activity of photostable pyrethroids is normally less dramatic but PBO is included in several commercial formulations, e.g. with permethrin, cypermethrin, deltamethrin and fenvalerate.

PBO has also found application in the laboratory as a tool in helping to understand the modes of action of insecticides in both susceptible and resistant insects. Additionally it has been used in monitoring for the presence of specific mechanisms of resistance in field-collected samples of arthropod pests, but as Raffa and Priester (1985) pointed out '. . . their [i.e. synergists] contribution as research tools and as control agents are quite different, and at times these approaches are conflicting.' This chapter attempts to review aspects of the mode of action of PBO with particular reference to its role in studies on resistance and

PIPERONYL BUTOXIDE
ISBN 0-12-286975-3

its place in laboratory studies, and highlights areas which require deeper investigation.

2. BIOASSAY METHODOLOGY

Any proposed explanation for the mode of action of insecticides or synergists has to accommodate all biological evidence. Studies must be based on data from tests that use sound toxicological techniques and so the starting point has to be a review and appraisal of *in vivo* toxicological evidence. Such data must then be correlated with evidence from biochemical, neurophysiological and other tests to explain the role played by PBO. Before discussing this it is necessary to be aware of pertinent bioassay techniques, their limitations and how methodology can influence observed results.

Laboratory toxicological studies of contact poisons often employ a method of topical application of measured drops over a range of concentrations, analysing the results by probit analysis (Finney, 1971) to estimate the LD_{50} (the dose that gives 50% mortality at kill endpoint, i.e. the stage beyond which there is no further kill other than control mortality) values and other parameters. Comparisons are made between such values for the insecticide without and with the additive to derive a factor of synergism (FOS) if this ratio is greater than 1, or a factor of antagonism if it is less than 1. In studies involving resistant strains the LD_{50} values and factors of synergism should be compared with those for a standard susceptible strain. However the bioassay protocols and techniques employed by various authors often make comparisons between results from different sources difficult, and interpretation of data can be confused.

Published results from bioassays, particularly against strongly resistant strains, may be suspect when the authors fail to recognize that the assumption on which probit analysis is based, namely that an incremental increase in log dose leads to an incremental increase in probit mortality, may not hold over the full response range, especially when applying extremely large doses. Although the sample population may be genetically homogeneous, this breakdown in the assumption is likely because of the physicochemical properties of the applied compound and hence its performance at the interface with the insect. This in turn is likely to lead to a distortion of the response curve and hence to a grossly inaccurate overestimate of the LD_{50} value. This can be further distorted by failing to use kill endpoint as the criterion of effect. Consequently these effects can, either singly or in combination, result in a nonlinear response similar to that which would be achieved if the sample population was heterogeneous, thus leading to misleading estimates of synergistic and/or resistance factors. The dose at which the probit response deviates from linearity will vary from insecticide to insecticide and additionally can be affected by solvent or formulation. For accurate estimates of LD_{50} values for comparative purposes the relevant probit lines for the susceptible and resistant populations should ideally be parallel with slopes greater than 2.0 and each based on a range of doses that extends from one giving greater than 85% mortality (preferably LD_{99}) and one giving less than 15% (preferably LD_1). This dose range should reveal any heterogeneity in the lethal response to the toxicant. Even when using a sensible protocol care must be exercised when discussing the consequent degrees of either resistance or synergism.

Another factor that influences results of bioassays for joint action concerns the choice of using either a constant ratio of insecticide to additive or a constant amount of additive whilst varying the dose of the insecticidal component of the mixture. Both methods have their place, the former being most suitable when the results of bioassays have relevance to field application. It is assumed that PBO interferes with processes within the insect which normally degrade the poison, thus enhancing the activity of the insecticidal component. By applying a constant dose these processes will be equally affected. The amount of additive applied is usually the maximum dose that avoids control mortality in the susceptible standard strain when applied in the absence of the insecticide.

Two important factors have to be recognized when performing fixed ratio bioassays. Firstly, little has been published on the minimum amount of additive needed to influence the insecticidal effect. A.W. Farnham (unpublished data) has investigated this using various pyrethroids with PBO against susceptible female house flies. In experiments in which it was applied topically approximately 1 hour before the insecticide, he found that at extremely small doses of PBO ($< 0.05 \mu g$ per fly) there was no discernible improvement in performance (FOS < 2) (Fig. 12.1). As the dose of PBO was increased, synergism increased linearly but at a different rate for each pyrethroid until above 5 µg per fly when the PBO was toxic. A similar result was reported by Levot (1994), who tested a strain of sheep body louse, *Bovicola ovis,* that had low-level resistance to cypermethrin and PBO. He found that at concentrations above 200 mg L^{-1} PBO was toxic, but up to that concentration synergism increased linearly with increasing concentrations of PBO. Thus it appears that for house flies at least 0.05 µg per fly PBO is required to elicit a synergistic effect. However, some pyrethroids alone are extremely potent with LD_{50} values of 0.005 µg per fly or less (e.g. deltamethrin, LD_{50} 0.000 33 µg per fly) and for such compounds bioassayed using a constant ratio of, say, 10 : 1 PBO : insecticide, the degree of synergism observed is likely to be negligible. Because synergism of insecticides by PBO is dose dependent, this has important implications not only when studying the synergism of a candidate compound against a susceptible strain, but also when these data are compared with those from a resistant one. Obviously this has to be taken into consideration when discussing results in the context of the impact of additives on degrees of synergism or levels of resistance.

The second point is the failure to recognize that PBO is toxic in its own right. For susceptible house flies the published LD_{50} values for PBO range from *c.* 10 to 40 µg per fly. Thus in bioassays employing the constant ratio as before of, say, 10 : 1, any quoted LD_{50} of greater than 1 µg per fly of the insecticidal component in the mixture could include a lethal contribution from the PBO. Again, this is pertinent when analysing results, particularly from assays with resistant strains.

Yet another aspect of the bioassay techniques that can influence observed mortality in tests where PBO is used at a constant dose is the time interval

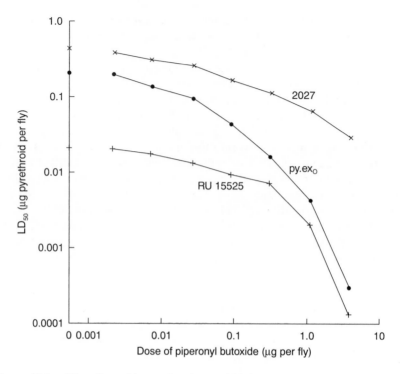

Figure 12.1.   The effect of increasing doses of PBO on the observed LD_{50} of three pyrethroids against susceptible house flies. 2027, 3-(α-cyanobenzyl)benzyl($1R$)-*trans*-chrysanthemate; RU15525, ($1R$)-*cis*-2,2-dimethyl-3-(tetrahydro-2-oxo-3-thienyl)cyclo-propanecarboxylate.

between application of each compound. Using susceptible house flies, Farnham (1977) showed that maximum enhancement of topically applied pyrethroid occurred when 2 μg PBO per fly was applied 0–2 hours earlier (Fig. 12.2). After 2 hours the potentiation of the pyrethroid declined until by 36 hours between treatments the PBO did not influence the mortality exhibited by flies treated with the pyrethroid alone. This decline shows that there is a loss of activity of PBO, but whether this is due to differential rates of penetration of the two chemicals, to degradation of the PBO in the fly, or to excretion of the intact molecule with time is not known. When the pyrethroid was applied before the PBO potentiation continued for up to 16 hours, suggesting that the activity of the pyrethroid did not diminish rapidly with time. This indicates that the pyrethroid was not rapidly denatured in susceptible flies. But if the loss of activity of the pyrethroid is comparatively slow as estimated by this method, it poses the question as what role PBO plays in enhancing activity when it can continue to do so for so long after the application of the pyrethroid.

Another influence on absolute LD_{50} values is the post-treatment temperature. This has a marked effect on the response of insects to insecticides, particularly pyrethroids with their negative temperature coefficients. However, the influence of PBO on this effect in either susceptible or resistant insects is little known.

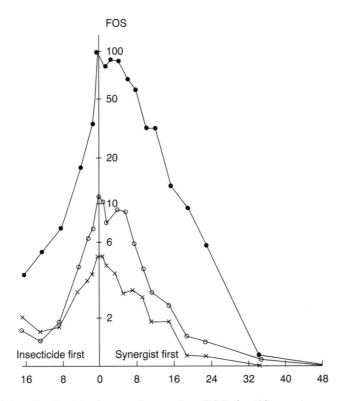

Figure 12.2.   Graph of the factors of synergism (FOS) for (1*R*, *trans*)-tetramethrin at LD$_{50}$ against time interval between treatments for three synergist applications. ⊷ , 1.0 µg NIA 16388 n-propyl-2-propynylphenylphosphonate; ⊶ , 0.09 µg NIA 16388; ×, 1.0 µg PBO. Reproduced with permission from Farnham (1977).

Comparing data from constant dose assays against different insect species becomes more difficult because of the tolerance of each species to PBO itself. For example, by topical application a dose of 1–5 µg per fly is normal for the house fly, but for the bollworm, *Helicoverpa armigera,* Forrester *et al.* (1993) applied 50 µg per larva, whilst Brewer and Keil (1989) used about 0.012 µg per insect to the sciarid fly, *Lycoriella mali.* These gross differences in amounts of PBO applied cannot be expected to elicit identical effects in each species when the response of the target species to the pyrethroids alone is of the same order of magnitude, e.g. for permethrin the LD$_{50}$ values in µg per insect are 0.014 for house fly adults, 0.016 for *H. armigera* larvae and *c.* 0.025 for *L. mali* adults.

Interpretation of results from bioassay techniques where observations are made on the knockdown effect is also questionable. When compounds are sprayed directly onto the insect and a difference in response occurs between the toxicant alone and the mixture of toxicant plus PBO, how quickly should one expect the enhancement effect to be revealed? This must depend on the physico-chemical properties of the components of the mixture, and on the immediate mode of action of the toxicant and how this can be perturbed by PBO. Cochran (1987) used a residue technique for studying the responses of cockroach (*Blattella germanica*) populations to the carbamate, bendiocarb, and observed no synergism in his susceptible strain whose LT_{50} (time to 50% knockdown for a fixed dose) was *c.* 20 minutes either without or with PBO in the residual treatment. He compared these results with those for his resistant strain, Mandarin, whose LT_{50} was > 1000 minutes for the same residual concentration of insecticide alone but was much less (LT_{50} *c.* 490 min) when PBO was included at twice the concentration of the bendiocarb in the residue. The rates of uptake of each constituent in the mixture from a treated surface are unlikely to be identical and so any interpretation of these data in the context of mode of action of either the carbamate or of PBO has to be circumspect.

These examples of the limitations of bioassay methodology show that comparative interpretations of data from various sources can be confused and so lead to ill-reasoned conclusions of both the extent and description of resistance and synergism. As Yamamoto (1973) wrote, 'Synergistic effect is greatly influenced by the test insect, formulation, pyrethroid–synergist ratio, method of administration and choice of response, either kill or knockdown; these experimental differences complicate [the] interpretation of results.' However, it is only on the basis of sound toxicological evidence that any theories concerning the mode of action of PBO, either alone or when in combination with another compound, must be founded.

3. MODE OF ACTION OF PIPERONYL BUTOXIDE ALONE

The mode of action of any active toxicant is normally studied biochemically in distinct phases to explain its lethal effects observed from *in vivo* bioassays. These involve aspects of penetration, metabolism and excretion as well as studies of the perturbation at the site of action. The vast majority of conventional contact insecticides are nerve poisons with their sites of action at the synapse or along the nerve axon. Much information derived from studies involving synergists concerning modes and sites of action of insecticides has been published and reviewed comprehensively (Haley, 1978; Raffa and Priester, 1985; B-Bernard and Philogène, 1993). Whilst the impact of PBO as an additive has been included in relevant experiments, particularly in studies on the metabolism of insecticides, little has been published on the penetration, metabolism and excretion of PBO itself, or on its site of action when it acts as a toxicant in its own right. Schunter *et al.* (1974) reported on the toxicity of PBO to the cattle tick, *Boophilus microplus,* and suggested that the lethal effect was due to its disrupting a vital oxidative but unidentified life process. As with other arthropod species, the site of action was not determined. All this despite an appeal by Camougis (1973) '. . . it is possible that synergists may have neuroexcitatory actions of their own. This should be investigated.'

Penetration of PBO into insects has been little studied. Ware (1960) showed that the rate of penetration was of the same order as that of either sevin or malathion. The presence of these insecticides made little difference to the rate of metabolism of PBO, although the metabolites were not identified. Casida (1970) reviewed published data on metabolism and excretion. House flies metabolize PBO to yield a variety of alcohol, catechol and carboxylic acid metabolites which are excreted, either without conjugation or as glucosides. Much other evidence comes from biochemical assays done using mammals with PBO applied either orally or intravenously; little has been done on insect preparations. It appears that neither the site of action of PBO nor the active moiety that causes the enhancement of performance of another co-applied poison are understood, but Casida concluded that methylenedioxyphenyl (MDP) synergists inhibit oxidative metabolism of co-applied toxicants.

Studies on the metabolism of any toxicant often involve techniques of comparison between susceptible and resistant strains. Although there are reports of resistance to topically applied PBO alone, e.g. in the house fly (Sawicki, 1974) and diamondback moth (Sun *et al.*, 1985), there appear to be no reports of relevant studies on either the mode of action or the mechanism(s) of resistance to PBO and how they might have evolved and been selected.

4. JOINT ACTION OF PIPERONYL BUTOXIDE WITH INSECTICIDES

Much has been done to expand our understanding of the mode of action of insecticides against both susceptible and resistant insects by including assays with PBO or its related MDP compound, sesamex. A large contribution to this knowledge has been through applying biochemical techniques, but it is important to interpret data generated from biochemical assays in the light of toxicological evidence.

Our understanding of the mode of action of pyrethroid insecticides has been expanded in a series of experiments designed to examine the influence of the resistance mechanism, reduced cuticular penetration, *pen* (Sawicki and Farnham, 1968). Burt and Goodchild (1974) compared the course of poisoning symptoms by pyrethrin-I (py-I) in two house fly strains, one fully susceptible (Cooper) and the other with only the genetically isolated factor *pen* (strain 314). They found that the topically applied py-I caused rapid knockdown in the Cooper strain but much slower knockdown in strain 314 (Fig. 12.3). However, after 48 hours the LD_{50} for each strain was statistically the same. When injected, the time-courses of poisoning were the same, py-I causing very rapid knockdown in both strains, but the ultimate LD_{50} values for both strains were again the same and not significantly different from those for the topically applied py-I. Hence the mode of application, whether topical application or injection, had little influence on the ultimate LD_{50} values for py-I. When the authors compared results for Cooper from topical application and injection assays for py-I and bioresmethrin (Fig. 12.4) they found slow knockdown by bioresmethrin in topical assays but very rapid knockdown by injection of each pyrethroid. Again the mode of application had little influence on the LD_{50} values for either pyrethroid.

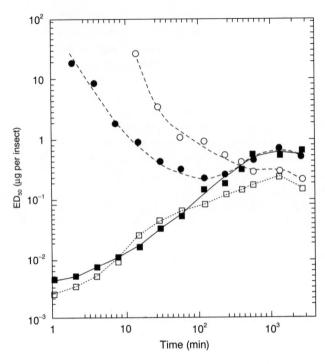

Figure 12.3. Plot of ED_{50} against time: susceptible house flies and house flies with factor *pen* treated with pyrethrin I. Topically applied to acetone: susceptible ‑‑•‑‑ ; factor *pen* ‑‑○‑‑. Injected: susceptible ‑‑•‑‑ ; factor *pen* ‑‑□‑‑ . (Burt and Goodchild, 1974 with permission).

Using the same two strains, Farnham (1973) showed that the rate of penetration of a dose of ^{3}H-py-I close to the LD_{50} for both strains was much faster into the Cooper strain (70% irretrievable in surface washings after 3 hours) than into strain 314 (5% irretrievable). The degree of potentiation of py-I following pre-treatment with 2 µg sesamex per fly for both strains was large (FOS = 350 for Cooper, 160 for strain 314). From these data and those of Burt and Goodchild (1974) it may be hypothesized that, to account for the strong synergism observed when py-I and sesamex are applied together, there must be rapid and efficient metabolism of py-I on its own following its rapid penetration into the Cooper strain, and that this metabolic system can be grossly disrupted by sesamex. However, if penetration of py-I is markedly slowed by the factor *pen*, theoretically this should permit the rapid and efficient metabolic system to be even more effective. This ought to be reflected in some increase in observed resistance, but this was not so. Using separated doses of toxicant and synergist, Wilson (1949) concluded that external effects such as increased penetration or protection of pyrethrum from oxidative destruction could not play more than a small part in the synergistic action of PBO or piperonyl cyclonene. This poses the question as to what role the MDP synergists play in the susceptible strain to so enhance the performance of natural pyrethrins including py-I.

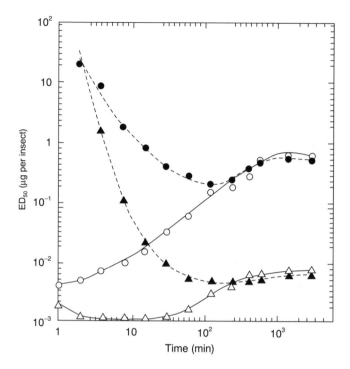

Figure 12.4. Plot of ED$_{50}$ against time: house flies treated with pyrethrin I or bioresmethrin. Pyrethrin I: topically applied in acetone ⸱⸱•⸱⸱ ; injected ⸰— . Bioresmethrin: topically applied in acetone ⸱⸱▲⸱⸱; injected △—. (Burt and Goodchild, 1974, with permission).

Various mechanisms of resistance to pyrethroids have been described. The major factor in many species including house fly is knockdown resistance, *kdr*, or one of its alleles. In house flies this factor provides strong to very strong resistance to all pyrethroids, but this resistance is little affected by MDP synergists, i.e. the factors of synergism following treatment with a fixed dose of the MDP compound to strains with *kdr* are similar to those for the susceptible strain treated likewise. This may not appear surprising when it is known that the mode of action of this resistance mechanism concerns modifications at the nerve target site and thus is not a factor of increased metabolism. What is less explicable is why at LD$_{50}$, *kdr*, a mechanism that delays the onset of poisoning but unlike *pen* requires a much larger dose of pyrethroids to be effectively lethal (up to 600-fold for deltamethrin in house flies with the allele *super-kdr*), is only potentiated to the same extent as the small amount of pyrethroid effective against the susceptible strain (LD$_{50}$ for deltamethrin is 0.00035 μg per fly), even though the quantity of additive is the same for each strain.

Because of the strong potentiation of pyrethroids by PBO in susceptible strains and the strong resistance provided by *kdr* and its alleles, it can be very difficult to establish the presence of those additional resistance factors of increased metabolism in field strains by *in vivo* bioassays that include PBO.

Such bioassays have to demonstrate that there is significantly greater synergism than that which occurs in the susceptible strain. This has been demonstrated in pyrethroid-resistant populations of *Helicoverpa armigera* in Australia (Forrester *et al.*, 1993), where the major resistance mechanism now appears to be PBO-suppressible rather than the *kdr*-type factor that had predominated in the samples collected until the mid-1980s (Gunning *et al.*, 1991). There was very much greater synergism in field-collected samples than in their standard susceptible strain following treatment with 50 μg per larva PBO, but the authors were unable to determine whether or not there was any resistance to PBO itself because of the natural tolerance level of the susceptible strain.

Scott and Georghiou (1986) selected for strong resistance to permethrin in a strain of house flies, Learn-PyR. This strain showed resistance to PBO which was applied at one-third of the maximum sublethal doses, i.e. at 2 μg per fly to the NAIDM-susceptible strain and 10 μg per fly to Learn-PyR, before pyrethroid treatment. There was much stronger synergism (FOS = *c*. 800) of permethrin in the resistant strain than in NAIDM (FOS = *c*. 4.4) and a consequent reduction in the level of resistance from > 5000-fold to 32-fold following PBO treatment. The difference in degrees of synergism between these two strains cannot be accounted for simply on the basis of the difference in the amount of PBO applied. This is positive evidence for the presence of a permethrin resistance factor that can be affected by PBO. The authors also showed that reduced cuticular penetration contributed to the observed resistance and that PBO could further reduce the penetration of permethrin into Learn-PyR flies, but it had little influence on the penetration of permethrin into flies of the NAIDM strain.

In an examination of the effect of strong selection with pyrethroids on cross-resistance spectra, Farnham (1971) studied a population of house flies (213ab) from Sweden which had originally developed resistance in the field to natural pyrethrins plus PBO (Davies *et al.*, 1958) and had subsequently been selected irregularly with natural pyrethrins plus PBO in the laboratory (J.D. Keiding, personal communication). Two substrains were established through selection with either natural pyrethrins alone to give strain NPR or resmethrin to give strain 104. After many generations of regular treatment, strain NPR became immune to the largest dose that could be sensibly applied i.e. 50 μg per fly of the selecting agent (< 10% kill of either sex, a level of resistance ⩾ 60-fold). This strain was equally immune to either py-I or pyrethrin-II (py-II), and exhibited strong resistance to bioresmethrin (resistance factor RF = 120) (Table 12.1), as did strain 104 (RF = 40). However this latter strain, in contrast to strain NPR, was only weakly resistant to py-I (RF = 1.6) and py-II (RF = 18). Pretreatment with 2 μg sesamex per fly led to similar responses to pyrethroids of the two strains to py-I, py-II and bioresmethrin (RFs = 120, 110 and 86 respectively in strain NPR and 69, 120 and 77 in strain 104). The outstanding consequence of this was the very small synergistic factors for strain 104, particularly with py-I (FOS = 6.8) and py-II (FOS = 4.8), by far the smallest for any house fly strain assayed with pyrethrolone esters and sesamex. On the accepted concept that MDP synergists inhibit oxidative metabolism (Casida, 1970) and that the selection regimes have only increased the gene frequencies of factors already present in strain 213ab, the question arises as to what mechanism(s) were selected for such strong resistance to synergized pyrethrins to arise in strain 104 when no synergist was included in the selecting agent, and what role the sesamex played when applied in combination with natural pyrethrins.

In a further study Farnham (1974) genetically analysed the NPR strain. He isolated and identified pyrethroid resistance factors on autosomes III (*kdr* and *pen*) and V (*py-ses*). This latter factor gave low resistance to pyrethrolone esters (*c.* 2- to 3-fold), stronger synergism of them by sesamex (*c.* 2- to 4-fold greater) than that found in the susceptible strain, but no resistance to bioresmethrin. He also identified a factor on autosome II which appeared to give slightly smaller synergistic factors than those for the susceptible strain. Neither of the factors on chromosomes II or V could alone account for the grossly reduced synergism of py-I or py-II in strain 104. Although not significantly different, there was a trend towards reduced synergism of both pyrethrins in the strain with *pen* by comparison with the susceptible, and in the strain with both *kdr* and *pen* by comparison with the strain with *kdr* alone, but the degree of synergism across all strains for bioresmethrin was remarkably similar (Table 12.1). Thus it appears that the performance of sesamex can be affected specifically by the presence of *pen* reducing the efficacy of the pyrethroid/synergist mixture.

Bridges (1957) has reported that PBO slows the penetration of [^{14}C]allethrin into adult susceptible house flies, and in a recent study Kennaugh *et al.* (1993) showed that PBO could also influence the penetration of isomers of permethrin into the larvae of *H. armigera*.

Another group of insecticides for which strong synergism of resistance by PBO has been demonstrated is the *N*-alkylamides. During a broad examination of cross-resistance in house fly populations, Farnham *et al.* (1986) tested a number of strains from several countries with 5-bromo-2(6-(1,2-dimethylpropyl-amino)-6-oxo-hexa–2(*E*),4(*E*)-dienyl)naphthalene (BTG 502) both without and with PBO. The authors showed that seven out of 20 strains resisted BTG 502 but only one of these resisted the combination of BTG 502 plus 1 μg PBO per fly. BTG 502 is synergized by PBO in susceptible house flies but there was much stronger synergism correlated with the level of resistance to BTG 502 (FOS up to 600) in the six resistant strains that had the PBO suppressible factor (Fig. 12.5). However, this factor did not correlate with the presence of any previously identified resistance mechanism in any of the strains, although the two strains most resistant to BTG 502 were also strongly resistant to several other insecticides including dimethoate.

5. CONCLUSIONS

Whilst there is irrefutable evidence to support the theory that PBO inhibits oxidative metabolism, the fact that it appears capable of performing other less well-defined functions casts a shadow over the widely accepted concept that because a synergistic effect can be demonstrated following application of PBO,

Table 12.1. LD_{50} values, resistance factors (RF) and factors of synergism (FOS) for various strains tested with pyrethroids alone or following pretreatment with 2 µg sesamex per fly

Strain	Pyrethrin I			Pyrethrin II			Bioresmethrin		
	LD_{50} (µg per fly)	RF	FOS	LD_{50} (µg per fly)	RF	FOS	LD_{50} (µg per fly)	RF	FOS
NPR	>50	#		>50	#		0.66	120	
+ sesamex	0.13	120		0.48	110		0.049	86	13
104	0.52	1.6		2.5	18		0.21	39	
+ sesamex	0.076	69	6.8	0.52	120	4.8	0.044	77	4.8
kdr	0.57	1.7		0.86	6.1		0.066	11	
+ sesamex	0.0043	4.0	130	0.0084	2.0	100	0.0073	13	8.2
pen	0.26	0.79		0.14	1.0		0.0071	1.3	
+ sesamex	0.0019	1.7	140	0.0075	1.7	19	0.0010	1.8	7.0
kdr + pen	1.3	3.9		1.4	10		0.11	20	
+ sesamex	0.015	14	89	0.059	14	24	0.013	23	8.3
Cooper	0.33			0.14			0.0054		
+ sesamex	0.0011		300	0.0043		49	0.00057		9.5

Too large to estimate

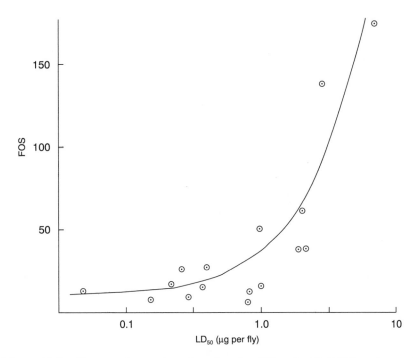

Figure 12.5. Factors of synergism plotted against LD_{50}s for BTG 502 against various strains. Reproduced with permission from Farnham *et al.* (1986).

however small, there must be mixed function oxidase activity that is being inhibited. This has major implications, especially where PBO is used as a diagnostic indicator for the presence of specific resistance factors.

PBO is usually applied at a dose that on its own is sublethal to the target species, in combination with a known toxicant whose performance is so enhanced that it too is applied at a rate which on its own would be sublethal. The assumption is that PBO interferes with processes in the organism, enabling the toxicant to be less readily degraded prior to reaching its site of action. But PBO on its own can exhibit toxic effects, and so at sublethal doses is likely to exert some stress on the insect. This in turn would improve the potency of the mixture without needing to affect metabolism of the insecticide directly. In the insecticidal mixture, when both components are applied at intrinsically sublethal rates, could each be contributing equally to the observed lethal effect, or could that effect be attributable to PBO enhancement by the known insecticide rather than vice versa?

Interference with penetration of co-applied toxicants, its own potency related to its unknown target site, the ability of some insect populations to resist PBO possibly through a metabolic process, and the likelihood that PBO can interfere with processes within the target species other than those concerned with oxidative metabolism all point to areas of investigation that should be undertaken to help us to understand the mode of action of this very valuable and important chemical.

REFERENCES

B-Bernard, C. and Philogène, B.J.R. (1993). Insecticide synergists: role, importance, and perspectives. *J. Tox. Environ. Health* **38**, 199–223.

Brewer, K.K. and Keil, C.B. (1989). A mixed function oxidase factor contributing to permethrin and dichlorvos resistance in *Lycoriella mali* (Fitch) (Diptera: Sciaridae). *Pestic. Sci.* **26**, 29–39.

Bridges, P.M. (1957). Absorption and metabolism of [^{14}C]allethrin by the adult housefly, *Musca domestica* L. *Biochem. J.* **66**, 316–320.

Burt, P.E. and Goodchild, R.E. (1974). Knockdown by pyrethroids: its role in the intoxication process. *Pestic. Sci.* **5**, 625–633.

Casida, J.E. (1970). Mixed-function oxidase involvement in the biochemistry of insecticide synergists. *J. Agric. Food Chem.* **18**, 753–772.

Camougis, G. (1973). Mode of action of pyrethrum on arthropod nerves. In: '*Pyrethrum*' (Casida, J., ed.), pp. 211–222. Academic Press, New York.

Cochran, D.G. (1987). Effects of synergists on bendiocarb and pyrethrins resistance in the German cockroach (Dictyoptera: Blattellidae). *J. Econ. Entomol.* **80**, 728–732.

Davies, M., Keiding, J. and von Hofsten, C.G. (1958). Resistance to pyrethrins and to pyrethrins-piperonyl butoxide in a wild strain of *Musca domestica* L. in Sweden. *Nature* **182**, 1816–1817.

Farnham, A.W. (1971). Changes in cross-resistance patterns of houseflies selected with natural pyrethrins or resmethrin (5-benzyl–3-furylmethyl(±)*cis-trans*-chrysanthemate). *Pestic. Sci.* **2**, 138–143.

Farnham, A.W. (1973). Genetics of resistance of pyrethroid selected houseflies, *Musca domestica* L. *Pestic. Sci.* **4**, 513–520.

Farnham, A.W. (1974). Genetics of resistance to insecticides in a pyrethrum resistant strain of houseflies. PhD Thesis, London University.

Farnham, A.W. (1977). The effect of various intervals between treatments on the synergism of a pyrethroid, (1R-*trans*)-tetramethrin, in houseflies. In: *Proceedings of the 1977 British Crop Protection Conference – Pests and Diseases* **1**, 149–153.

Farnham, A.W., Sawicki, R.M. and White, J.C. (1986). The response of resistant houseflies to an unsynergised and synergised N-alkylamide. In: *Proceedings of the 1986 British Crop Protection Conference – Pests and Diseases* **2**, 645–649.

Finney, D.J. (1971). '*Probit Analysis*', 3rd edn. Cambridge University Press, Cambridge.

Forrester, N.W., Cahill, M., Bird, L.J. and Layland, J.K. (1993). Management of pyrethroid and endosulfan resistance in *Helicoverpa armigera* (Lepidoptera: Noctuidae) in Australia. *Bull. Entomol. Res. Suppl. Series* (supplement 1), 132.

Gunning, R.V., Easton, C.S., Balfe, M.E. and Ferris, I.G. (1991). Pyrethroid resistance mechanisms in Australian *Helicoverpa armigera*. *Pestic. Sci.* **33**, 473–490.

Haley, T. (1978). PBO, [α-(2-butoxyethoxy)ethoxy]-4,5-methylenedioxy-2-propyltoluene: a review of the literature. *Ecotox. Environ. Safety* **2**, 9–31.

Kennaugh, L., Pearce, D., Daly, J.C. and Hobbs, A.A. (1993). A piperonyl butoxide synergisable resistance to permethrin in *Helicoverpa armigera* which is not due to increased detoxification by cytochrome P450. *Pestic. Biochem. Physiol.* **45**, 234–241.

Levot, G.W. (1994). Pyrethroid synergism by piperonyl butoxide in *Bovicola ovis* (Schrank) (Phthiraptera: Trichodectidae). *J. Austr. Entomol. Soc.* **33**, 123–126.

Raffa, K.F. and Priester, T.M. (1985). Synergists as research tools and control agents in agriculture. *J. Agric. Entomol.* **2**, 27–45.

Sawicki, R.M. (1974). Genetics of resistance of a dimethoate-selected strain of houseflies (*Musca domestica* L.) to several insecticides and methylenedioxyphenyl synergists. *J. Agric. Food Chem.* **22**, 344–349.

Sawicki, R.M. and Farnham, A.W. (1968). Examination of the isolated autosomes of the SKA strain of houseflies (*Musca domestica* L.) for resistance to several insecticides with and without pre-treatment with Sesamex and TBTP. *Bull. Entomol. Res.* **59**, 409–421.

Schunter, C.A., Roulston, W.J. and Wharton, R.H. (1974). Toxicity of PBO to *Boophilus microplus*. *Nature* **249**, 386.

Scott, J.G. and Georghiou, G.P. (1986). Mechanisms responsible for high levels of permethrin resistance in the house fly. *Pestic. Sci.* **17**, 195–206.

Sun, C.N., Wu, T.K., Chen, J.S. and Lee, W.T. (1985). Insecticide resistance in diamondback moth. In: *Proceedings of the 1st International Workshop on the Diamondback Moth,* AVRDC Publ. no. 86-248 (Talekar, N.S., ed.), pp. 359–371. AVRDC (Asian Vegetable Research and Development Centre), Shanhua, Taiwan.

Ware, G.W. (1960). The penetration of PBO as a synergist and as an antagonist in *Musca domestica* L. *J. Econ. Entomol.* **53**, 14–16.

Wilson, C.S. (1949). Piperonyl butoxide, piperonyl cyclonene and pyrethrum applied to selected parts of individual flies. *J. Econ. Entomol.* **42**, 423–428.

Yamamoto, I. (1973). Mode of action of synergists in enhancing the insecticidal activity of pyrethrum and pyrethroids. In: *Pyrethrum* (Casida, J., ed.), pp. 195–210. Academic Press, New York.

13

Inhibition of Resistance-related Esterases by Piperonyl Butoxide in *Helicoverpa armigera* (Lepidoptera: Noctuidae) and *Aphis gossypii* (Hemiptera: Aphididae)

ROBIN V. GUNNING, GRAHAM D. MOORES and
ALAN L. DEVONSHIRE

1. INTRODUCTION

The cotton bollworm *Helicoverpa armigera* (Hübner) is a serious pest of cotton and other summer crops in Australia, where it has a long history of insecticide resistance to DDT (Wilson, 1974; Goodyear *et al.*, 1975; Goodyear and Greenup, 1980), pyrethroids (Gunning *et al.*, 1984), carbamates (Gunning *et al.*, 1992), organophosphates (Goodyear and Greenup, 1980; Kay *et al.*, 1983; Gunning and Easton, 1993) and endosulfan (Kay, 1977; Kay *et al.*, 1983; Gunning and Easton, 1994).

Chemical insecticides are currently essential for the control of *H. armigera* on cotton and are likely to remain an important component of control strategies for the foreseeable future. However, insecticide resistance is a major threat to the economic production of cotton in Australia. The development of resistance had been delayed by the insecticide resistance management strategy for *H. armigera* but the frequency of pyrethroid resistance has gradually increased over recent years (Forrester *et al.*, 1993; Gunning *et al.*, 1991; Gunning, 1994).

Pyrethroid resistance in some strains of *H. armigera* is largely due to over-production of specific esterase isoenzymes (R_m 0.24–0.33) which are thought to sequester and hydrolyse pyrethroids. The resistance factor is positively correlated to esterase titre, so that increasing resistance is accompanied by increasing esterase activity (Gunning *et al.*, 1996). Unfortunately, conventional esterase inhibitors, such as profenofos, have failed to synergize pyrethroids

PIPERONYL BUTOXIDE
ISBN 0-12-286975-3

against resistant *H. armigera* (Gunning *et al.*, 1991), even though they can be very effective against resistant strains of the related species *Heliothis virescens*. Piperonyl butoxide (PBO) is partially effective as a pyrethroid synergist in *H. armigera* and its synergistic properties have been used in tank mixes to improve pyrethroid efficacy in the field (Forrester *et al.*, 1993). PBO is known to facilitate pyrethroid penetration through the cuticle of resistant *H. armigera* (Gunning *et al.*, 1995); however, there is no evidence of a monooxygenase mediated resistance mechanism (Kennaugh *et al.*, 1993; Gunning *et al.*, 1995). It therefore seemed possible that PBO might act as an inhibitor of the pyrethroid-resistance-related esterase in *H. armigera*.

Both *Aphis gossypii* and *Myzus persicae* are major agricultural pests, and insecticide resistance in these species is a significant problem. Organophosphate and carbamate tolerance in these aphids has been related to detoxification via increased carboxylesterase activity (Suzuki *et al.*, 1993; Devonshire and Moores, 1977).

In this chapter we present the results from studying the possible inhibitory effects of PBO on the hydrolysis of 1-naphthyl acetate by putative pyrethroid resistance-related esterases in *H. armigera* and the resistance-related esterases in *A. gossypii* and *M. persicae*.

2. EXPERIMENTAL MATERIALS AND METHODS

2.1. Insects – *H. armigera*

A susceptible *H. armigera* strain was originally obtained from Dr R. Teakle (Department of Primary Industries, Queensland) (Gunning *et al.*, 1996).

H. armigera were collected as eggs from cotton at Warren in the Macquarie Valley, New South Wales, and at Emerald, Queensland. These *H. armigera* were reared to the F_1, as previously described (Gunning *et al.*, 1984). Fenvalerate-selected strains were obtained by retaining the survivors of discriminating doses ($LD_{99.9}$ or higher of the susceptible strain; Gunning *et al.*, 1984); the more highly resistant strains (Goodyear and Greenup, 1980) (30-fold) were achieved by selection with fenvalerate synergized with PBO. The resistant strains were approximately 10-, 30- and 60-fold resistant to fenvalerate.

2.2. Insects – Aphids

The *Aphis gossypii* strain used in the experiments, clone 968E, was derived from a sample obtained in Greece in 1991. It has high esterase activity, around $10\times$ the levels of the standard susceptible strains, and an insensitive AChE (Moores *et al.*, 1996). This strain shows very high resistance to pirimicarb (>20 000-fold) and to organophosphates (demeton-*S*-methyl, 400-fold). The *Myzus persicae* strain used was 794J, a well-defined resistant R_3 clone derived from a sample collected from a glasshouse at Evesham in 1982, possessing amplified esterase activity.

2.3. Insecticides and Bioassay – *H. armigera*

The insecticides and synergist used were technical grade: fenvalerate (98%, Shell), profenofos (99%, Ciba Geigy) and PBO (99%, Sigma).

Larval bioassays to determine pyrethroid resistance status in the test insects were similar to those recommended by the Entomological Society of America (Anon., 1970). Technical grade insecticide was dissolved in acetone and five serially diluted concentrations prepared. For each concentration, ten 3rd instar larvae (30–40 mg) were treated with 1 μL of solution applied by microapplicator or micropipette to the dorsal thorax. Each test was replicated three times and every replicate included acetone-treated controls which confirmed no control mortality. Synergists PBO and profenofos were applied 30 minutes before the fenvalerate at 10, and 0.1 μg per larva, respectively (Gunning *et al.*, 1991). After dosage, the larvae were held individually at 25 ± 1°C with adequate food. Mortality was assessed 48 hours after treatment. Larvae were considered dead if unable to move in a coordinated way when prodded with a blunt probe. The data were analysed by probit, and resistance factors calculated as the ratio of resistant LD_{50} to susceptible LD_{50}.

2.4. Enzyme Activity – *H. armigera*

Esterase activity was detected using 1-naphthyl acetate as a substrate, using kinetic assays, as previously described (Gunning *et al.*, 1996). Groups of fifty individual adult (3–4 mg) resistant and susceptible *H. armigera* were homogenized in Eppendorf tubes in 1.0 mL of 0.02 mol L^{-1} phosphate buffer (pH 7.0) containing 0.05% Triton X-100. Homogenates were centrifuged for 3 minutes, at 3000 rpm and decanted into clean Eppendorf tubes. Synergists dissolved in acetone (1 mL) were added to the *H. armigera* homogenates at various concentrations (PBO 1–592 μmol L^{-1}: profenofos 1–10 μmol L^{-1}). Homogenates were incubated at 25°C, as were controls (insect homogenate alone and homogenate plus acetone). At various times after synergist addition (10–180 min), aliquots (10 μL) were transferred to a microplate containing 240 μL of 0.2 mol L^{-1} phosphate buffer (pH 6.0) with 0.6% Fast Blue RR Salt and 1 mmol L^{-1} 1-naphthyl acetate. Kinetic assays for hydrolysis of 1-naphthyl acetate were immediately performed on a Bio-Rad 3550 microplate reader (Bio-Rad Laboratories, utilizing the Kinetic Collector 2.0 software run on Macintosh SE microcomputer), taking absorbance readings (450 nm) automatically at 14 second intervals for 10 minutes. The rate was calculated by the online computer as the slope of the fitted regression line, using an absorbance limit of 2.0.

An extract of the partially purified esterases of interest (electrophoretic mobility, R_m 0.25–0.34; Gunning *et al.*, 1996), was isolated using the following procedure. Forty 3–4 mg larvae were homogenized in 200 μL of 0.02 mol L^{-1} phosphate buffer (pH 7.0) containing 0.5% Triton X-100. The tube was centrifuged at 3000 rpm for 5 minutes and the supernatant loaded on to a polyacrylamide gel prepared with a flat surface. After electrophoresis, a strip of the gel was stained to locate the esterases of interest and the corresponding region of the remaining unstained gel was excised and minced. The esterase was extracted from the gel using an electro-eluter (Bio-Rad Laboratories) and concentrated by centrifugation (Millipore MC filter unit) to a convenient volume. The inhibition of hydrolysis of 1-naphthyl acetate by this enzyme extract was examined, utilizing enzyme kinetic techniques, as described above.

The nature of enzyme inhibition by profenofos and PBO was further studied in the 60-fold resistant strain and enzyme, examining enzyme kinetic activity toward 12.5, 25, 50, 100 and 200 mmol L^{-1} concentrations of 1-naphthyl acetate.

2.5. Enzyme Activity – Aphids

Mass homogenates of fresh aphid material were prepared immediately before the inhibition analysis by crushing 60 aphids in 5.5 mL of phosphate buffer (0.02 mol L^{-1}, pH 7.0) on ice. The homogenates were then spun at 5000 rpm for 10 minutes, and the supernatant kept for further analysis.

Dilutions of PBO in buffer (as above) were prepared from an acetone stock to give concentrations of 2×10^{-4}, 10^{-3}, 2×10^{-3} and 4×10^{-3} mol L^{-1}, respectively, plus a control. Acetone was then added to these dilutions to give 10% v/v final concentration. 20 μL aliquots of these dilutions were then placed into the wells of a microtitre plate. At zero time, 20 μL of the aphid homogenate was added to each well and mixed.

At times 1, 5, 10, 20, 30 and 40 minutes, 200 μL of buffer (as before) containing 1 mmol L^{-1} 1-naphthyl acetate and 1.5 mmol L^{-1} Fast Blue RR Salt were added to stop the inhibition and monitor the esterase activity remaining. This was performed by using a T_{max} microplate reader (Molecular Devices) monitoring the assays at 450 nm for 10 minutes (Grant *et al.*, 1989).

3. RESULTS

3.1. *H. armigera*

Bioassay data for the *H. armigera* strains are shown in Table 13.1. Strains were 10-, 30- and 60-fold resistant to fenvalerate. Profenofos was ineffective as a pyrethroid synergist, but PBO rendered the 10-fold-resistant strain almost completely susceptible to fenvalerate. In the more highly resistant populations (30- and 60-fold), PBO was much less effective.

The inhibitory effect of profenofos on esterase activity towards 1-naphthyl acetate in the crude homogenates after a 30 min incubation, is shown in Fig. 13.1. There was some inhibition of esterase activity in the susceptible and all the resistant strains (~20%); however, there was no inhibition with 1-naphthyl acetate in the partially purified resistant enzyme extract. Increasing the profenofos beyond a threshold concentration of 0.64 μmol L^{-1} did not have any additional effect.

The results of similar experiments using PBO as an inhibitor (after a 30 min incubation) are shown in Fig. 13.2. While PBO had no inhibitory effects on

Table 13.1. Toxicity of topically applied fenvalerate and fenvalerate + synergist to pyrethoid-resistant and susceptible larvae of *Helicoverpa armigera*

Insecticide/synergist	Strain	Slope (±SE)	LD$_{50}$ (µg fenvalerate per larva)	RF[a]
Fenvalerate	Susceptible	3.9 (0.3)	0.032 (0.028–0.036)	
Fenvalerate + PBO[b]	Susceptible	4.1 (0.4)	0.033 (0.030–0.037)	
Fenvalerate + profenofos[c]	Susceptible	4.1 (0.3)	0.030 (0.026–0.034)	
Fenvalerate	Warren	4.8 (0.2)	0.29 (0.25–0.32)	10
Fenvalerate + PBO	Warren	4.4 (0.4)	0.033 (0.028–0.037)	1
Fenvalerate + profenofos	Warren	4.5 (0.4)	0.30 (0.26–0.35)	10
Fenvalerate	Warren	4.2 (0.3)	0.92 (0.85–0.98)	30
Fenvalerate + PBO	Warren	3.8 (0.4)	0.83 (0.77–0.87)	28
Fenvalerate + profenofos	Warren	3.9 (0.5)	0.94 (0.82–1.04)	31
Fenvalerate	Emerald	2.9 (0.4)	1.75 (1.27–2.48)	59
Fenvalerate + PBO	Emerald	3.2 (0.4)	1.61 (1.33–1.95)	54
Fenvalerate + profenofos	Emerald	3.0 (0.3)	1.81 (1.35–2.36)	60

[a] Resistance factors were calculated as the ratio of the LD$_{50}$ of the resistant strains to that of the susceptible strain (which was unaffected by the additives).
[b] PBO, 10 µg per larva.
[c] Profenofos, 0.06 µg per larva.

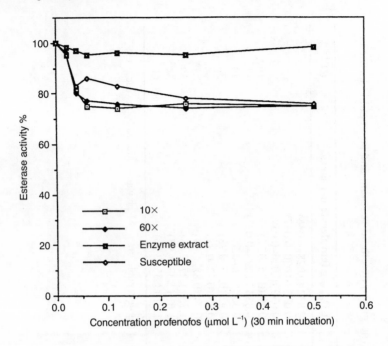

Figure 13.1   Effects of incubating resistant and susceptible *H. armigera* homogenates, and a partially purified resistant enzyme extract with profenofos, on esterase activity toward 1-naphthyl acetate.

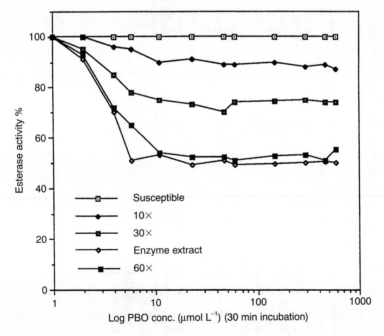

Figure 13.2   Effects of incubating resistant and susceptible *H. armigera* homogenates and a partially purified resistant enzyme extract with PBO, on esterase activity toward 1-naphthyl acetate.

esterase activity in the susceptible strain, it effectively inhibited esterase activity towards 1-naphthyl acetate in the resistant strains, at concentrations of approximately 10^{-5} mol L^{-1}. Inhibition was greater in the strains with increased resistance factor. In the 60-fold-resistant strain, and with the extract of pyrethroid-detoxication-related enzymes, approximately 50% of enzyme activity toward 1-naphthyl acetate was inhibited. The differences between resistant strains are clearly related to the amount of resistance related esterase present.

The effects of incubation time on PBO inhibition of *H. armigera* esterase activity using a concentration of 10^{-5} mol L^{-1} PBO were investigated (Fig. 13.3). Results showed that incubation of esterase for 30 minutes with PBO was necessary for maximum inhibition of enzyme activity in resistant homogenates and the esterase extract. No esterase inhibition occurred in the susceptible strain.

3.2. Aphids

Nonspecific esterase activity of crude homogenate prepared from *A. gossypii* (clone 968E) after incubation with PBO is shown in Fig. 13.4. There was clear inhibition of esterase activity in these incubations over 40 minutes, and this inhibition increased with PBO concentration. However, the limited solubility of PBO precluded use of concentrations exceeding 2 mmol L^{-1}.

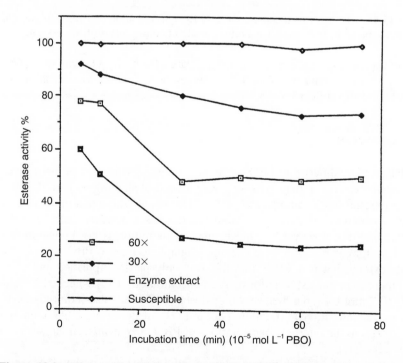

Figure 13.3. Effects of incubation time on PBO inhibition of esterase-mediated hydrolysis of 1-naphthyl acetate in resistant and susceptible *H. armigera* homogenates and a partially purified enzyme from the resistant strain.

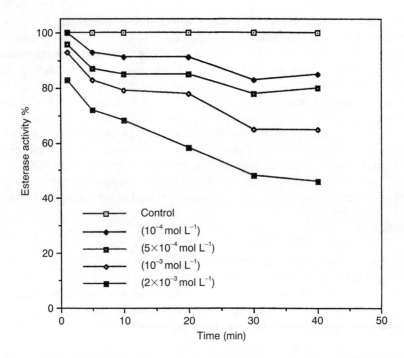

Figure 13.4.   Effects of incubation time and PBO concentration on PBO inhibition of esterase-mediated hydrolysis of 1-naphthyl acetate in homogenates of *A. gossypii.*

In contrast, using crude homogenates prepared from *M. persicae*, no inhibition of esterase activity was found over 40 minutes, even at concentrations of 2 mmol L^{-1} PBO (Fig. 13.5).

4. DISCUSSION

The data, which shows that PBO apparently interacts with *H. armigera* resistance-related esterases, may go a long way toward explaining its partial synergism of pyrethroids (Gunning *et al.*, 1991). Our results show that up to 70% of the esterase activity related to pyrethroid resistance is apparently inhibited in the presence of PBO; however, the mechanism of inhibition is not known at this stage. It is not clear why 30% of enzyme is unable to bind to PBO, even in the enzyme extract, but it is likely that this was only partially purified and comprised more than one esterase form. The concentration of PBO needed is quite high (~10^{-5} mol L^{-1}), but then very high doses of PBO are also required to synergize fenvalerate *in vivo* (10–50 µg per larva) (Forrester *et al.*, 1993; Gunning *et al.*, 1991), so the internal concentrations of PBO in *H. armigera* larvae could also be very large.

There was a lack of inhibition of esterase activity observed in the *H. armiger*-susceptible enzyme compared with that in the resistant enzyme. Also, no inhibition of the esterase (E4) was observed in *M. persicae*; furthermore, inhibition

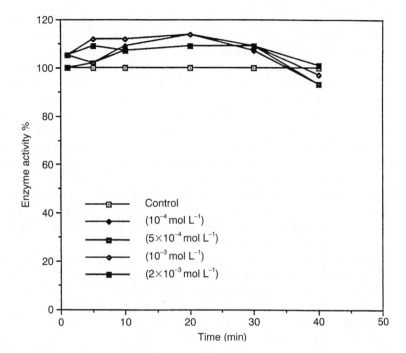

Figure 13.5.   Effects of incubation time and PBO concentration on PBO inhibition of esterase-mediated hydrolysis of 1-naphthyl acetate in homogenates of *M. persicae.*

of the *A. gossypii* esterase did not occur if the aphids had been kept frozen before incubation with PBO (G. Moores, unpublished data). These observations suggest that whilst PBO may be acting as a conventional esterase inhibitor, it is possible that certain esterases will enter a physicochemical reaction with the PBO, leaving them unavailable to the substrate. It is unlikely that PBO was interacting with the substrate directly, as no differences would then be expected between fresh/frozen material, or indeed between the different species.

Some inhibition of general esterase activity by profenofos was present, but there was probably no binding to the *H. armigera* esterases associated with pyrethroid resistance. This is consistent with bioassay data, which showed no synergism of pyrethroids by profenofos.

PBO has been assumed to inhibit mono-oxygenase-based detoxication of pyrethroids in resistant *H. armigera* (Daly, 1993; Forrester *et al.*, 1993), although there is no indication of cytochrome P450-mediated metabolism of pyrethroids (Kennaugh *et al.*, 1993; Gunning *et al.*, 1995). Instead there is compelling evidence to support an esterase-mediated metabolic resistance mechanism (Gunning *et al.*, 1996) and the present data show that PBO can inhibit these *H. armigera* resistance-associated esterases. In addition, PBO can facilitate fenvalerate penetration through the cuticle of resistant *H. armigera* (Gunning *et al.*, 1995). It is probable that the combination of these two effects explains the partial synergism of fenvalerate by PBO in pyrethroid-resistant

Australian *H. armigera*. Experiments (unpublished) have also been performed to investigate the synergistic effect of PBO on *A. gossypii*, which are consistent with the observed esterase inhibition.

Our data show that PBO only partially inhibited *H. armigera* resistance-associated esterases and this is consistent with bioassay data. Our previous studies have shown that only the more highly resistant individuals, with a greatly increased esterase titre, survive PBO/fenvalerate mixtures (Gunning *et al.*, 1996). In less resistant individuals, PBO can inhibit sufficient enzyme to prevent pyrethroid detoxification. The ability of PBO to facilitate pyrethroid penetration through the cuticle (Gunning *et al.*, 1995) would assist here in producing very high internal concentrations of insecticide.

Other studies have also shown that PBO was a moderately effective inhibitor of esterase isoenzyme activity in insects. It caused some inhibition of permethrin hydrolysis by esterases in *Pseudoplusia includens* (Dowd and Sparkes, 1987) and inhibited permethrin-hydrolysing esterases in pyrethroid-tolerant *Wiseana cervinata* larvae (Chang and Jordan, 1983). Concentrations of PBO required for noticeable inhibition of esterase activites were in the order of 10^{-4} to 10^{-5} mol L^{-1}, similar to those required here to inhibit *H. armigera* esterase activity.

Our finding that the esterases involved in pyrethroid resistance in *H. armigera* are partially inhibited by PBO has interesting implications for resistance management on cotton in Australia. PBO does not have a limitless capacity to inhibit these resistance-associated enzymes and our bioassay data only showed high synergism in the lower-order resistant populations. These less resistant individuals have only a moderate increase in esterase titre compared with susceptible insects (Gunning *et al.*, 1996). However, highly resistant *H. armigera*, which are not synergized by PBO, can have enormous quantities of esterase (Gunning *et al.*, 1996), probably more than enough to detoxify pyrethroids even after some binding with PBO has occurred. PBO has been used as a tank additive to pyrethroid sprays against *H. armigera* in Australia (Forrester *et al.*, 1993), but the present data suggests that PBO should be used in the field with caution, like any other insecticide, as it will select the more highly resistant individuals with an increased esterase activity, and so lead to control failures. Resistance factors of Australian *H. armigera* range from 2- to 300-fold (Gunning *et al.*, 1996), and those insects with resistance of less than 30-fold can still be controlled with commercial rates of pyrethroids, provided sprays are applied against eggs and small larvae (R.V. Gunning, unpublished). Thus since pyrethroid resistance is fixed in *H. armigera* populations (Gunning, 1994), the emphasis should be placed on maintaining relative susceptibility of the lower-order resistant populations, which fortunately still predominate in Australia (R.V. Gunning, unpublished).

Many investigators regard synergism by PBO as a specific inhibitor of mixed function oxidases. The findings that nonspecific esterase activity can also be severely decreased by the same compound should serve as a warning against such assumptions, and suggests the need for careful reappraisal of earlier investigations.

5. ACKNOWLEDGEMENTS

The *H. armigera* work was initiated at IACR-Rothamsted, UK whilst R.V. Gunning was a visiting scientist there in 1989 and 1992. IACR-Rothamsted receives grant-aided support from the Biotechnology and Biological Sciences Research Council of the United Kindom. We are grateful to the late R.M. Sawicki, A.W. Farnham and F.J. Byrne, all of Rothamsted, and V.E. Edge of NSW Agriculture for support and useful discussions. It is a pleasure to acknowledge the laboratory assistance of M.E. Balfe, N.A. Coleman and B.C. Craswell (all of NSW Agriculture). *H. armigera* work was partially funded by the Winston Churchill Memorial Trust of Australia, John Swire and Sons Australia Pty. Ltd, The Australian Academy of Science/Royal Society Scientific Exchange Programme and the Cotton Research and Development Corporation, Australia.

REFERENCES

Anon. (1970). Standard method for detection of insecticide resistance in *Heliothis zea* (Boddie) and *H. virescens* (F.). *Bull. Entomol. Soc. Am.* **16**, 147–150.

Chang, C.K. and Jordan, T. W. (1983). Inhibition of permethrin-hydrolysing esterases from *Wiseana cervinata* larvae. *Pestic. Biochem. Physiol.* **19**, 190–196.

Daly, J.C. (1993). Ecology and genetics of insecticide resistance in *Helicoverpa armigera:* interactions between selection and gene flow. *Genetica* **90**, 217–226.

Devonshire, A.L. and Moores, G.D. (1977). The properties of a carboxylesterase from the peach-potato aphid, *Myzus persicae* (Sulz.), and its role in conferring insecticide resistance. *J. Biochem.* **167**, 675–683.

Dowd, P.F and Sparkes, T.C. (1987). Inhibition of trans-permethrin hydrolysis in *Pseudoplusia includens* (Walker) and use of inhibitors as pyrethroid synergists. *Pestic. Biochem. Physiol.* **27**, 237–242.

Forrester, N.W., Cahill, M., Bird, L.J. and Layland, J.K. (1993). Management of pyrethroid and endosulfan resistance in *Helicoverpa armigera* (Lepidoptera: Noctuidae) in Australia. *Bull. Entomol. Res.* (supplement 1).

Goodyear, G.J. and Greenup, L.R. (1980). A survey of insecticide resistance in the cotton bollworm *Heliothis armigera* (Hübner) (Lepidoptera: Noctuidae) in New South Wales. *Gen. Appl. Entomol.* **12**, 37–39.

Goodyear, G.J., Wilson, A.G.L., Attia, F.I. and Clift, A.D. (1975). Insecticide resistance in the cotton bollworm *Heliothis armigera* (Hübner) (Lepidoptera: Noctuidae) in the Namoi Valley of New South Wales, Australia. *J. Aust. Entomol. Soc.* **14**, 171–173.

Grant, D.F., Bender, D.M. and Hammock, B.D. (1989). Quantitative kinetic assays for glutathione *S*-transferase and general esterase in individual mosquitoes using an EIA reader. *Insect Biochem.* **19**, 741–751.

Gunning, R.V. (1994). Pyrethroid resistance increasing in unsprayed populations of *Helicoverpa armigera* (Hübner) (Lepidoptera: Noctuidae) in New South Wales 1987–90. *J. Aust. Ent. Soc.* **33**, 381–383.

Gunning, R.V. and Easton, C.S. (1993). Resistance to organophosphate insecticides in *Helicoverpa armigera* (Hübner) (Lepidoptera: Noctuidae) in Australia. *Gen. Appl. Entomol.* **25**, 27–35.

Gunning, R.V. and Easton, C.S. (1994). Endosulfan resistance in *Helicoverpa armigera* (Lepidoptera: Noctuidae) in Australia. *J. Aust. Entomol. Soc.* **33**, 9–12.

Gunning, R.V., Easton, C.S., Greenup, L.R. and Edge, V.R. (1984). Pyrethroid resistance in *Heliothis armigera* (Hübner) (Lepidoptera: Noctuidae) in Australia. *J. Econ. Entomol.* **77**, 1283–1287.

Gunning, R.V., Easton, C.S., Balfe, M.E. and Ferris, I.G. (1991). Pyrethroid resistance mechanism in Australian *Helicoverpa armigera*. *Pestic. Sci.* **33**, 473–490.

Gunning, R.V., Balfe, M.E. and Easton, C.S. (1992). Carbamate resistance in *Helicoverpa armigera* (Hübner) (Lepidoptera: Noctuidae) in Australia. *J. Aust. Entomol. Soc.* **31**, 97–103.

Gunning, R.V., Devonshire, A.L. and Moores, G.D. (1995). Metabolism of es-fenvalerate in pyrethroid susceptible and resistant Australian *Helicoverpa armigera* (Lepidoptera: Noctuidae). *Pestic. Biochem. Physiol.* **51**, 205–208.

Gunning, R.V., Moores, G. D. and Devonshire, A.L. (1996). Esterases and fenvalerate resistance in Australian *Helicoverpa armigera* (Hübner) Lepidoptera: Noctuidae. *Pestic. Biochem. Physiol.* **54**, 12–23.

Kay, I.R. (1977). Insecticide resistance in *Heliothis armigera* (Hübner) (Lepidoptera: Noctuidae) in areas of Queensland, Australia. *J. Aust. Entomol. Soc.* **16**, 43–45.

Kay, I.R., Greenup, L.R. and Easton, C.S. (1983). Monitoring *Heliothis armiger* (Hübner) strains from Queensland for insecticide resistance. *Queensland J. Agric. Anim. Sci.* **40**, 23–26.

Kennaugh, L., Pearce, D., Daly, J.C. and Hobbes, A.A. (1993). A PBO synergisable resistance to permethrin in *Helicoverpa armigera* which is not due to increased detoxification by cytochrome P450. *Pestic. Biochem. Physiol.* **45**, 234–241.

Moores, G.D., Gao, X., Denholm, I. and Devonshire, A.L. (1996). Characterisation of insensitive acetylcholinesterase in insecticide-resistant cotton aphids, *Aphis gossypii* Glover (Homoptera: Aphididae). *Pestic. Biochem. Physiol.* **56**, 102–110.

Suzuki, K., Hama, H. and Konno, Y. (1993). Carboxylesterase of the cotton aphid, *Aphis gossypii* Glover (Homoptera: Aphididae), responsible for feniththion resistance as a sequestering protein. *Appl. Entomol. Zool.* **28**, 439–441.

Wilson, A.G.L. (1974). Resistance of *Heliothis armigera* to insecticides in the Ord irrigation area, North Western Australia. *J. Econ. Entomol.* **67**, 256–258.

14

Potential of Piperonyl Butoxide for the Management of the Cotton Whitefly, *Bemisia tabaci*

GREG DEVINE and IAN DENHOLM

1. INTRODUCTION

As piperonyl butoxide (PBO) is effective at inhibiting the oxidative metabolism of certain toxins by insects, it is often used to synergize insecticides (e.g. Haley, 1978; B-Bernard and Philogène, 1993). In this role it increases the efficacy of some compounds, and is a valuable tool for the control of a range of dipteran and coleopteran pests in medical, veterinary, household and grain-store situations. In the arable crop sector, however, there has been little attempt to integrate PBO with chemical control measures. An exception to this has been the use of PBO for overcoming supposed mono-oxygenase mediated resistance in the bullworm *Helicoverpa armigera* in Australian cotton fields since 1991. The major constraint to the use of PBO under arable or greenhouse conditions is the perception that, because of its lack of photostability, its efficacy is compromised under conditions of strong sunlight.

Despite the lack of use of PBO on food and fibre crops, evidence is accumulating for effects mediated by PBO alone, that may have a significant role in the control of insects, particularly ones that have developed resistance to conventional insecticides. This chapter reviews such effects and summarizes research into the potential use of PBO in controlling the cotton whitefly, *Bemisia tabaci*. Results of this work will be published formally in due course.

Bemisia tabaci is a highly polyphagous pest of cotton, vegetables and ornamentals that has developed resistance to almost all the insecticides used for its control (Denholm *et al.*, 1996). It transmits viruses, causes direct feeding damage and contaminates crops with honeydew, thereby promoting secondary infection with sooty moulds (*Capnodium* spp.). It was estimated to have caused $500 million of damage to crops in the USA in 1991 (Perring *et al.*, 1993), even though the first control failures in that country were reported as recently as the mid-1980s (reviewed by Brown *et al.*, 1995).

PIPERONYL BUTOXIDE
ISBN 0-12-286975-3

Problems in containing the cotton whitefly have been exacerbated by the toxicity of insecticides to natural enemies, especially parasitoids that occur worldwide and have been recorded as infesting up to 80% of nymphs in cotton monocultures in Israel (Gerling, 1986; Horowitz, 1993).

2. TOXICITY OF PBO TO *B. TABACI*

Although PBO can inflict mortality in its own right, its toxicity varies greatly between and within species. Table 14.1 summarizes some available reports, both positive and negative, on the lethal effects of PBO against several insect and mite pests.

In leaf dip bioassays, PBO was lethal at doses of between 60 and 600 ppm to early instars of four strains of insecticide-resistant and susceptible *Bemisia tabaci* (Table 14.2). PBO also killed whitefly adults, but only when insects were exposed to fresh residues of over 1000 ppm. After 24 hours the toxicity of these residues declined markedly (G. Devine, unpublished).

A comparison of the lethality of PBO with that of conventional insecticides is difficult, since few bioassays on whiteflies measure effects against nymphs, but Cahill *et al.* (1995) found that pyrethroids and organophosphates often need to be applied at doses of 100 to 1000 ppm to give reasonable kill of resistant adults. Products effective against nymphs and eggs (e.g. buprofezin, a chitin synthesis inhibitor; and pyriproxifen, a juvenile hormone analogue), tend to be active at doses much lower than those needed for control of nymphs by PBO (Horowitz and Ishaaya, 1994; Ishaaya and Horowitz, 1995).

Table 14.1.   Reported lethal effects of PBO against insect and mite pests

Species	Order	Strain/Life stage	Dose and effect	Reference
Musca domestica	Diptera	Field/Adults	>100 µg per fly, LD_{50}	Sawicki (1974)
Musca domestica	Diptera	Dimethoate selected/ Adults	0.05 µg per fly, LD_{50}	Sawicki (1974)
Plutella xylostella	Lepidoptera	Field/Larvae	4520 ppm, LC_{50}	Sun *et al.* (1986)
Plutella xylostella	Lepidoptera	Fenvalerate selected/ Larvae	11 700 ppm, LC_{50}	Sun *et al.* (1986)
Boophilus microplus	Acari	Field/Nymphs	450 ppm, LC_{50}	Schunter *et al.* (1974)
Aphis fabae	Homoptera	Field/Larvae	2000 ppm, no effect	Lal (1972)
Aphis gossypii	Homoptera	Field/Adults	5000 ppm, no effect	Gubran *et al.* (1993)
Psylla pyricola	Homoptera	Field/Adults	1200 ppm, ~ LC_{40}	Burts *et al.* (1989)
Aleurothrixus floccosus	Homoptera	Field/Larvae and adults	1200 ppm, > LC_{90}	Castaner *et al.* (1989)

Table 14.2. Toxicity of PBO against early instar *B. tabaci* nymphs

Strain[a]	No. tested	LC$_{50}$ (ppm)	95% CLs[b]		Slope
SUD-S	532	184	146–229	1	3.0
PYR-R	488	66	45–93	3	1.0
ISR-R	578	227	150–416	1,2	2.5
USA-B	695	518	391–614	2	5.6

[a] SUD-S originated from the Sudan in 1974. It is fully susceptible to insecticides. PYR-R and ISR-R were insecticide-resistant strains from Israel and USA-B a resistant strain from Arizona.
[b] Confidence limits on fitted LC$_{50}$s. Numbers indicate confidence limits overlap ($P < 0.05$).

3. JUVENOID EFFECTS

The developmental effects of PBO on insects have been described by a number of authors. It has been found to retard development in house fly (*Musca domestica*: Mitlin and Konecky, 1955; Yu and Terriere, 1974) and blowfly larvae (*Lucilia cuprina*: Kotze and Sales, 1994) and in *Aphis gossypii* and *B. tabaci* nymphs (Satoh and Plapp, 1993; Satoh *et al.*, 1995). The effects are typically manifest as a delay of 1 or 2 days to adulthood. Bowers (1968) and Terriere and Yu (1973) suggested that PBO acts as a functional (rather than structural) mimic of juvenile hormone (JH) by altering oxidase levels and thus inhibiting the metabolism or activation of these hormones, but the mechanisms driving the developmental effects of PBO remain generally unclear.

Yu and Terriere (1974) reported PBO to induce oxidase activity in young house fly larvae, but, despite this, there was a delay in pupation. Oxidase activity was also induced in insecticide-susceptible adults, but inhibited in resistant adults. However, when fed to either strain, PBO inhibited ovarian development, egg production and egg hatch.

When applied against nymphs developing on whole cotton plants, PBO caused large delays (up to 4 days) in the development of a resistant (USA-B) strain of *B. tabaci* (Fig. 14.1). One possible explanation of this effect is that PBO acts as an antifeedant, thereby slowing insect development by starvation. This was tested in the laboratory by quantifying honeydew production by *B. tabaci* nymphs treated previously with PBO. Honeydew production was only significantly reduced for those nymphs that had been severely affected and which subsequently died within 2 days of treatment. Surviving nymphs produced the same amount of honeydew as untreated whitefly. This strongly suggested that juvenoid effects seen in surviving whitefly were not the result of feeding disruption (G. Devine, unpublished).

Feeny (1976) and Dyte (1967) suggested that increased generation times should leave an insect more exposed to its predators and parasitoids. This is likely to be of particular importance for parasitoid species which prefer specific juvenile stages, e.g. for *Eretmocerus* spp. that favour early instar *B. tabaci* (Gerling and Sinai, 1994). Another consequence of increased generation times is that the number of generations per season may be reduced. A 5-day increase

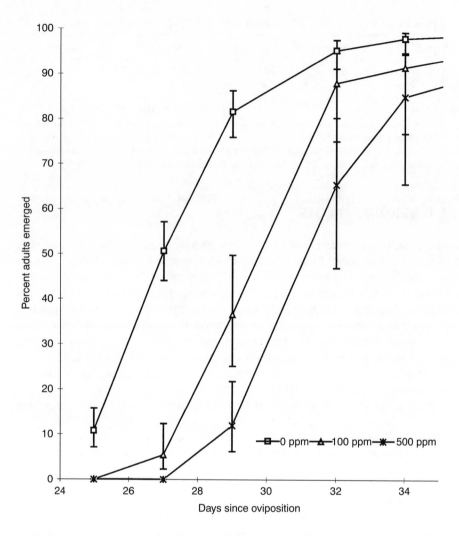

Figure 14.1.   The effect of increasing concentrations of PBO on the pre-adult development of *B. tabaci* (mean and confidence limits).

in generation time could, in the tropics, result in one or two fewer generations over a 5- to 6-month cotton season. As unchecked population growth is exponential, this may have substantial implications for population size at the end of a season.

4. INTERACTIONS WITH PLANT SECONDARY COMPOUNDS

Many xenobiotics affect development in insects, and it is possible that by combining plant-mediated effects and chemical manipulation, such developmental effects can be increased. When applied to 3rd instar Colorado potato beetles (*Leptinotarsa decemlineata*), PBO reduced the percentage of larvae developing to 4th instars by 60% on a standard potato cultivar. However, in combination with a cultivar which had a high solanin content, it completely suppressed development (Carter, 1987).

In comparison with whiteflies (USA-B strain) being reared on a standard cotton cultivar, populations on a cultivar with high gossypol content exhibited delays in development. When nymphs on the latter cultivar were also treated with PBO at 250 ppm, additional developmental effects among surviving nymphs were apparent (Figure 14.2). The delay caused by the combination of high gossypol content and PBO was more than twice that caused by the application of PBO to a standard cultivar (3.7 days as opposed to 1.8 days). This again has two possible implications for pest control: to reduce the pest population by

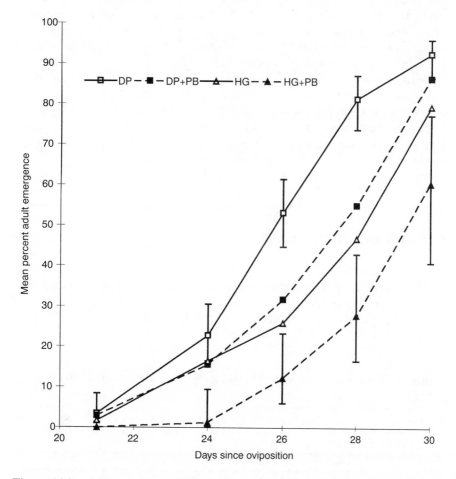

Figure 14.2. Additive effects of PBO and a cotton cultivar with high gossypol content on pre-adult development of *B. tabaci*. DP, cv. deltapine (standard cultivar); HG, cv. HG063 (bred for high gossypol content).

decreasing the number of generations, and/or to expand the time interval within which parasitoids or predators can exert their influence.

5. EFFECTS ON INSECT PARASITOIDS

Contact with surfaces treated with 10 000 ppm of PBO did not affect the adult parasitoid *Diglyphus begini* (Rathman *et al.*, 1992) whereas concentrations of 1200 ppm killed 74% of the *Cales noacki* pupae developing in *Aleurothrixus flocossus* hosts (Castaner *et al.*, 1989; Garrido *et al.*, 1990). The latter was probably a result of the incidental kill of the hosts.

PBO concentrations of *c.* 50 ppm (presented as fresh foliar residues) were toxic to adult *Eretmocerus mundus* and *Encarsia formosa*, both important parasitoids of *B. tabaci* (Table 14.3). After 2 days, however, such toxicity became negligible. As shown in Fig. 14.3, PBO applied at 150 ppm had no pronounced effect when applied to a population of developing parasitoid larvae (ISR-R strain) in the laboratory. PBO was applied to 11-day-old whitefly populations previously exposed to 10 female parasitoids for 24 hours. In both untreated (Fig. 14.3(a)) and treated (Fig. 14.3(b)) cages, the ratio of peak adult whitefly numbers to peak adult parasitoid numbers was similar (0.08 and 0.11, respectively), despite the fact that PBO had caused 50% mortality of treated whitefly hosts. Hence the proportion of parasitoid larvae dying on treated plants was no greater than would be expected as a result of host mortality, and the percentage parasitism under these conditions remained at approximately 10%. Despite its immediate toxicity to adult parasitoids, PBO does not appear to have an adverse effect on the survival of immature stages.

6. FIELD TRIALS IN ISRAEL

In a cotton field in Israel, weekly applications of PBO at a high concentration (5000 ppm) gave equivalent control of whitefly to a single treatment with buprofezin at the recommended field rate of 375 ppm. In contrast, weekly sprays of cypermethrin at 300 ppm caused whitefly numbers to increase markedly above those in untreated plots (Fig. 14.4). Work currently in progress suggests that the

Table 14.3. Toxicity of fresh PBO residues to adult parasitoids of *Bemisia tabaci*

Species[a]	No. tested	LC$_{50}$ (ppm)	95% CLs[b]	Slope
Eretmocerus mundus	139	40	17.8–56.0	2.5
Encarsia formosa	172	55	40.2–73.3	2.0

[a] *Eretmocerus mundus* originated from Pakistan in 1990. *Encarsia formosa* was supplied by Ciba Bunting (UK).
[b] Confidence limits on fitted LC$_{50}$s.

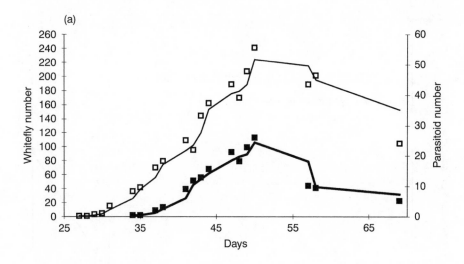

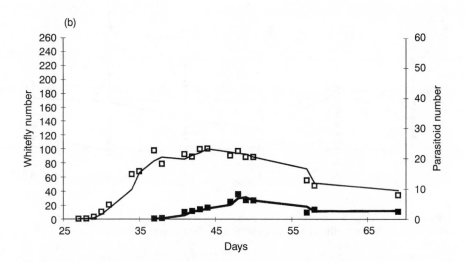

Figure 14.3. Numbers of adult whiteflies and parasitoids emerging from cotton plants left untreated or treated with PBO. (a) Untreated. (b) Treated. □, adult whitefly; ■, adult parasitoids.

concentration of PBO can be decreased substantially to give similar levels of population suppression.

In the same trial (Fig. 14.5), PBO at 5000 ppm had little discernible effect on parasitism of *B. tabaci* while cypermethrin reduced parasitism throughout the experiment. Hence the acute toxicity of PBO against parasitoid adults appeared to have little impact on eventual levels of parasitism.

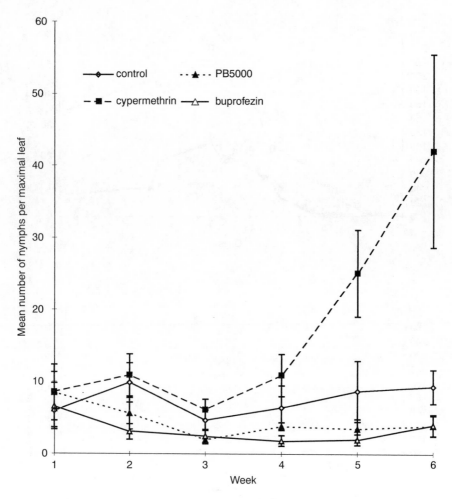

Figure 14.4. Mean numbers (with confidence limits) of unparasitized, late instar whitefly nymphs surviving chemical treatments in Israeli cotton plots.

7. SUMMARY

As the registration of new compounds becomes increasingly difficult and costly, researchers are likely to reassess the utility of compounds such as PBO in pest management programmes. This prospect is particularly appealing in the light of sporadic reports in the literature which, over the past four decades, have shown that this compound has effects beyond the synergism of other insecticides.

Firstly, it is intrinsically lethal to some pest species, but is not particularly harmful to beneficials. Secondly, alone or in conjunction with other toxins, it can affect insect development. These effects have potentially profound implications for the control of insect pests, particularly when allowed to express them-

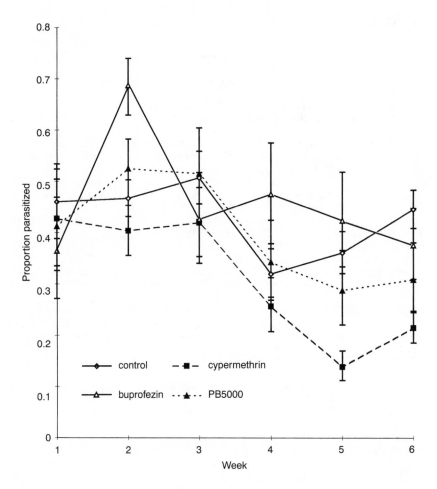

Figure 14.5.   Mean parasitism (with confidence limits) of whitefly nymphs collected from treatments in Israeli cotton plots.

selves in combination with the pest's natural enemies, or the natural defence mechanisms of plants.

For *B. tabaci*, the combination of lethal and sublethal effects shows considerable potential to be exploited within the framework of an integrated pest management strategy.

ACKNOWLEDGEMENTS

This project is being funded by Endura Spa, Bologna, Italy. The authors thank Drs Isaac Ishaaya and Rami Horowitz of the Agricultural Research Organisation, Israel for their advice on the field work, and Dr Greg Constable, CSIRO, Australia, for providing the high gossypol cultivar.

REFERENCES

B-Bernard, C. and Philogène, B.J.R. (1993). Insecticide synergists: role, importance and perspectives. *J. Toxicol. Environ. Health* **38**, 199–223.

Bowers, W.S. (1968). Juvenile hormone: activity of natural and synthetic synergists. *Science* **168**, 895 – 897.

Brown, J.K., Frohlich, D.R. and Rosell, R.C. (1995). The sweetpotato or silverleaf whiteflies: biotypes of *Bemisia tabaci* or a species complex. *Ann. Rev. Entomol.* **40**, 511–534.

Burts, E.C., van de Baan, H.E. and Croft, B.A. (1989). Pyrethroid resistance in pear psylla, *Psylla pyricola* (Foerster) (Homoptera: Psyllidae) and synergism of pyrethroids with PBO. *Canad. Entomol.* **121**, 219–223.

Cahill, M., Byrne, F.J., Gorman, K., Denholm, I. and Devonshire, A.L. (1995). Pyrethroid and organophosphate resistance in the tobacco whitefly *Bemisia tabaci* (Homoptera: Aleyrodidae). *Bull. Entomol. Res.* **85**, 181–187.

Carter, C. (1987). Difference in pyrethroid efficacy among Solanum clones resistant and susceptible to Colorado potato beetle. *Am. Potato J.* **64**, 432.

Castaner, M., Garrido, A., Busto, T. and Del Malagon, J. (1989). Effecto de diversos insecticidas en laboratorio sobre la mortilidad de los estados inmaduros de la mosca blanca algodonsa *Aleurothrixus floccosus* (Mask.) e incidencia sobre el insecto util *Cales noacki* (How). *Investigacion agraria, Produccion y proteccion vegetales* **4**, 420.

Denholm, I., Cahill, M., Byrne, F.J. and Devonshire A.L. (1996). Progress with documenting and combating insecticide resistance in *Bemisia*. In: *Bemisia 1995: Taxonomy, Biology, Damage Control and Management* (Gerling, D. and Mayer, R.T., eds), pp. 577–603. Intercept, Andover.

Dowling, D. (1993). Back to the drawing board for resistance strategy. *Australian Cottongrower* **14**, 10–12.

Dyte, C.E. (1967). Possible new approach to the chemical control of plant feeding insects. *Nature* **216**, 298.

Feeny, P.P. (1976). Plant apparency and chemical defense. *Rec. Adv. Phytochem.* **10**, 1–40.

Garrido, A., Castaner, M., Del Busto, T. and Malagon, J. (1990). Toxicidad de diversos plaguicidas sobre los estados immaduros de *Aleurothrixus floccosus* (Mask.) e incidencia sobre el insecto util *Cales noacki* (How). *Boletin de Sanidad Vegetal Plagas* **16**, 173–181.

Gerling, D. and Sinai, P. (1994). Buprofezin effects on two species of whitefly (Homoptera: Aleyrodidae). *J. Econ. Entomol.* **87**, 842–846.

Gerling, D. (1986). Natural enemies of *Bemisia tabaci*; biological characteristics and potential as biological control agents: a review. *Agric. Ecosyst. Environ.* **17**, 99–110.

Gubran, E.M.E., Delorme, R., Auge, D. and Moreau, J.P. (1993). Pyrethroids and organochlorine resistance in cotton aphid *Aphis gossypii* (Glov.) (Homoptera: Aphididae) in the Sudan Gezira. *Int. J. Pest. Manag.* **39**, 197–200.

Gunning, R.V., Easton, C.S., Balfe, M.E. and Ferris, I.G. (1991). Pyrethroid resistance mechanisms in Australian *Helicoverpa armigera*. *Pestic. Sci.* **33**, 473–490.

Haley, T.J. (1978). PBO, α[2-(2-butoxyethoxy)ethoxy]-4,5-methylenedioxy-2-propyltoluene: a review of the literature. *Ecotoxicol. Environ. Safety* **2**, 9–31.

Horowitz, A.R. (1993). Control strategy for the sweetpotato whitefly, *Bemisia tabaci*, late in the growing season. *Phytoparasitica* **21**: 281–291.

Horowitz, A.R. and Ishaaya, I. (1994). Managing resistance to insect growth regulators in the sweetpotato whitefly (Homoptera: Aleyrodidae). *J. Econ. Entomol.* **87**, 866–871.

Howie, B. (1992). A view on cotton insecticide strategies. *Australian Cottongrower* **13**, 70–71.

Ishaaya, I. and Horowitz, A.R. (1995). Pyriproxifen, a novel insect growth regulator for controlling whiteflies: mechanisms and resistance management. *Pestic. Sci.* **43**, 227–232.

Kotze, A.C. and Sales, N. (1994). Cross resistance spectra and effects of synergists in insecticide resistant strains of *Lucilia cuprina* (Diptera: Calliphoridae). *Bull. Entomol. Res.* **84**, 355–360.

Lal, O.P. (1972). Synergism of terramycin to inhibit the reproductivity of *Aphis fabae*. *Experientia* **28**, 717–718.

Mitlin, N. and Konecky, M.S. (1955). Inhibition of development in the housefly by piperonyl butoxide. *J. Econ. Entomol.* **48**, 93–94.

Perring, T.M., Cooper, A.D., Rodriguez, R.J., Farrar, C.A. and Bellows, T.S. Jr. (1993). Identification of a whitefly species by genomic and behavioural studies. *Science* **259**, 74–77.

Rathman, R.J., Johnson, M.W., Rosenheim, J.A., Tabashnik, B.E. and Purcell, M. (1992). Sexual differences in insecticide susceptibility and synergism with piperonyl butoxide in the leafminer parasitoid *Diglyphus begini* (Hymenoptera; Eulophidae). *J. Econ. Entomol.* **85**, 15–20.

Satoh, G.T. and Plapp, F.W. Jr (1993). Use of juvenoids for management of cotton aphids and sweet-potato whitefly populations. In: *Proceedings 1993 Beltwide Cotton Conferences,* National Cotton Council, Memphis, TN.

Satoh, G.T., Plapp, F.W. Jr and Slosser, J.E. (1995). Potential of juvenoid insect growth regulators for management of cotton aphids (Homoptera: Aphididae). *J. Econ. Entomol.* **88**, 254–258.

Sawicki, R.M. (1974). Genetics of resistance of a dimethoate selected strain of housefly (*Musca domestica L.*) to several insecticides and methylenedioxyphenyl synergists. *J. Agric. Food. Chem.* **22**, 344–349.

Schunter, C.A., Roulston, W.J. and Wharton, R.H. (1974). Toxicity of piperonyl butoxide to *Boophilus microplus. Nature* **249**, 386.

Sun, C.N., Wu, T.K., Chen, J.S. and Lee, W.T. (1986). Insecticide resistance in diamondback moth. In: *Diamondback Moth Management* (Talekar, N.S. ed), pp. 359–371. Asian Vegetable Research and Development Center. Shanhua, Taiwan.

Terriere, L.C. and Yu, S.J. (1973). Insect juvenile hormones. Induction of detoxifying enzymes in the housefly and detoxication by housefly enzymes. *Pestic. Biochem. Physiol.* **3**, 96–107.

Yu, S.J. and Terriere, L.C. (1974). Possible role for microsomal oxidases in metamorphosis and reproduction in the housefly. *J. Insect. Physiol.* **20**, 1901–1912.

15

The Use of Synergized Pyrethroids to Control Insect Pests in and Around Domestic, Industrial and Food-handling Premises

JOHN C. WICKHAM

1. INTRODUCTION

Piperonyl butoxide (PBO) is primarily used as a synergist in combination with natural pyrethrins or pyrethroids in space spray, residual and admixture products for the control of insect pests in or around domestic and commercial premises, especially food preparation areas. The suitability of PBO for this purpose has been proven over many decades of study and widespread practical use. The main attributes of the compound which have ensured this long-lasting utilization in urban pest control procedures are summarized below. The beneficial effect that PBO has on insects showing resistance to insecticides is referred to in Chapters 12, 13 and 19 of this volume.

1.1. Physiological Effects of Adding a Synergist

An insect reacts to an insecticide by attempting to excrete the chemical or to degrade it to nontoxic metabolites, principally by oxidative processes, although hydrolysis and esterase actions are also involved (Demoute, 1989). The relative importance of such mechanisms gives rise to variations in synergistic effect between species of insects. PBO inhibits mixed function oxidative mechanisms so that insecticide breakdown is reduced and more insecticide becomes available internally at the site of action (Casida, 1970). This results in an increased mortality, or alternatively, the same effect may be observed by using decreased amounts of insecticide, i.e. *synergism* (Hewlett, 1968).

It is important to note that the degree of inhibition of these oxidation processes is related to the *amount* of synergist added.

The decrease in metabolic breakdown produced by the action of PBO narrows

PIPERONYL BUTOXIDE
ISBN 0-12-286975-3

the range of responses by individuals within an insect population (Holborn, 1957) so that the overall effect is further enhanced, particularly at high performance levels. In bioassay tests, this is demonstrated by a steeper regression line and the contraction of the confidence limits given to estimates of performance. In practical terms, this improves the reliability of a treatment.

Some insects do not possess oxidative degradative processes and thus for these insects PBO cannot act as a synergist. Also, with some toxicants, an oxidative process is necessary to produce an insecticidally active metabolite, and use of PBO in these instances reduces performance (Palmer *et al.*, 1990).

Mixtures of PBO with pyrethroid or carbamate insecticides are often more effective against insect strains resistant to these compounds where oxidative metabolism is responsible for the decreased effect (Davies *et al.*, 1958; Wilkinson, 1968; Glynne-Jones, 1983; MacDonald *et al.*, 1983; Funaki *et al.*, 1986). There is no evidence to indicate that PBO increases the low toxicity of pyrethrins and pyrethroids to mammals. The acceptable daily intake (ADI) of PBO for humans has been established at 0.2 mg per kg body weight (JMPR, 1995).

1.2. Physical effects of PBO

The presence of PBO may prolong the effectiveness and possibly enhances the stability of products containing natural pyrethrins and other light-sensitive pyrethroids when exposed as thin films on glass to daylight or in water (Nash, 1954; Head *et al.*, 1968; Ahmed *et al.*, 1976; Brooke, 1958). Other investigators report that no such protective action occurs (Nasir, 1953; Phipers and Wood, 1957; Blackith, 1952a,b,c; Brooke, 1967). An illustration of the stability of talc-based powders containing pyrethrins or bioallethrin with or without PBO when stored at 28°C is given in Table 15.1 (Davies *et al.*, 1970).

PBO is an effective solvent of many substances and this property assists in the preparation of many insecticide formulations. The volatility of PBO is low and in common with many oils having this characteristic, the rate of penetration through the insect cuticle by the active ingredient in such a solution is reduced (Hewlett, 1968; Hayashi *et al.*, 1968). Whilst rates of penetration play little part

Table 15.1. Stability of powders based on pyrethrins or bioallethrin (BA) stored in the dark at 28°C using *Sitophilus granarius* as test insect (Davies *et al.*, 1970)

Compound	Initial concentration (%)	% Loss of efficiency			
		3 months	6 months	12 months	18 months
Pyrethrins	0.5	40	100	–	–
Pyrethrins + PBO	0.1 + 1.0	<13	< 17	<26	19–30
Bioallethrin	1.5	–	65	–	–
BA + PBO	0.15 + 1.5	Nil	Nil	<10	10

in determining the ultimate toxicities of pyrethroids (Burt and Goodchild, 1974), the presence of PBO may give a slightly slower knockdown effect.

2. PRINCIPLES OF INSECTICIDE USE

Insecticides have a long history, although it is only within the last 50 years that they have come into widespread use, as relatively inexpensive synthetic products have been manufactured. These materials have provided reliable, efficient and effective means to overcome insect infestations. Insecticide applications have often been made without due consideration of their merits and limitations or their 'ecology' (Wright *et al.*, 1984; Bennett *et al.*, 1984), i.e. their positioning in relation to the insect pest and its distribution in the environment. Similarly, correction of the environmental conditions or other circumstances which have allowed and supported an infestation are often ignored. Integrated pest management systems (IPM) which take these factors into account need to be developed for each major pest control situation.

When using insecticides it is essential to use an appropriate formulation which makes the active ingredient available to the insect in the most cost-effective manner.

Manufacturers of insecticide products are strongly motivated towards providing formulations which show maximum advantages with minimum side effects, and governments throughout the world monitor efficacy and safety by requiring the registration of insecticide products. Internationally accepted legislation covers the safety of the operator applying the treatment, the conditions under which the treatment may be applied, and the precautions to be taken to prevent contamination of food or the general environment. Maximum residue levels (MRLs), which define the finite quantity allowed in and around foods, have been set for every registered insecticide after consideration has been given to all the relevant effects on nontarget organisms arising from the methods of use applicable to the product.

These considerations provide a safeguard against any reasonably predictable and most unforeseen effects arising from use of insecticide products. Since all data concerning methods and rates of application, safety, etc., are summarized on the product label, it is important to:

ALWAYS READ THE LABEL BEFORE USING ANY INSECTICIDE

3. PIPERONYL BUTOXIDE: THE PREFERRED SYNERGIST

In the early 1940s a search began for compounds which would extend the usefulness of natural pyrethrins (see Chapter 1 in this volume by Antonio Tozzi). Wachs (1947), supported by Dove (1947), demonstrated that 4,5-methylenedioxyphenoxy-2-propyl benzyl butyl diethylene glycol ether, now known as piperonyl butoxide (PBO), was highly effective and it soon came into commercial production in the USA. On the initiative of an English company (Cooper, McDougall and Robertson), the manufacture and distribution of PBO was subsequently extended worldwide. Other methylenedioxy compounds which were effective as synergists also became available during the next 10 years. Although some of these, such as sulfoxide (Synerholm *et al.*, 1947), safroxan (Sawicki, 1961), bucarpolate (Mitchell, 1959) and sesamex (Beroza, 1956; Beroza and Barthel, 1957), showed good activity in some situations, this was not sufficient in terms of cost-effectiveness to bring them into widespread use.

As an illustration of the wide-ranging synergistic activity of PBO, the data in Table 15.2 show factors of synergism recorded by several workers in laboratory tests against various insect species by different methods of application.

4. THE OPTIMUM FORMULA USING PBO AS A SYNERGIST

As noted previously, it is the *amount* of PBO present which determines the degree of synergistic effect and not the ratio *per se*. Increasing the amount of PBO generally increases the factor of synergism, which is limited only at extreme high dosages by the toxic effects of PBO or at low dosages by being below the threshold of synergistic activity. For a given level of performance (e.g. LD_{50}), the relationship between amount of pyrethroid and amount of PBO can be shown to be logarithmic (Wickham *et al.*, 1974), as demonstrated by the logarithmic isobole illustrated in Chapter 11 in this volume by Duncan Stewart.

As an example, using the data derived from house fly studies by Nagasawa and Shibuya (1987) to obtain estimates of LD_{50} levels, logarithmic isoboles can be constructed. These make it possible to make comparisons between any of the five pyrethroids evaluated, at any level of PBO. The data in Table 15.3 show how the amounts of pyrethroid and PBO interrelate with ratio and factor of synergism. This table includes findings derived from logarithmic isoboles at arbitrarily selected levels of PBO of $1.0 \, \mu g \, mL^{-1}$ and $3.2 \, \mu g \, mL^{-1}$ (the latter is an extrapolation to an arbitrary maximum to show the theoretical maximum factor of synergism); and a further selection where the insecticide : PBO ratio is set at 1 : 10. Since all the values are for a single performance level it is clear that selection of a mixture could be made at a variety of ratios or factors of synergism at will and hence choice can be based on other criteria. Comparisons made on the basis of identical ratios or identical factors of synergism are unlikely also to be equal in cost or availability, so product selection must take such factors into account.

The prime value of laboratory-based findings, such as those quoted here, is to indicate those formulations that are worthy of field evaluation or to isolate factors associated with a specific material or method of presentation which may influence performance. All recommendations for use of products must be determined as a result of *field assessment with products* and not simply by extrapolation from laboratory studies (Schneider and Bennett, 1985; Owens, 1990). As far as possible, the recommendations made in subsequent sections are based on field evaluations or the product label claims made by the manufacturer.

Table 15.2. Factors of synergism recorded by several workers on some insect species using different methods of application

Reference and method of application	Insecticide	Ratio insecticide : PBO	Factor of synergism	Reference and method of application	Insecticide	Ratio insecticide : PBO	Factor of synergism
Musca domestica							
Goodwin-Bailey (1960)	Pyrethrins	1:1	2.3	Nagasawa and Shibuya (1987) (calc. from data)	Allethrin	1:0.7	3.0
		1:5	5.1			1:7.7	8.0
		1:10	7.0			1:57	14.9
Topical		1:20	9.5	Topical	Tetramethrin	1:0.3	2.0
		1:40	13.0			1:1.1	3.1
Spray	Pyrethrins	1:5	5.0			1:7.7	5.4
		1:10	8.0		Fenvalerate	1:1	1.6
Hadaway *et al.* (1963)	Pyrethrins	1:10	8.4			1:2.8	2.3
	Allethrin	1:10	2.9			1:91	7.5
Davies *et al.* (1970)	Pyrethrins	1:5	6.8	Topical	Permethrin	1:0.9	2.4
	Allethrin	1:5	3.6			1:3.3	3.5
	Bioallethrin	1:5	4.2			1:9.1	4.8
				Topical	Phenothrin	1:0.6	1.6
						1:2.6	2.5
						1:25	4.3
Blattella germanica (males)							
Chadwick (1971)	Pyrethrins	1:5	1.1	Chadwick (1971)	Pyrethrins	1:5	2.3
		1:10	1.6		Allethrin	1:5	1.7
		1:20	1.6	Dusting	Bioallethrin	1:5	1.5
		1:100	2.6		Bioresmethrin	1:5	1.1
Direct spray	Allethrin	1:5	1.8		Tetramethrin	1:5	2.1
	Bioallethrin	1:5	0.7		D-Tetramethrin	1:5	1.6
		1:10	0.9				
		1:20	1.2				
		1:100	3.2				
	s-Bioallethrin	1:5	0.9				

Table 15.2. Factors of synergism (*cont.*)

Reference and method of application	Insecticide	Ratio insecticide : PBO	Factor of synergism	Reference and method of application	Insecticide	Ratio insecticide : PBO	Factor of synergism
	Bioresmethrin	1:5	1.0	Goodwin-Bailey (1960)			
		1:10	1.0				
		1:20	1.0				
		1:100	1.4				
	Tetramethrin	1:5	1.9	Dust deposit	Pyrethrins	1:10	3.0
		1:10	1.9		Pyrethrins	1:10	3.0
		1:20	3.3				
	D-Tetramethrin	1:5	1.4				
Tribolium castaneum							
Carter *et al.* (1975)	Pyrethrins	1:5	2.6[a]				
	Allethrin	1:5	3.7[a]				
Topical	Bioallethrin	1:5	5.0[a]				
	Bioresmethrin	1:5	1.5[a]				
	Tetramethrin	1:5	c. 400[a]				
Stegobium paniceum							
Carter *et al.* (1975)	Pyrethrins	1:5	1.0				
	Bioallethrin	1:5	1.4				
Topical	Bioresmethrin	1:5	1.0				
Lasioderma serricorne							
Goodwin-Bailey (1960)	Pyrethrins	1:5	1.5	Lloyd (1961)			
		1:10	1.7	Topical	Pyrethrins	1:10	2.0
Topical							
Deposit	Pyrethrins	1:5	2.3	Direct spray	Pyrethrins	1:10	2.5
		1:10	3.8				
		1:40	13.0				
Carter *et al.* (1975)	Pyrethrins	1:5	2.0				
	Bioallethrin	1:5	0.5				

Method	Reference	Insecticide	Ratio	Factor	Reference	Insecticide	Ratio	Factor
Topical		Bioresmethrin	1:5	0.5				
Periplaneta americana								
Topical	Goodwin-Bailey (1960)	Pyrethrins	1:2.5	1.2	Chadwick (1979) Topical	Pyrethrins	1:5	1.1
			1:5	1.8		Bioresmethrin	1:5	3.4
Dust			1:10	2.0		D-Phenothrin	1:5	1.5
			1:20	2.5				
Sitophilus granarius								
Topical	Goodwin-Bailey (1960)	Pyrethrins	1:2.5	9.0	Parkin and Lloyd (1960) Topical	Pyrethrins	1:10	4.6
			1:5	11.0				
Dust			1:10	15.0	Lloyd and Parkin (1963) Topical	Pyrethrins	1:10	5.2
			1:20	19.0		Allethrin	1:10	12.9
Deposit		Pyrethrins	1:10	9.0				
Prostephanus truncatus								
Topical	Makundi (1986)	Deltamethrin	1:5	1.85	Makundi (1986) Dust	Deltamethrin	1:10	4.0
		Cyfluthrin	1:5	1.87		Cyfluthrin	1:10	8.0
Anopheles stephensi								
Topical	Hadaway et al. (1963)	Pyrethrins	1:10	2.0				
		Allethrin	1:10	1.7				
Aedes aegypti								
Topical	Hadaway et al. (1963)	Pyrethrins	1:10	2.7	Brooke (1958) Larvae in water	Pyrethrins	1:10	3.0
		Allethrin	1:10	2.0			1:20	3.6

[a] In mineral oil solvent.

NB. Comparisons between products should *not* be made on the basis of the factor of synergism; relative effectiveness should be judged on the amount of insecticide/synergist required (not given here – see references and note comments in text).

Table 15.3. The interrelationships between amounts of pyrethroid and PBO with ratio and factor of synergism at a single performance level

Compound	LD_{50} ($\mu g \, mL^{-1}$) Pyrethroid + PBO	Ratio[a]	Factor of synergism[b]	Source[c]
Allethrin	0.516	–	–	Regression from data[d]
	0.176 + 0.125	0.7	2.9	Regression from data[d]
	0.065 + 0.5	7.7	7.9	Regression from data[d]
	0.035 + 2.0	57.1	14.7	Regression from data[d]
	0.050 + 1.0	20.0	10.3	From log isobole
	0.025 + 3.2	128.0	20.6	From log isobole
	0.064 + 0.64	10.0	8.1	From log isobole
Tetramethrin	0.706	–	–	Regression from data
	0.360 + 0.125	0.3	2.0	Regression from data
	0.230 + 0.25	1.1	3.1	Regression from data
	0.130 + 1.0	7.7	5.4	Regression from data
	0.125 + 1.0	8.0	5.6	From log isobole
	0.073 + 3.2	43.8	9.7	From log isobole
	0.117 + 1.17	10.0	6.0	From log isobole
Fenvalerate	0.082	–	–	Regression from data
	0.050 + 0.5	1.0	1.6	Regression from data
	0.036 + 0.1	2.8	2.3	Regression from data
	0.011 + 1.0	90.9	7.5	Regression from data
	0.011 + 1.0	90.9	7.5	From log isobole
	0.006 + 3.2	533.3	13.7	From log isobole
	0.023 + 0.23	10.0	9.1	From log isobole
Permethrin	0.053	–	–	Regression from data
	0.022 + 0.02	0.9	2.4	Regression from data
	0.015 + 0.05	3.3	3.5	Regression from data
	0.011 + 0.1	9.1	4.8	Regression from data
	0.004 + 1.0	250.0	13.2	From log isobole
	0.0025 + 3.2	1280.0	21.2	From log isobole
	0.0107 + 0.107	10.0	5.0	From log isobole
Phenothrin	0.068	–	–	Regression from data
	0.043 + 0.025	0.6	1.6	Regression from data
	0.027 + 0.07	2.6	2.5	Regression from data
	0.016 + 0.4	25.0	4.2	Regression from data
	0.011 + 1.0	90.9	6.2	From log isobole
	0.007 + 3.2	457.1	9.7	From log isobole
	0.020 + 0.2	10.0	3.4	From log isobole

[a] Ratio of PBO:1 of pyrethroid.
[b] LD_{50} of insecticide alone/LD_{50} of insecticide in pyrethroid + PBO mixture.
[c] Log isobole constructed from LD_{50}s derived from regression (column 2).
[d] From log dose/probit mortality regression derived from data in Nagasawa and Shibuya (1987), against house flies, *Musca domestica*.

5. APPLICATION METHODS

There are four major methods of insecticide application:

- space treatment
- residual (surface) treatment
- admixture treatment
- baits/traps.

The last of these categories incorporates specialized methods of control which are outside the scope of this chapter.

It is essential that thorough treatment is applied if good control is to be achieved with insecticides *and application must be repeated at suitable intervals* as determined by monitoring inspections.

5.1. Space Treatment

In space treatment the insecticide is dispersed as a cloud of fine droplets. This is the main method by which flying insects are controlled, as the chemical is rapidly brought into contact with the insect. As a result of this immediate and intimate contact only small quantities of non-residual active ingredient are necessary. It is important to apply compounds which not only produce a rapid knockdown effect (mainly pyrethrins and some synthetic pyrethroids), but also provide subsequent mortality.

The principal group of insecticides having knockdown activity are those known collectively as the pyrethroids, which include the naturally occurring pyrethrins. The performance of a space spray formulation can be adjusted to choice by using different proportions of knockdown agent, killing agent and synergist.

Synthetic pyrethroid compounds were derived from long-term studies on the structural activity of pyrethrins extracted from the dried flower heads of the pyrethrum plant, *Chrysanthemum cinerariaefolium*. In general, all pyrethroids are biodegradable, although with some this process may be slow. Many are unstable in the presence of sunlight. These features contribute to the safety and ecological suitability that is manifest by most pyrethroids (Demoute, 1989). The group is also characterized by the varying effects on biological performance shown by different isomeric forms of the active ingredient. This has led to the development of many compounds showing considerable differentiation in levels of activity and methods of manufacture. The pyrethroids listed in Table 15.4 are among those registered in the UK and most other countries under many product names and most are available in combination with PBO for use in food premises. The chemical structures of all the pyrethroids mentioned herein may be found in Worthing and Hance (1991).

This variety of pyrethroid compounds from many sources and having different levels of activity has allowed a wide range of products to be developed which have the ability to control flying insect pests in most situations. For space

Table 15.4.   Pyrethroids used as insecticides

Frequently used	Less frequently used
Knockdown agents	
Bioallethrin	Allethrin
S-Bioallethrin	D-Allethrin
Tetramethrin	D-Tetramethrin
Pyrethrins	Cyfluthrin
Killing agents	
Bioresmethrin	Resmethrin
Cypermethrin	Alphacypermethrin
Deltamethrin	Lambdacyhalothrin
Permethrin	D-Phenothrin

treatments such products are usually formulated either as oil solutions of the insecticide(s) or as emulsifiable preparations where a solvent solution of the active ingredients is mixed with emulsifiers so that the product can subsequently be diluted with water.

Space spray treatments have their limitations in that they will only control insects present at time of treatment and the insecticide will not penetrate deeply into cracks and crevices where many insects live. Consequently, frequent treatments are necessary for reliable control. Also, although insecticide will be deposited from space sprays onto horizontal surfaces and have some immediate effect, this will not be adequate to function as a residual insecticide. As a general hygiene precaution all food materials or surfaces upon which food will be handled are liable to deposition and must be covered during spraying.

5.2. Residual Treatments

These are applied to surfaces, cracks and crevices to provide a deposit that will be effective for a given period termed 'residual life'. Rapid action (to the extent that effects must be observed within minutes) is not a major requirement of residual sprays, although it is desirable with space sprays. With residual treatments it is the ability to provide a deposit which is effective over a period of time which is important. This seeks to take advantage of the crawling habit of many insect pests and also their strong preference to be in contact with surfaces other than that upon which they stand, known as the thigmotactic response, when selecting resting areas or 'harbourages' (Berthold and Wilson, 1967). The selection of surfaces to be treated during control procedures should include all such harbourages.

Formulations which make the insecticide readily available and easy to detach from the surface, such as suspensions, wettable powders and microencapsulated preparations, provide the most active residual deposits, although oil solutions or emulsions are effective and frequently used. Application is usually through standard hydraulic pumps or compression sprayers (knapsack sprayers), where a tank pressure of about 20 psi (~ 138 kPa) can be maintained.

5.3. Admixture Treatments

Admixture treatments, in which the insecticide becomes an integral part of, or is in close association with a commodity, are used to provide long-term protection of bulk materials in store. Application is often by spraying, in much the same way as residual treatments, when the material is entering a store (e.g. grain protectant or dried fruit applications) or by dipping or brushing to preserve timber, wool, skins and hides. A long residual life is needed, hence products similar to those prepared for residual applications are used through mechanical sprayers and powder applicators, or adapted to specific situations by brushing or immersion in baths containing liquid formulations.

6. THE VALUE OF PBO IN SPACE SPRAY TREATMENTS

The pre-eminent position of PBO in space sprays arises from the assurance that products containing this synergist will provide a high mortality of insects when in combination with an effective level of knockdown compound. In the absence of PBO, a much higher level of active ingredient would be necessary, at a considerably increased cost. Such products are safe to use in or around domestic and commercial premises, especially food preparation areas.

The production and dispersion of spray clouds for space treatments can in the main be carried out using any one of five methods:

- pressure packed aerosols (low or high pressure)
- handsprayers
- mechanical mist sprayers
- thermal sprayers/foggers
- ULV (ultra-low volume) sprayers.

An alternative, more specialized procedure for small-scale use involves the use of vaporizing devices in which a high concentration of an insecticide is dispersed on a solid carrier from which during heating (mosquito mats) or burning (smokes or mosquito coils), or sometimes at room temperature (impregnated plastic strips), an insecticidal vapour is produced. When used in such products PBO acts more as a solvent or evaporation retardant which evens out the rate of volatilization than as a synergist.

Low-pressure aerosols which are self-contained, or handsprayers which can be refilled, are the most suitable products to use in relatively confined spaces such as are found in domestic premises, offices and small stores. Windows and doors should be closed before treatment. In retail outlets, small storage facilities and dairy barns (Wright, 1977), specially formulated aerosols can be dispensed from wall-mounted units which automatically discharge measured doses at regular intervals. Such devices can keep premises with volumes up to 170 m^3 (6000 ft^3) insect free for 3 months.

Aerosol formulations may contain the combination of a knockdown pyrethroid in concentrations ranging from 0.05% to 0.5%; a killing pyrethroid from 0.005% to 0.2%; and a synergist, PBO, at not less than 0.4% up to 3.0%. All formulations are w/w. Ready-for-use sprays suitable for application through handsprayers will contain mixtures prepared at about one-third to one-half the above concentrations.

For larger rooms, warehouses and stores, or for outdoor treatments where the volume of spray is much greater, machines are employed to produce mists, fogs or mechanically generated aerosols (ULV). A rather more specialized but extremely effective method is the use of high-pressure CO_2 cylinders containing synergized pyrethrins, in which the CO_2 acts as both solvent and propellant.

Mists are composed of relatively coarse droplets (mean diameter 80–150 μm) generated by the shearing action of an air blast or a spinning disc to atomize the spray. Although very useful in many circumstances, such droplets fall from the air much faster than the finer fogs and aerosols, making mists potentially less efficient, especially in outdoor situations.

Fogs (mean diameter 50–100 μm) are produced from thermal fogging machines which generate copious quantities of hot air into which is injected the insecticide formulation. This vaporizes, and on reaching the cooler air outside the machine it condenses to form a fog. Oil solutions produce dense fogs; water-based formulations produce a haze.

Mechanical aerosol generators commonly utilize a high volume of fast moving air directed to shear the insecticide formulation into an aerosol (mean diameter (>10 to <50 μm). Small droplet size allows good distribution to be attained which has led to the use of lower volumes of insecticide at higher concentrations (ULV) thus providing the desired biological effect with maximum economy.

High-pressure CO_2 cylinders can be used as self-contained mobile units on small or large trolleys or as part of a fixed installation having a piped distribution system with strategically placed spray heads which may be controlled manually or automatically to a specific treatment regime. Because the mean diameter of the droplets from this procedure is <15 μm, deposition of the spray is much delayed. This lengthens the exposure time and improves performance against flying insects in enclosed spaces (Groome and Martin, 1994).

6.1 Rates of Application

Products for use in all these types of equipment (except aerosols filled at low or high pressure) are prepared as concentrates. Because rate of knockdown in these circumstances is not as critical as in domestic situations, these may be diluted with oil or water as desired with little overall alteration in performance. There are many products available with a large variety of different mixtures of pyrethroids and PBO. The general rates of application and levels required to give good performance can be defined within limits; these are indicated in Table 15.5 below. It is of the greatest importance that the spray cloud is distributed as uniformly as possible throughout the area to be treated. Indoors this is easily related to volume, but outdoors efficacy will depend upon how much the density of obstacles, the wind speed and the effective swathe width affect the ease of distribution of the spray cloud.

Table 15.5.   Standard rates of application

Type	Area	Metric dose	US dose
Mist	Indoors	0.3–1.5 L/1000 m^3	= US 0.3–1.5 fl oz/1000 ft^3
Fog	Indoors	0.3–1.5 L/1000 m^3	= US 0.3–1.5 fl oz/1000 ft^3
	Outdoors*	10–50 L/ha	= US 1.0–5.3 gal/acre
	*spot treatment	25 L/ha	= US 2.7 gal/acre
ULV	Indoors	0.4–1.25 L/10 000 m^3	= US 0.4–1.2 fl oz/10 000 ft^3
	Outdoors*	0.5–2.0 L/ha	= US 6.8–27.4 fl oz/acre
	*spot treatment	2.0 L/ha	= US 27.4 fl oz/acre
CO_2 cyl	Indoors	150–850 g/1000 m^3 of total product	= US 4–24 g/1000 ft^3 of total product

To illustrate the relationship between the rates of application and levels of treatment in metric and US measures, consider a mist treatment with a product containing 0.1% pyrethrins + 1.0% PBO:

Metric – application of 1 litre/1000 m^3 gives a dose level of 1.0 + 10.0 g/1000 m^3.

US – application of 1 fl oz/1000 ft^3 gives a dose level of 29.6 + 296 mg/1000 ft^3 (which is equivalent to 1.0 + 10.0 g/1000 m^3).

Control of flies, mosquitoes, stored products moths, etc. [and tobacco beetles], is achieved by treatment levels of synergized pyrethroid:

Indoors – 2 g (or less)/1000 m^3; [5 g/1000 m^3 [for tobacco beetles]
Outdoors – 10 g (or less)/ha.

Typical recommendations are given below in Table 15.6 (Indoors) and Table 15.7 (Outdoors).

7. THE VALUE OF PBO IN RESIDUAL TREATMENTS

The pre-eminent position of PBO as a synergist in residual sprays is because it allows the use of safer products containing pyrethroids (natural and synthetic) in and around areas where food is prepared and processed, at cost-effective and economic levels. As with space sprays, this arises mainly from the assurance that products containing this synergist will provide high mortality of insects in circumstances where the effective level of knockdown compound is insufficient to provide high killing activity. Otherwise a much higher level of active ingredient would be necessary at a considerably increased cost. An additional feature is that PBO significantly prolongs the effective life of deposits of light-sensitive insecticides.

The effective residual life of any insecticide deposit rarely exceeds 6–8 weeks, depending upon such circumstances as the persistent nature of the active ingredient, the formulation, the type of surface, the efficiency of treatment and

Table 15.6. Typical recommendations for insecticide application indoors

Mixture	Flies, moths, wasps, etc. (g/1000 m^3)			Tobacco beetles (g/1000 m^3)		
	Mist/fog	ULV	Aerosol	Mist/fog	ULV	Aerosol
Pyrethrins	0.75–1.5	1.0–2.0	0.9–1.4	5.0	4.0	4.3
+ PBO	7.5–15.0	8.0–16.0	7.2–11.2	50.0	64.0	34.4
S-Bioallethrin		0.1–0.2				
+ Permethrin		0.6–1.2				
+ PBO		0.7–1.4				
Tetramethrin	0.1–0.2	0.15				
+ Cypermethrin	0.25–0.5	0.15				
+ PBO	0.3–0.6	0.45				
Bioresmethrin	1.0	1.0				
+ PBO	1.0	1.0				

Table 15.7. Application rates for outdoor control of flies and mosquitoes – WHO – Anon. (1984)

Mixture	Fog (g/ha)	ULV (g/ha)
S-Bioallethrin	2.5–10.0	0.25–1.0
+ Permethrin	5.0–15.0	5.00–7.5
+ PBO	12.5–50.0	5.25–5.75
Tetramethrin	1.5–16.0	2.0–2.5
+ Phenothrin	4.0–7.0	5.0–12.5
+ PBO	2.0–48.0	5.0–10.0
Deltamethrin		0.5–1.0

the length of time the deposit remains undisturbed. Because of these factors, no insecticide formulation can maintain a uniform and consistently effective deposit over time. Effective monitoring of an infestation will greatly assist in making the decision as to when control measures must be re-applied.

It is now common practice, especially with the more recently introduced residual pyrethroids, to recommend that treatment be given in two phases. An initial higher treatment level to provide 'clean out' of an infestation is followed by a 'maintenance phase' of repeated applications at suitable intervals, where the dosage applied at each re-treatment is often half that of the initial treatment.

The product label describes the proper application rate for any specific residual insecticide as quantity of active ingredient (a.i.) per unit area – usually mg (a.i.) m^{-2} or mg (a.i.) ft^{-2}. However, it is general practice to quote the concentration of product which should be applied to a surface either 'to point of runoff' or 'as a coarse spray'. The 'point of runoff' approach compensates for the lower effects from deposits applied to absorbent surfaces. Even though considerable variation may be expected, as shown in Table 15.8, it is generally implied that 'to point of runoff' is based on the assumption that 5 L of spray is applied to 100 m^2 of surface. Hence a spray concentration of 1.0% would be recommended where a deposit of 500 mg m^{-2} (50 mg ft^{-2}) is intended. The 'coarse spray' instruction is often linked with a compensatory dilution rate dependent upon the surface to be treated. The data in Table 15.9 illustrate the levels of residual treatment arising from applications at given spray concentrations.

In domestic, industrial and food-handling premises, where safety is the prime

Table 15.8. Volume of sprays giving runoff – metric and US measures

Spray/surface	Volume	
	L/100 m^2	= US gal/1000 ft^2
Oil solution on glass	1.0–1.5	0.25–0.4
Oil solution on gloss-painted wood	1.5–2.0	0.5
Water spray on emulsion-painted wood	~ 5.0	1.3
Water spray on limewash	6.5	1.7
Water spray on concrete	8.0–10.0	2.0–2.5
Water spray on building block	> 10.0	> 2.5

Table 15.9. Treatment levels (as active ingredient per unit area) arising from residual applications at given spray concentrations. (a) Application rate, (b) effective treatment levels using typical application rate.

(a) Application to	Volume applied	
	Litre/100 m^2	US gal/1000 ft^2
Metal, glass or enamel	1.0–2.0, say 1.5[a]	0.25–0.5, say 0.375[a]
Painted surface on wood or other absorbent material	4.0–6.0, say 5.0[a]	1.0–1.5, say 1.25[a]
Concrete or building block	8.0–10.0, say 9.0[a]	2.0–2.5, say 2.25[a]

[a] median value of range

(b) Spray concentration	Metal, glass or enamel		Painted surface on wood or other absorbent material		Concrete or building block	
	mg/m^2	mg/ft^2	mg/m^2	mg/ft^2	mg/m^2	mg/ft^2
1.0	150.0	14.0	500	46.5	900	84.0
0.5	75.0	7.0	250	23.0	450	42.0
0.2	30.0	2.8	100	9.3	180	16.7
0.1	15.0	1.4	50	4.6	90	8.4
0.05	7.5	0.7	25	2.3	45	4.2
0.02	3.0	0.3	10	0.9	18	1.7
0.01	1.5	0.1	5	0.5	9	0.8

consideration, the pyrethroids are again the principal group of insecticides used. Whilst PBO can improve the killing properties of most of these compounds against beetles and cockroaches (Carter *et al.*, 1975; Chadwick, 1979), it may not be included in certain residual products because its strong solvent properties can assist absorption into a porous surface with the result that the deposits have a shorter residual life (Carter and Chadwick, 1978). It is often cost-effective to include PBO where direct spraying of the pest is an important aspect of the treatment (Chadwick and Evans, 1973) and in some residual products, such as water-based sprays and microencapsulated preparations (Bennett and Lund, 1977). Low and high pressure (CO_2) aerosols containing pyrethroid + PBO mixtures are convenient and highly effective both for direct spraying and residual (especially crack and crevice) application.

The use of pyrethrins in heavy (nonvolatile) oil films to control moths in stored products was practised prior to the discovery of synergists (Potter, 1938), typically using solutions of 1.3% pyrethrins. These have now been replaced by synergized formulations, which are more cost-effective, e.g. 0.3% pyrethrins + 3.0% PBO; 0.25% bioresmethrin + 0.25% PBO. Applications for this purpose are from mist, fog or ULV sprayers adjusted to give a coarse droplet size (>80 μm) which provides a surface deposit on stacked commodities when applied at dosage levels equivalent to 4 mg synergized pyrethrins per m^3.

The treatment of walls and the general fabric of food storage buildings by water-based sprays of synergized pyrethroids applied from knapsack sprayers fitted with standard or 'crack and crevice' nozzles (Mampe, 1976) provides effective control of beetles and cockroaches for up to 2–3 weeks. All of these treatments need to be repeated several times until newly emerging adults of the pest population are brought under control. This is especially true with cockroach infestations because of the survival of some oothecae after the initial treatment with say, diazinon and pyrethrins + PBO (Heuvel and Shenker, 1965).

A further aspect contributing to effective cockroach control is the ability of many pyrethroids to agitate the insects so that they leave their harbourages and consequently receive a greater dose of insecticide than would penetrate into their hiding places (Fuchs, 1988). This phenomenon is known as 'flush out' (Cornwell, 1976), and is related to the change in cockroach behaviour in response to light after exposure to pyrethroids (Miall and Le Patourel, 1989). Pyrethrins + PBO, even at low deposit levels, can exert a repellent action capable of protecting the treated area from reinfestation.

The effectiveness of various pyrethroid + PBO mixtures for cockroach control has been shown in laboratory and field trials reported by Chadwick and Evans (1973), McNeal and Bennett (1976) and Moore (1977). Rates of application to produce a residual deposit (always indoors) through mist, fog or ULV machines are up to 10 times those recommended for space spraying, e.g. mist or fog, 0.3–1.3 L per 100 m^3 (US 0.3–1.3 fl oz per 100 ft^3); and ULV, 0.4–1.25 L per 1000 m^3 (US 0.4–1.2 fl oz per 1000 ft^3).

Expected levels of synergized pyrethroids from these methods of treatment would be 4–12 mg m^{-3}. Since only about one-third of this will reach the horizontal surfaces, the effective deposit level is 1.5–4.0 mg/m^2. This is sufficient initially to control most beetles and cockroaches.

A somewhat specialized but highly effective form of cockroach control is the treatment of sewers and refuse chutes by the introduction of fogs or ULV sprays of synergized bioresmethrin (0.15% BRM + 0.15% PBO) into manholes and access points (Chadwick and Shaw, 1974; Chadwick *et al.*, 1977).

In another study, 0.25% pyrethrins + 1.0% PBO applied from domestic aerosol cans protected woollen cloth from attack by webbing clothes moths for up to 27 months and by carpet beetles for shorter periods when the treated articles were stored in the dark (Bry *et al.*, 1977).

Other formulations which are of value as residual treatments are powders and smokes. In powder formulations, the technical grade of insecticide is dispersed on an inert carrier (talc or clay). Distribution is best achieved though a hand-held shaker 'pepper pot' device or by application from a mechanical duster and usefully applied in places that are difficult to reach or where spraying liquids is hazardous and in areas where the presence of a dust can be tolerated. Again, synergized pyrethroids are among the safest and most effective active ingredients and a 0.2% pyrethrins + 1.0% PBO powder applied at the rate of 2–10 g m^{-2} is a good all-round product. Another, more specialized, form of powder is based on amorphous silica gel, which is insecticidal in its own right by virtue of its ability to absorb the waxy layer of the cuticle leading to insect death by dehydration. This material is extremely light and 'fluffy' so is used at lower dosage levels (1–5 g m^{-2}; US 0.5–2 oz per 100 ft^2) with 1.0% pyrethrins + 10.0% PBO to provide long-lasting control of crawling insects (Tarshis, 1967; Cunningham and Kelly, 1989).

8. THE VALUE OF PIPERONYL BUTOXIDE IN ADMIXTURE TREATMENTS

The pre-eminent position of PBO as a synergist in admixture treatments is because it allows safer products containing pyrethroids (natural and synthetic) to be used on food and in and around areas where food is prepared and processed, at cost-effective and economic levels.

Admixture is used only in a limited number of circumstances because of the need for specialized handling procedures. Equally, the conditions governing the use of insecticides in the treatment of food are of necessity very stringent. However, although the number of circumstances is small, such treatments can occur widely within any particular pest control context, e.g. grain, dried fruit or fish protection; timber, wool and hide preservation; or the treatment of food packaging. Sometimes there is also direct mixing of insecticide with an insect food material to form a bait, with an inert substrate to form a slow-release strip, or into paints or other surface coating materials.

Grain protection by liquid and powder applications of synergized pyrethroids is discussed in detail elsewhere in this volume (see Chapter 16 by Mervyn Adams). The protection of dried fruit from infestation by pyrethrins + PBO mixtures used in the dressing oil of sultanas at <10 ppm pyrethrins has been reported by Amos *et al.* (1978). Similarly, the inclusion of pyrethrins + PBO in the pea flour used to dress sides of bacon during curing gave protection at $4 + 40$ mg ft^{-2} against attack by blow flies (Davies, 1957).

Wet fish may be protected during air drying by dipping in pyrethrins + PBO mixtures at 0.125% pyrethrins (McLellan, 1963; Somme and Gjessing, 1963; Morris and Andrews, 1968; Meynell, 1978). Dried fish may be protected against infestation by *Dermetses frischii* by dipping at the same level (Green, 1967).

Packages of foodstuffs may be protected from insect attack by mixtures of pyrethrins + PBO. The synergized mixture may be incorporated into the varnish layer overlying the printed surface of packets (at $10 + 100$ mg ft^{-2}), into a wax layer on paper lining a box in which a commodity (e.g. currants) is wrapped, or into the thread and/or crepe paper which is used to seal paper or fabric sacks. Work on such procedures was reported by Gray (1952), Incho *et al.* (1953) and Goodwin-Bailey and Brooke (1957), and was subsequently reviewed by Brooke (1961), Brooke and Lomax (1967) and Langbridge (1970). Similarly, the use of synergized pyrethrins in laminated wrappings (Highland *et al.*, 1977) and textile food bags (Yeadon *et al.*, 1971; Highland *et al.*, 1975) produced highly insect-resistant packages without contaminating the contents. Treatments may also be carried out by dipping empty sacks in emulsifiable preparations of pyrethroids + PBO and other pyrethroids (Ramzan *et al.*, 1987).

9. THE VALUE OF PIPERONYL BUTOXIDE IN 'NONINSECTICIDAL' PREPARATIONS

Whilst the use of insecticides is the only means to provide immediate freedom from pests as well as providing longer-lasting protection from reinfestation, there are other procedures coming into increasing use whereby compounds which have chronic effects on insect metabolism may be employed. The most prominent groups are those affecting the hormonal systems of insects, principally insect growth regulators (IGRs). PBO, which is an inhibitor of mixed function oxidases, may be expected to interfere with several aspects of juvenile hormone or anti-juvenile hormone activity and is structurally related to 1,3-benzodioxole compounds, among which are several potent inhibitors of terminal juvenile hormone epoxidation *in vitro* (Pratt and Finney, 1977). In laboratory tests Bowers (1968) showed that several pyrethroid synergists had an IGR action in their own right. However, *in vivo* examination of the effect of PBO on several farnesyl juvenile hormone mimics showed only low factors of synergism (Solomon *et al.*, 1973), whereas more significant synergistic effects were shown against adult stages of the imported fire ant by Bigley and Vinson (1979). Further studies along the lines of those reported by Osmani *et al.* (1987) are needed to investigate the potential of mixtures of PBO with the latest series of commercially available IGRs.

PBO has shown an enhancing effect on antifeeding substances such as extracts from *Azadiracthta indica*, known as neem (Lange, 1983).

10. CONCLUSION

It is beyond any doubt that the safe control of insect pests in and around domestic, industrial and food-handling premises would not have been so readily achieved over a period of nearly 50 years, were it not for the availability of pyrethroids synergized by PBO. There is every indication that this state of affairs will continue for some time.

REFERENCES

Ahmed, S.M., Ravindranath, G.M. and Bhavanagary, H.M. (1976). Stabilisation of pyrethrins for prolonged residual toxicity. Part II: Development of new formulations. *Pyreth. Post* **13**, 119–123.

Amos, T.G., Evans, P.W.C. and Johns, R.E. (1978). Use of synergised pyrethrins to protect processed sultanas from insect attack. *Pyreth. Post* **14**, 76.

Anon. (1984). *Chemical Methods for the Control of Arthropod Vectors and Pests of Public Health Importance.* WHO, Geneva.

Bennett, G.W. and Lund, R.D. (1977). Evaluation of encapsulated pyrethrins (Sectrol™) for German cockroach and cat flea control. *Pest Control* **45**, 44–50.

Bennett, G.W., Runstrom, E.S. and Bertholf, J. (1984). Examining the where, why and how of cockroach control. *Pest Control* **52**, 42–50.

Beroza, M. (1956). Morton Beroza Reports Sesamex as synergist *J. Agric. Food Chem.* **4**, 49.

Beroza, M. and Barthel, W.F. (1957). Chemical structure and activity of pyrethrin and allethrin synergists for control of the housefly. *J. Agric. Food Chem.* **5**, 855–859.

Berthold, R. and Wilson, B.R. (1967). Resting behaviour of the German cockroach, *Blattella germanica. J. Econ. Entomol.* **60**, 347–351.

Bigley, W.S. and Vinson, S.B. (1979). Effects of piperonyl butoxide and DEF on the metabolism of methoprene by the imported fire ant, *Solenopsis invicta* Buren. *Pestic. Biochem. Physiol.* **10**, 14–22.

Blackith, R.E. (1952a). Stability of contact insecticides. I. Ultra-violet photolysis of the pyrethrins. *J. Sci. Food Agric.* **3**, 219–224.

Blackith, R.E. (1952b). Stability of contact insecticides. II. Protection of the pyrethrins against ultra-violet photolysis. *J. Sci. Food Agric.* **3**, 224–230.

Blackith, R.E. (1952c). Stability of contact insecticides. III. Allethrin, DDT and BHC in ultra-violet light. *J. Sci. Food Agric.* **3**, 482–487.

Bowers, W.S. (1968). Juvenile hormone activity of natural and synthetic synergists. *Science* **161**, 895–897.

Brooke, J.P. (1958). The stabilisation of pyrethrins by piperonyl butoxide in water with special reference to their use as mosquito larvicides. *Ann. Appl. Biol.* **46**, 254–259.

Brooke, J.P. (1961). The treatment of food packaging materials with pyrethrins and PBO for the protection of packaged materials from insect infestation. *Pyreth. Post* **6**, 14–22.

Brooke, J.P. (1967). The effect of five methylenedioxyphenyl synergists upon the stability of the pyrethrins. *Pyreth. Post* **9**, 18–37.

Brooke, J.P. and Lomax, P.H. (1967). Protection of packaged foods from insects. *Pyreth. Post* **9**, 36–39.

Bry, R.E., Boatwright, R.E., Lang, J.H. and Simonaitis, R.A. (1977). Long-term protection of woollen fabrics with synergised pyrethrins. *Pyreth. Post* **14**, 26–29.

Burt, P.E. and Goodchild, R.E. (1974). Knockdown by pyrethroids: its role in the intoxication process. *Pestic. Sci.* **5**, 625–633.

Carter, S.W. and Chadwick, P.R. (1978). Permethrin as a residual insecticide against cockroaches. *Pestic. Sci.* **9**, 555–565.

Carter, S.W., Chadwick, P.R. and Wickham, J.C. (1975). Comparative observations on the activity of pyrethroids against some susceptible and resistant stored products beetles. *J. Stored Prod. Res.* **11**, 135–142.

Casida, J.E. (1970). Mixed function oxidase involvement in the biochemistry of insecticide synergists. *J. Agric. Food Chem.* **18**, 753–772.

Chadwick, P.R. (1971). Activity of some new pyrethroids against *Blattella germanica* L. *Pestic. Sci.* **2**, 16–19.

Chadwick, P.R. (1979). The activity of some pyrethroids against *Periplaneta americana* and *Blattella germanica*. *Pestic. Sci.* **10**, 32–38.

Chadwick, P.R. and Evans, M. (1973). Laboratory and field tests with some pyrethroids against cockroaches. *Int. Pest Control* **15**, 11–16.

Chadwick, P.R. and Shaw, R.D. (1974). Cockroach control in sewers in Singapore using bioresmethrin and piperonyl butoxide as a thermal fog. *Pestic. Sci.* **5**, 691–701.

Chadwick, P.R., Martin, M. and Marin, J. (1977). Use of thermal fogs of bioresmethrin and cismethrin for control of *Periplaneta americana* (Insecta: Blattidae) in sewers. *J. Med. Entomol.* **13**, 625–626.

Cornwell, P.B. (1976). *The Cockroach*, vol. II: Insecticides and Cockroach Control. The Rentokil Library, Associated Business Programmes, London.

Cunningham, B. and Kelly, M.P. (1989). Drione dust against *Blattella orientalis*. *Int. Pest Control* **31**, 90–92.

Davies, M. (1957). Blowfly damage to bacon: a new method of protection. *Pyreth. Post* **4**, 28–33.

Davies, M.S., Chadwick, P.R., Holborn, J.M., Stewart, D.C. and Wickham, J.C. (1970). Effectiveness of the (+)-*trans*-chrysanthemic acid ester of (±)-allethrolone (Bioallethrin) against four insect species. *Pestic. Sci.* **1**, 225–227.

Davies, M., Keiding, J. and Von Hofstein, C.G. (1958). Resistance to pyrethrins and to pyrethrins/piperonyl butoxide in a wild strain of *Musca domestica* L. in Sweden. *Nature* **182**, 1816–1817.

Demoute, J-P. (1989). A brief review of the environmental fate and metabolism of pyrethroids. *Pestic. Sci.* **27**, 375–385.

Dove, W.E. (1947). Piperonyl butoxide, a new safe insecticide for the household and field. *Am. J. Trop. Med.* **27**, 339–345.

Fuchs, M.E.A. (1988). Flushing effects of pyrethrum and pyrethroid insecticides against the German cockroach (*Blattella germanica* L.). *Pyreth. Post* **17**, 3–7.

Funaki, E., Tabaru, Y. and Motoyama, N. (1986). Synergistic effect of piperonyl butoxide on pyrethroid-resistant houseflies. *J. Pestic. Sci.* **11**, 415–420.

Glynne-Jones, G.D. (1983). The use of piperonyl butoxide to increase susceptibility of insects which have become resistant to pyrethroids and other insecticides. *Int. Pest Control* **25**, 14–15, 21.

Goodwin-Bailey, K.F. (1960). Synergism of piperonyl butoxide with pyrethrins against several species of insect. *Chem. & Ind.* 18 June, 700–702.

Goodwin-Baily, K.F. and Brooke, J.P. (1957). The treatment of wrapping materials for foodstuffs with pyrethrins and piperonyl butoxide for protection against insect infestation. In: *Proceedings of the IVth International Congress on Crop Protection*, Hamburg, vol. 2, pp. 1761–1767.

Gray, H.E. (1952). Packaging of cereals and some chemical treatments to increase resistance to penetration by insects. *Trans. Am. Cereal Chem.* **10**, 50–58.

Green, A.A. (1967). The protection of dried sea fish in South Arabia from infestation by *Dermestes frischii* Kug. *J. Stored Prod. Res.* **2**, 331–350.

Groome, J.M. and Martin, R. (1994). A new approach for the control of insects in industrial premises. *Proceedings of the Brighton Crop Protection Conference 1994*, vol. III, pp. 1031–1038.

Hadaway, A.B., Barlow, F. and Duncan, J. (1963). Effects of piperonyl butoxide on insecticidal potency. *Bull. Entomol. Res.* **53**, 769–777.

Hayashi, A., Saito, T. and Iyatomi, K. (1968). Studies on the increment of efficiency of insecticides (Part VIII). Metabolism of ^{3}H-pyrethroids in the adult housefly, *Musca domestica vicina*. Macq. *Botyu-Kagaku* **33**, 90–95.

Head, S.W., Sylvester, N.K. and Challinor, S.K. (1968). The effect of piperonyl butoxide on the stability of films of crude and refined pyrethrum extract. *Pyreth. Post* **9**, 14.

Heuvel, M.J. van den and Shenker, A.M. (1965). Cockroach control using non-persistent insecticides. *Int. Pest Con.* **7**, 10–11.

Hewlett, P. S. (1968). Synergism and potentiation in insecticides. *Chem. & Ind.*, 1 June, 701–706.

Highland, H.A., Secreast, M. and Yeadon, D.A. (1975). Insect-resistant textile bags: new construction and treatment techniques. *U.S.D.A Tech. Bull. No. 1511*, 1975.

Highland, H.A., Cline, L.D. and Simonaitis, R.A. (1977). Insect-resistant food pouches made from laminates treated with synergised pyrethrins. *J. Econ. Entomol.* **70**, 483–485.

Holborn, J. M. (1957). The susceptibility to insecticides of laboratory cultures of an insect species. *J. Sci. Food Agric.* **8**, 182–188.

Incho, H.H., Incho, E.S. and Matthews, N.W. (1953). Insect proofing paper. Laboratory evaluation of pyrethrins/piperonyl butoxide formulations. *J. Agric. Food Chem.* **1**, 1200–1203.

JMPR (1995). *Piperonyl butoxide*. A monograph prepared by the Joint FAO/Who Meeting on Pesticide Residues, Geneva.

Lange, W. (1983). Piperonyl butoxide: synergistic effects on different neem seed extracts and influence on degradation of an enriched extract by ultra-violet light. In: *Proceedings of the 2nd International Neem Conference*, Rauischholzhausen, pp. 129–140.

Langbridge, D.M. (1970). The protection of packaged foods from insect attack (trials using pyrethrum, synergised pyrethrins and carbaryl as paper coatings). *Pyreth. Post* **10**, 6–9, 14.

Lloyd, C.J. (1961). The effect of piperonyl butoxide on the toxicity of pyrethrins to the cigarette beetle. *Pyreth. Post* **6**, 3–4.

Lloyd, C.J. and Parkin, E.A. (1963). Further studies on a pyrethrum-resistant strain of the grain weevil, *Sitophilus granarius* (L.). *J. Sci. Food Agric.* **9**, 655–663.

MacDonald, R.S., Surgeoner, G.A., Solomon, K.R. and Harris, C.R. (1983). Effect of four spray regimes on the development of permethrin and dichlorvos resistance in the laboratory by the house fly (Diptera: Muscidae). *J. Econ. Entomol.* **76**, 417–422.

Makundi, R.H. (1986). The toxicity of deltamethrin and cyfluthrin to the Larger Grain Borer, *Prostephanus truncatus* (Horn) (Coleoptera: Bostrychidae). *Int. Pest Control.* **28**, 79–81.

Mampe, C.D. (1976). Roach control in the food handling business. *Pest Control* **44** (6), 15–18; (7), 27–34; (8), 44–50.

McLellan, R.H. (1963). The use of pyrethrum dip as a protection for drying fish in Uganda. *Pyreth. Post* **7**, 8–10.

McNeal, C.D. and Bennett, G.W. (1976). Utilisation of ULV aerosols for control of the German cockroach. *J. Econ. Entomol.* **69**, 506–508.

Meynell, P.J. (1978). Reducing blowfly spoilage of sun drying fish in Malawi using pyrethrum. *Proc. Indo-Pacific Fishery Commission* 18, 347–353.

Miall, S.M. and Le Patourel, G.N.J. (1989). Response of the German cockroach *Blattella germanica* (L.) to a light source following exposure to surface deposits of insecticides. *Pestic. Sci.* **25**, 43–51.

Mitchell, W. (1959). Bucarpolate pyrethrum synergist. *Pyreth. Post* **5**, 19.

Moore, R.C. (1977). Field tests of pyrethrins and resmethrin applied with ULV generators or total release aerosols to control the German cockroach. *J. Econ. Entomol.* **70**, 86–88.

Morris, R.F. and Andrews, D. (1968). Investigations into the use of pyrethrum and other insecticides for the control of the blowfly *Calliphora terraenovae* (Macq) infesting light-salted cod fish in Newfoundland. *Pyreth. Post* **9**, 9–12, 36.

Nagasawa, S. and Shibuya, S. (1987). Toxicity to the housefly, *Musca domestica* L. (Diptera; Muscidae) of mixtures of synthetic pyrethroids and piperonyl butoxide. *Jap. J. Appl. Entomol. Zool.* **31**, 150–155.

Nash, R. (1954). Studies on the synergistic effect of piperonyl butoxide and isobutyl-undecyleneamide on pyrethrins and allethrin. *Ann. Appl. Biol.* **41**, 652–663.

Nasir, N.M. (1953). Stability of contact insecticides, IV. Relationship between the ultra violet absorption spectrum and the photolysis of DDT and the pyrethrins. *J. Sci. Food Agric.* **4**, 374–378.

Osmani, Z., Anees, I. and Sighamony, S. (1987). Potentiation of action of JHA by pepper seed extracts and JH mimic action of the same. *Int. Pest Control.* **29**, 128–129.

Owens, J.M. (1990). Problems associated with evaluating cockroach control chemicals. Laboratory test methods. In: *Proceedings of the National Conference on Urban Entomology – 1990* (Robinson, W.H., ed.), pp. 25–29. Blacksburg, VA, USA.

Palmer, C.J., Smith, I.H., Moss, M.D.V. and Casida, J.E. (1990). 1-[4-[(Trimethyl-silyl)ethynyl]phenyl]-2,6,7-trioxabicyclo[2,2,2] octanes: a novel type of selective proinsecticide. *J. Agric. Food Chem.* **38**, 1091–1093.

Parkin, E.A. and Lloyd, C.J. (1960). Selection of a pyrethrum-resistant strain of the grain weevil, *Calandra granaria* L. *J. Sci. Food Agric.* **8**, 471–477.

Phipers, R.F. and Wood, M.C. (1957). An investigation into the reported stabilisation of pyrethrins by piperonyl butoxide. *Pyreth. Post* **4**, 11.

Potter, C. (1938). The use of protective films of insecticide in the control of indoor insects, with special reference to *Plodia interpunctella* Hb. and *Ephestia elutella* Hb. *Ann. Appl. Biol.* **25**, 836.

Pratt, G.E. and Finney, J.R. (1977). Chemical inhibitors of juvenile hormone biosynthesis in vitro. In: *Crop Protection Agents* (McFarlane, N.R., ed.), pp. 113–132 Academic Press, London.

Ramzan, M., Chahal, B.S., Judge, B.K. and Narang, D.D. (1987). Field trials on the impregnation of gunny bags with synthetic pyrethroids against storage loss of wheat due to stored grain pests. *Bull. Grain Technol.* **25**, 160–164.

Sawicki, R.M. (1961). The effect of safroxan on the knockdown and the 24-hour toxicity of commercial pyrethrum extract against houseflies (*Musca domestica* L.). *Pyreth. Post* **6**, 38.

Schneider, B.M. and Bennett, G.W. (1985). Comparative studies of several methods for determining the repellency of blatticides. *J. Econ. Entomol.* **78**, 874–878.

Solomon, K.R., Bowlus, S.B., Metcalf, R.L. and Katzenellenbogen, J.A. (1973). The effect of piperonyl butoxide and triorthocresyl phosphate on the activity of juvenile hormone mimics and their sulfur isosteres in *Tenebrio molitor* L., and *Oncopeltus fasciatus* (Dallas). *Life Sci.* **13**, 733–742.

Somme, L. and Gjessing, E.T. (1963). Insecticides for protection against blowflies in the stockfish industry. *Pyreth. Post* **7**, 3–7, 18.

Synerholm, M.E., Hartzell, A. and Arthur, J.M. (1947). *Contrib. Boyce Thompson Inst.* **15**, 35.

Tarshis, I. B. (1967). Silica aerogel insecticides for prevention and control of arthropods of medical and veterinary importance. *Angewandte Parasitologie* **8**, 210–237.

Wachs, H. (1947). Synergistic insecticides. Discovery of 4,5-methylenedioxyphenoxy-2-propyl benzyl butyl diethylene glycol ether. *Science* **105**, 530–531.

Wickham, J.C., Bone, A.J. and Stewart, D.C. (1974). The application of computer-based techniques in the evaluation of pesticide products. *Pestic. Sci.* **5**, 353–362.

Wilkinson, C.F. (1968). In: *Enzymatic Oxidation of Toxicants* (Hodgson, E., ed.) p. 113. North Carolina State University, Raleigh, or *World Review of Pest Control* **17**, 155–168.

Worthing, C.R. and Hance, R.J. (1991). *The Pesticide Manual.* Published by British Crop Protection Council, Farnham, Surrey.

Wright, R.E. (1977). Evaluation of Air Guard actuators and insecticide aerosols for the control of flies in dairy barns. *Pyreth. Post* **14**, 2.

Wright, C.G., Leidy, R.B. and Roper, E. (1984). Insecticide ecology. *Pest Management* **3**, 22–24.

Yeadon, D.A., Danna, G.F., Cooper, A.S. and Reeves, W.A. (1971). Insect-resistant packaging. Effectiveness of fabric treatment varies with oil content, particle size; is improved by PVC barrier. *Modern Packaging* Sept., 54–57.

16

The Use of Piperonyl Butoxide in Grain Protection

MERVYN ADAMS

1. INTRODUCTION

In the context of world trade, grain refers not only to cereals such as wheat, maize (corn), rice, barley and sorghum, but also to those oilseeds, legumes and pulses that are stored for long periods. These are known as durable products, i.e. they have a long storage life when stored in a dried state. Drying may be natural in the field or by artificial means after harvest. The intention is to produce a product which has a storage life of months to years.

The scale of the grain industry is evident from Table 16.1. Some of the stocks form part of food security plans to ensure adequate supplies in countries liable to suffer harvest failures and others are intervention stocks held for commercial reasons.

The extent of losses during storage and transport have been studied by many experts (Adams, 1977; Boxall, 1986) and estimates range from negligible to total. A conservative estimate of 5% loss to insects during storage and transport is equivalent to 26.4 million tonnes of wheat valued at US \$4.144 billion. It is this potential loss that makes the use of grain protectants so vital to the world's food supply and economy. In 1985, Webley concluded that the cost of using pest

Table 16.1. Worldwide grain situation 1994–1995 (in millions of tonnes)

Grain	Stock at start	Production	Demand	Stock at end	Trade (1993–1994)
Wheat (incl. flour)	127	528	550	104	92.6
Maize	72.3	555.1	539.8	87.6	54.3
Rice (milled)	49.8	360.5	361	49.3	16.34
Barley	31.7	160.9	166.6	26	17.1
Soybeans	17.3	137.6	130.7	24.2	28.08
Sorghum	4.1	56.8	55.8	5.1	6.6

Reproduced with permission from Anon. (1995).

PIPERONYL BUTOXIDE
ISBN 0-12-286975-3

control measures, such as a grain protectant ($0.65 per tonne), were considerably less than the costs of extra storage and handling ($9.75–13 per tonne), and the differential between milling grade and lower grades of wheat (($1.30–5.20 per tonne); making it essential to maintain grain quality at the highest level.

2. PEST PROBLEMS

Although there are a variety of pests of stored grain including rodents and birds, this chapter is confined to insect pests which can be controlled by the use of insecticides.

The insect problem is insidious. Grain may become infested in the field and the pest carried into a clean store (Giles and Ashman, 1971). Small amounts of infested grain from the previous year's crop, lodged in crevices, can effect a rapid infestation of the new crop. Insects thrive on a wide range of conditions (Howe, 1965) and constitute a threat in most parts of the world.

The nature of the infestation may be categorized as follows:

- **Primary pests** are the initial invaders capable of attacking completely sound grain. They may infest in the field, e.g. *Sitophilus zeamais* Motsch., *Sitotroga cerealella* (Olivier). Many species lay eggs on the commodity, their larvae boring into the individual grains, remaining hidden until they emerge as adults, e.g. *Sitophilus* species, bruchids. Others lay eggs loose and their larvae attack intact grains, e.g. *Rhizopertha dominica* (Fabricius), *Prostephanus truncatus* (Horn) and *Ephestia* species.
- **Secondary pests** can only attack grain damaged during handling or by other insects, e.g. *Tribolium castaneum* (Herbst), *Oryzaephilus surinamensis* L.
- **Scavengers and predators** live on other insects, moulds resulting from insect attack and general refuse, e.g. *Alphitobius* species, predatory mites, psocids and cockroaches.

Insects will not only survive under most storage conditions, but can modify their environment in ways that encourage both further infestation and an increase in their rate of reproduction.

3. CONTROL METHODS

The prevention of the widespread destruction of grain in store and the transport of insects to other locations requires specialized control techniques. These include:

- **fumigation** with a gas such as methyl bromide or phosphine. A gas-proof structure encloses the commodity and the gas permeates throughout it, killing the insects within. There is no residual effect.
- **space treatments** to prevent or reduce the number of insects entering stored grain.

- *residual insecticide* treatments of empty stores and applications onto the fabric of bags are used to prevent insects entering, but have no effect on grain that is already infested.
- *grain protectants* which are mixed with the grain, usually at the time of storage or prior to transport, to kill insects already present within the produce and to provide protection against further infestation. Because the insecticide is in contact with the grain, species with hidden stages will be killed as adults emerge. Long-term residual protection is provided.

Grain may be stored in bulk or in packaging such as sacks. Bulk is preferable for long-term storage and international trade because it enables protectant treatments to be applied at the time of movement, e.g. from harvest into store, from one silo to another, to ships, or to train wagons. The main disadvantages are equipment costs for storage and conveying, and the fact that it has to be moved from silo to silo for admixture treatments *in situ*. It is also possible to admix a protectant with the grain before bagging.

Residual pesticides provide the basis for the most flexible methods available for pest management in stored products pest control (Champ and Ryland, 1986).

4. SUITABILITY OF INSECTICIDES AS GRAIN PROTECTANTS

The direct application of insecticides to grain is the most efficacious solution to the insect problem, but it is also the most difficult to put into practice. The reasons are not technical, but regulatory. This is one of the few applications where a residual insecticide is deliberately added to food intended for human consumption. There are few insecticides with an acceptable acute and chronic toxicity profile that have sufficient efficacy over long periods to render them suitable candidates. Those applied to vegetables rely on their degradation during the post-application interval to minimize residues, whereas grain may be consumed at any time during the storage period.

Residue levels are regulated internationally by the use of maximum residue limits (MRLs) which define the maximum residue that would result from application of the pesticide according to good agricultural practice (GAP). They are not the highest levels that are toxicologically acceptable. The link with toxicity is via the acceptable daily intake (ADI), which is the amount of the insecticide that could be consumed every day for a lifetime without increasing the risk of an unfavourable effect. An insecticide with a high ADI will be the preferred choice for treating staple human foods.

Webley (1994) summarized the position on residues from grain protectants, concluding that residues found in practice are much lower than the theoretical figure given in trials.

Grain protectants have to be proved to possess negligible human toxicity at the dose rates applied, yet be stable enough to provide protection for up to a year or more. The ideal grain protectant possesses the following attributes:

- low mammalian toxicity
- high insecticidal efficacy and wide spectrum of activity

- good stability under storage conditions
- rapid degradation after the storage period and during processing of the commodity
- economic and easy to apply.

Very few insecticides meet these criteria and possess an internationally accepted toxicity package proving their safety.

5. ORGANOCHLORINE COMPOUNDS

The main protectants used after the 1939–1945 war were organochlorine compounds, following their success in the control of public health pests. The most commonly used compound was lindane, used mainly in village stores, partly because of its vapour activity which gave some advantages in crops with hidden insect stages. McFarlane (1969) demonstrated that lindane powder at 2.5 ppm resulted in 52% pre-emergence mortality of *Acanthoscelides obtectus* (Say.) compared with 23% for a 1 : 5 ratio of pyrethrins : piperonyl butoxide (PBO) at 1.25 and 2.5 ppm.

PBO played a very minor role in the possible extension of the life of lindane as a grain protectant. In trials, the addition of PBO enhanced the toxicity of lindane against susceptible and resistant strains of *T. castaneum* at synergistic ratios of 5 and over 40, respectively (Udeaan and Kalra, 1983), but this was of no commercial interest.

It was the concern over the long-term toxicity of organochlorine compounds that led to the search for safer alternatives.

6. THE PYRETHRINS

Although the organochlorine compounds were seen to be of use for village grain storage, products that were safer and easier to handle were sought for large-scale applications. Owen and Waloff (1946) reported the success of pyrethrins in the control of the stored products moth, *Ephestia elutella* (Hübner), leading the way for their further development. All that was required was a way of providing them with sufficient activity and persistence to fulfil the task, as pyrethrins alone required an uneconomically high application rate of 25 ppm to be effective (Le Pelley and Kockum,1954).

6.1. Pyrethrins Stability

The propensity of pyrethrins to degrade rapidly was the main handicap to their initial use in storage. Page and Blackith (1950) revealed that the addition of PBO stabilized pyrethrins and increased their insecticidal efficacy. Phipers and Wood (1957) using PBO with mineral oil as a solvent for thin films of pyrethrins, demonstrated a stabilization effect.

Hewlett (1951) examined the specific synergistic activity of pyrethrin films in relation to stored products pests and the most cost-effective ratios. During these experiments it was found that although PBO alone in Shell oil P31 sprays had no effect on *T. castaneum* as a direct spray treatment, it did when applied as a thin film.

6.2. Synergized Pyrethrins

Dove (1947a, b, 1949) and Dove and MacAllister (1947) introduced the idea that PBO could play a major role in synergizing pyrethrins for the control of stored products insects. This was commercially developed into a combination admixture grain protectant and a warehouse and mill spray in the early 1950s, as reported by Dove (1951, 1952a, b, 1953, 1958).

In the UK, Goodwin-Bailey and Holborn (1952) working at The Cooper Technical Bureau developed PBO as a synergist for pyrethrum insecticidal powders as grain protectants. Using a range of carriers, both in the laboratory and in field trials, they settled on 0.04% pyrethrins : 0.8% PBO, applied at 1.3 ppm pyrethrins : 27 ppm PBO. Protection remained satisfactory for 11 months with significant repellency occurring for the first 3 months.

Concurrently, Wilbur (1952) was conducting field trials on wheat in Kansas, USA, with a 0.08% pyrethrins : 1.1% PBO powder mixture applied at 1 ppm pyrethrins : 13.7 ppm PBO. The results indicated an outstanding protectant value for on-farm storage.

The varied species and life cycles of grain storage insects makes control quite complex. Indications of differential susceptibility between related species and size were described by Glynne Jones and Green (1959). They conducted comparative trials with pyrethrins alone and 0.25% pyrethrins plus 0.5% PBO powders against two major grain pests, *Sitophilus oryzae* L. and *Sitophilus granarius* L., rearing them on wheat and maize. There was a positive correlation between body weight and ability to survive the pyrethrins treatment, with the smaller *S. oryzae* from wheat more susceptible (2.15 times) than those raised on maize. *S. oryzae* was also more susceptible (2.34 times) than *S. granarius*. The synergistic effect of adding PBO at different ratios ranged from a factor of 2.1 at 1 : 1 for *S. oryzae* to 18.7 at 1 : 20 for *S. granarius*. There was little increase in synergism when the ratio rose above 1 : 8, which suggests that the quantity of PBO, rather than the ratio, was the limiting factor.

The success of PBO in reducing the cost of the pyrethrins treatments led to success in the marketplace with the replacement of DDT and lindane. It also inspired further developments in application techniques.

6.3. Application Developments

Work carried out by Dove and Schroeder (1955) compared the use of oil- or water-based emulsion of 0.2% pyrethrins synergized by 2% PBO. The latter were found to be excellent since the emulsions broke on the surface, leaving an insecticidal deposit of equivalent performance to powder formulations, whereas the oils either carried the insecticide into the grain kernels, making it unavailable as a contact toxicant, or formed a sticky deposit, binding the kernels together.

Goodwin-Bailey (1960) examined application methods in relation to synergism, with the results shown in Table 16.2.

Wettable powder formulations of pyrethrins synergized with PBO were found to be longer-lasting for surface sprays against Indian meal moth, *Plodia interpunctella* (Hübner), than water-diluted emulsions (Kantack and Laudani, 1957). This is because of absorption of the active material into surface substrates when emulsion concentrates are used, whereas wettable powders leave a residual surface deposit, even if the water penetrates the surface.

An exhaustive testing regime with synergized pyrethrums, covering a wide range of stored products insects, was carried out by Lloyd and Hewlett (1958), using films and sprays containing pyrethrins alone or synergized by PBO. A wide range of tolerances was found: adult moths and bruchids were the most susceptible; adult ptinids were the least susceptible; and adults were generally more susceptible than larvae. A 0.3% pyrethrins : 3.0% PBO treatment was found to be equivalent to a 1.3% pyrethrins only treatment. Field trials against *Ptinus villiger* (Reit.), the hairy spider beetle, demonstrated that a 0.1% pyrethrins : 1.0% PBO oil-based spray, applied monthly, could provide control as good as that of malathion in commercial flour storages, although inferior to that of lindane (Watters, 1961).

6.4. Progress in the USA

In 1959, Goodwin-Bailey reviewed trials reported in the USDA Marketing Research Report No. 322 (Walkden and Nelson, 1959) as an impressive achievement, with over half a million bushels (14 000 tonnes) of wheat and shelled corn treated and protected with pyrethrins and the synergists PBO, sulfoxide or MGK 264 between 1952–1957. The most commonly used combination was pyrethrins plus PBO, in some cases providing 2–3 years' protection. The effective control of some 13 species of insects encouraged the spread of the pyrethrins plus PBO combination.

One highly significant sector of the grain market is the provision of government-owned reserve grain stocks, to ensure food security and stabilize

Table 16.2.   Factors of synergism at LD_{50} at different pyrethrins : PBO ratios

Ratio	Sitophilus granarius		Lasioderma serricorne	
	Dust	Deposit	Topical	Deposit
1 : 2.5	9	–	–	–
1 : 5	11	–	1.5	2.3
1 : 10	15	9	1.7	3.8
1 : 20	19	–	–	–

Reproduced with permission from Goodwin-Bailey (1960).

markets where intervention policies are in force. In the 1950s some of these stocks were held in converted Liberty ships anchored in James River and Hudson River, USA (Phillips, 1959). Each held about 225 000 bushels of wheat in the lower hold with headspace above, creating ideal conditions for insect infestation. Synergized pyrethrum sprays at 0.3% pyrethrum : 3.0% PBO were applied at 2 US quarts per 1000 ft^2 ($\approx$ 2.05 L/100 sq metres) to the grain, the exposed surfaces of the ship and bulkheads above the grain and the underside of the between deck. The success of the treatment at Hudson River was outstanding, with the greatest number of insects found in any month being 98 live Indian meal moths and 4 Cadelle beetles (*Tenebroides mauritanicus* L.) in over 350 holds of wheat.

6.5. Developments in Other Countries

The increase in successful treatments with synergized pyrethrins in the USA led to applications by Bulk Handlers in Australia, and trials in South Africa (Joubert, 1965; Joubert and DeBeer, 1968; Joubert and du Toit, 1968). The South African work concluded that a water-based spray at a dosage of 1.25 ppm of pyrethrins : 12.5 ppm PBO afforded excellent protection to maize against insect infestation for a year.

Weaving (1970) in Kenya, working with the bruchid beetles, *A. obtectus* and *Zabrotes subfasciatus* (Boh.), demonstrated that a 0.2% pyrethrins powder in a 1 : 10 to 1 : 15 ratio with PBO was suitable for practical control when applied at 2 ppm pyrethrins.

6.6. Synergist Efficiency

For many years pyrethrins synergized by PBO was the only pesticide approved for post-harvest applications as a direct admixture to durable products such as grain and pulses in store. Soon the high cost of pyrethrins, plus their erratic supply, began to threaten this supremacy. This instigated more studies on synergism to identify the most cost-effective synergists and ratios.

Glynne Jones and Chadwick (1960) compared PBO with bucarpolate, S421, and sulfoxide. They found that for flour beetles, *T. castaneum,* efficacy ratings were sulfoxide > PBO > S421 and bucarpolate, at ratios of 1 : 1, 1 : 5, and 1 : 10. For the weevils *S. oryzae* and *S. granarius,* at only a 1 : 1 ratio PBO was similar to sulfoxide and both were approximately twice as active as bucarpolate and S421, but this was species dependent. This agreed with the previous work by Glynne Jones and Green (1959) using a powder formulation shown in Table 16.3.

Baker (1963) combined pyrethrins : PBO : MGK 264 in a ratio of 1 : 2 : 3.3 to attempt to obtain a cheaper formulation than a 1 : 5 ratio of pyrethrins : PBO for public health pests. However, work on *S. oryzae* and *S. granarius* by Chadwick (1963a) demonstrated that pyrethrins plus MGK 264 was only about 0.4 times as toxic as pyrethrins plus PBO. In the same experiment with *T. castaneum,* a 1 : 5 ratio of pyrethrins : MGK 264 was found to be 0.7 times as active as that for the same ratio with PBO. A similar result was found by Saxena (1967).

Table 16.3.   Degree of synergism at various ratios of pyrethrum : PBO
(Glynne Jones and Green, 1959, with permission)

Ratio pyrethrum : PBO	Degree of synergism	
	Sitophilus oryzae	*Sitophilus granarius*
1 : 1	2.1	5.5
1 : 8	5	15
1 : 20	6.5	18.7

The choice of synergists and combination was heavily influenced by cost, as reviewed by Chadwick (1963b). Hewlett (1969) examined the mathematical relationship between pyrethrins and PBO, utilizing *T. castaneum* as the test insect. The resulting isoboles, which describe the relationship between pyrethrins and PBO concentrations for fixed mortality levels, revealed that for indefinitely large doses of PBO, mixtures were 3.04 times as toxic as pyrethrins alone.

7. USE OF MALATHION

From 1960 to 1970 malathion, an organophosphorus insecticide, was shown to be a very cost-effective grain protectant and the usage of synergized pyrethrins diminished. Malathion was shown to have greater persistence in grain storage than synergized pyrethrins (Floyd, 1961), and soon became established as the standard grain treatment in Australia (Watt, 1962) and in the USA (LaHue 1965, 1966, 1967, 1969, 1970).

In Australia, Banks and Desmarchelier (1978) determined the half-life of malathion to be 3 months, supporting existing commercial experience. This led to revised thoughts on dosage rates.

7.1. Resistance to Malathion

The rapid spread of the use of malathion as a grain protectant continued during the 1960s and early 1970s, until resistance was reported in the USA by Gillenwater and Burden in 1973 and in Australia in 1974 (Greening *et al.*, 1974). The level of resistance was surveyed in 1975 (Bengston *et al.*, 1975). The problem was found to be serious with factors of × 9 for *S. oryzae,* × 6 for *R. dominica* and × 39 for *T. castaneum.*

7.2. Strategies for Overcoming Resistance

In South Africa, Joubert and du Toit (1968) recommended alternating synergized pyrethrins with cheaper insecticides, such as malathion, as a precaution against the selection of insecticide-resistant strains. The South African authorities, predicting the onset of resistance to malathion, insisted that synergized pyrethrins were used *every third year* and averaged the treatment cost. This action delayed the onset of acute malathion resistance.

In Australia the failure of malathion to maintain the nil insect tolerance required in Australian export grain led to revised strategies.

A mixture of malathion plus synergized pyrethrins was introduced into New South Wales, Australia. Its efficacy was studied in commercial stores by Desmarchelier *et al.* (1979), who found that treatment of stores in the South Dubbo region in 1977, with 18 g t^{-1} malathion plus 3 g t^{-1} of synergized pyrethrins resulted in 255 infestations (1 live insect or more) of *R. dominica* and *T. castaneum*, of which only 17 were heavy (more than 1 insect per 5 kg). This compared favourably with the 614 infestations, of which 34 were heavy, reported for the same stores in 1976, when 18 g t^{-1} malathion alone was used.

Greening (1980) also compared the efficacy of synergized pyrethrins with and without the addition of malathion when applied to newly harvested grain during 1976. The results after 17 months on-farm storage in New South Wales revealed that both wheat and barley were completely protected by 3 ppm pyrethrins : 27 ppm PBO. The addition of malathion was deemed unnecessary.

8. INTRODUCTION OF SYNTHETIC PYRETHROIDS

8.1. First Generation Pyrethroids

The first group of synthetic pyrethroids based on the structure of pyrethrins included bioallethrin, resmethrin, bioresmethrin and tetramethrin.

Lloyd (1973) carried out laboratory toxicity tests using a selection of candidate pyrethroids on *T. castaneum* and susceptible and resistant *S. granarius*. The results were bioresmethrin > resmethrin > bioallethrin > allethrin > tetramethrin. When synergized with PBO the same pattern emerged. The results are shown in Table 16.4.

8.2. Use of Bioresmethrin

Such data, coupled with its extremely low mammalian toxicity, made synergized bioresmethrin the ideal candidate for grain storage.

Work with bioresmethrin + PBO (1 : 5) in powder formulations was undertaken by Carter *et al.* (1975). This combination was more active than malathion against susceptible *O. surinamensis, S. granarius, S. oryzae and T. castaneum* and malathion-resistant *S. oryzae* and *T. castaneum*. Synergized bioresmethrin was also used in Brazil by Bitran and Campos (1975).

Trials in Australia by Ardley and Desmarchelier (1975) on stored wheat showed that a formulation of bioresmethrin 4 ppm plus PBO 20 ppm controlled most insects which had become resistant to malathion.

Concurrently, the use of alternative organophosphorus compounds such as fenitrothion were also being examined. Ardley and Sticka (1977) found that fenitrothion at 6 ppm was equal to 12 ppm of malathion, but the response of *R.*

Table 16.4.   Relative toxicities of five first generation pyrethroids with and without PBO

Toxicant	T. castaneum (flour beetle)		Pyrethrin-resistant S. granarius (grain weevil)	
	Factor of synergism	LD$_{50}$ mg (0.03 mL doses)	Factor of synergism	LD$_{50}$ mg (0.03 mL doses)
Pyrethrins	–	0.200[a]	–	30.2[b]
Pyrethrins				
+ PBO (1:10)	2	0.1	145.2	0.208[a]
Bioresmethrin	–	0.0372	–	1.85[a]
Bioresmethrin				
+ PBO (1:10)	1.4	0.0257	31.2	0.0593[a]
Resmethrin	–	0.129	–	9.12
Resmethrin				
+ PBO (1:10)	3.4	0.038	155.1	0.0588[a]
Bioallethrin	–	0.151	–	43.7[b]
Bioallethrin				
+ PBO (1:10)	2.5	0.0593[a]	151.7	0.288
Allethrin	–	0.388[a]	–	47.9[b]
Allethrin				
+ PBO (1:10)	4	0.0966[a]	66.2	0.724
Tetramethrin	–	20.3[a]	–	>25[b]
Tetramethrin				
+ PBO (1:10)	42.6	0.476[a]	>215.5	0.116[a]

[a] Weighted mean values.
[b] 0.05 ml doses.

Reprinted from *Journal of Stored Products Research* **9**, 77–92, © 1973, with permission from Elsevier Science.

dominica was variable. The solution to the relatively high cost of synergized bioresmethrin alone or the unreliable control of 6 ppm of fenitrothion was soon discovered. Desmarchelier (1977) proposed that although synergized bioresmethrin provided excellent protection against *R. dominica*, fenitrothion, pirimiphosmethyl or dichlorvos was required to control *T. castaneum*. His studies determined that there was no antagonistic effect between the different groups, therefore a combination would allow reduced use of insecticide leading to lower residues and cost. Combinations of synergized bioresmethrin and fenitrothion were tested (Ardley and Desmarchelier, 1978). A 2 : 2 : 2 and a 2 : 2 : 4 ratio were approved for further development.

In 1980/1981 a combination of 1 ppm of bioresmethrin synergized by 8 ppm of PBO plus 12 ppm of fenitrothion became the standard treatment for long-term storage of export quality grain in New South Wales, Queensland and Victoria, where malathion resistance was most prevalent, maintaining the nil insect tolerance policy (Standing Committee on Agriculture, 1981).

9. LIGHT-STABLE PYRETHROIDS

The instability of bioresmethrin to light, often resulting in an undesirable odour, was not a limiting factor in its use as a grain protectant, but it remained an expensive insecticide (manufacture required isomer resolution). The new light-stable pyrethroids (Elliott *et al.*, 1978) were rapidly evaluated by grain storage researchers because of their lower cost potential.

9.1. Use of Light-Stable Pyrethroids in Australia

Bengston (1979), working for the Queensland Department of Primary Industries, described the potential use of permethrin, fenvalerate, phenothrin, cypermethrin and deltamethrin, concluding that 2 ppm of deltamethrin synergized by 10 ppm of PBO showed the most promise.

This work was carried into large-scale field trials in wheat by Bengston *et al.* (1983b), which included residue analysis. The following treatments were successful:

- deltamethrin + PBO ($2 + 8$ mg kg^{-1})
- fenitrothion + fenvalerate + PBO ($12 + 1 + 8$ mg kg^{-1})
- fenitrothion + phenothrin + PBO ($12 + 2 + 8$ mg kg^{-1})
- pirimiphos-methyl + permethrin + PBO ($4 + 1 + 8$ mg kg^{-1})

The degradation of these insecticides was found to be very slow in the relatively dry grain found in Queensland. Further work on their stability on stored wheat was carried out by Noble *et al.* (1982), who demonstrated half-lives (in weeks), at 25°C, 12% moisture and 35°C, 15% moisture as shown in Table 16.5.

Table 16.5.   Pyrethroid half-lives in wheat stored under different conditions (Noble *et al.*, 1982, with permission)

Pyrethroid	Half-life in weeks	
	25°C, 12% moisture	35°C, 15% moisture
Deltamethrin	114	35
Fenvalerate	210	74
Permethrin	252	44
Phenothrin	72	29

Sorghum is another important crop in Australia which is normally stored for 6 months at 25°C and 13.5% moisture. Bengston *et al.* (1983a) carried out further field trials with grain protectants and sorghum. The combination that provided the best cost-effective ratio for commercial use on sorghum was fenitrothion + bioresmethrin synergized by PBO, but the search for alternatives amongst the light-stable pyrethroids continued.

Bengston *et al.* (1984), using laboratory assays taken from 500 t silos of bulk-treated sorghum, classed the following combinations as generally effective:

- deltamethrin + PBO (2 + 8 mg kg^{-1}).
- fenitrothion + fenvalerate + PBO (12 + 1 + 8 mg kg^{-1}).
- fenitrothion + phenothrin + PBO (12 + 2 + 8 mg kg^{-1}).

All the residues at the completion of the experiment at 26 weeks were below the Codex MRL.

Laboratory studies undertaken by Samson and Parker (1989) in Australia, compared nine protectants on maize against four species of Coleoptera, including resistant strains. Maize is stored at higher moisture levels than wheat, and degradation should be greater, especially with organophosphorus compounds. Trial results were:

- Bioresmethrin at 1 mg kg^{-1} synergized by 1 mg kg^{-1} PBO failed against *R. dominica* at 3 months.
- Unsynergized deltamethrin applied at 0.04 mg kg^{-1} failed to control *Sitophilus* species, provided 9 months protection against *R. dominica*, but was felt to be inadequate for maize storage.
- Permethrin (0.5–1 mg kg^{-1}) plus 1 mg kg^{-1} of PBO also provided 9 months protection against *R. dominica*, as did;
- Fenvalerate (0.25 mg kg^{-1}) plus 1 mg kg^{-1} of PBO.

The addition of PBO as a synergist was thought to be able to provide better protection with deltamethrin but there were concerns over potentiating irritation in certain formulations.

In the following year (Samson *et al.*, 1990), laboratory tests with *S. oryzae* and *S. zeamais* were carried out on maize, using deltamethrin in combination with varying amounts of PBO, rather than a fixed ratio. These proved the superior efficacy provided by the inclusion of PBO.

The data in Table 16.6 demonstrates the relationship between intended storage period, level of PBO, and the required rate of deltamethrin for effective

Table 16.6. Minimum effective rates of deltamethrin (mg kg^{-1}) for different periods of protection against *S. zeamais* (QSZ102) in combination with PBO (after Samson *et al.*, 1990)

PBO (mg kg^{-1})	Period of protection (weeks)						
	0	6	12	18	24	36	48
0	>4	4–8	4–8	4–8	4–8	4–8	4–8
2	0.25–0.5	0.5–1	1–2	1–2	2–4	2–4	2–4
4	0.25–0.5	0.5–1	1–2	1–2	1–2	2–4	1–2
8	0.25–0.5	0.25–0.5	0.5–1	0.5–1	0.5–1	1–2	1–2
16	0.25–0.5	0.125–0.25	0.125–0.25	0.25–0.5	0.25–0.5	0.5–1	0.5–1

control. In this experiment an effective treatment was taken as preventing living F_2 generations in at least two out of three replicates. The ranges shown in the table cover the highest ineffective to the lowest effective application rate (Samson *et al.*, 1990).

As a result of these trials, a minimum field application rate of 1 mg kg^{-1} of deltamethrin plus 8 mg kg^{-1} PBO was recommended for 24 weeks protection, or 1 mg kg^{-1} of deltamethrin plus 16 mg kg^{-1} PBO for up to 48 weeks.

Similar work was carried out with paddy rice (Samson *et al.,* 1989) because degradation was also believed to be considerably more than on wheat. Initial results indicated:

- Approved organophosphorus compounds had to be applied at levels above their MRLs (10 mg kg^{-1}), and were then only effective against *S. oryzae* for 6 weeks.
- The minimum effective rate for unsynergized deltamethrin was 2 mg kg^{-1}. However, synergism by PBO was considered as a viable option. Against *R. dominica* unsynergized deltamethrin was very effective at low rates for 9 months.

Other pyrethroids synergized by 8 mg kg^{-1} of PBO and applied below their respective MRLs (shown in parentheses) were also very effective:

- Permethrin (2 mg kg^{-1}).
- Bioresmethrin (5 mg kg^{-1}).
- D-Phenothrin (5 mg kg^{-1}).
- Fenvalerate (5 mg kg^{-1}).

A summary of the synergistic effect of PBO (8 mg kg^{-1}) on a range of pyrethroids against *R. dominica* on wheat is provided in Table 16.7.

The problem of irritancy with synergized deltamethrin was solved by reformulation, so this mixture became the candidate of choice to replace fenitrothion + synergized bioresmethrin in areas where this was felt likely to fail, or the cost of the mixture was considered too high.

Table 16.7. Synergism of pyrethroid insecticides with PBO (8 mg kg^{-1}) on wheat for *Rhyzopertha dominica* using treated grain bioassays

Insecticide	Synergism factor[a]
Bioresmethrin	4.3
Deltamethrin	3.6
Fenvalerate	1.9
Permethrin	2.1
D-Phenothrin	1.6

[a] Relative potency of pyrethroid + PBO compared with pyrethroid alone.
Reproduced with permission from Bengston *et al.* (1990).

9.2. Non-Australian research

In Brazil, Bitran *et al.* (1983a) concluded that flucythrinate (2–8 ppm), fenvaler-ate (2–8 ppm), and permethrin (2–8 ppm) synergized at 1 : 5 by PBO controlled *S. zeamais* for 9 months on maize, but were not as effective when unsynergized. Further trials with synergized deltamethrin (0.5 : 5.0 and 0.75 : 7.5 ppm) on maize (Bitran *et al.*, 1983b), demonstrated effective control of *S. zeamais* for 9 months, the combination being superior to malathion at 10 ppm.

During the 1980s in East Africa the maize crop was attacked by an imported pest, *P. truncatus* (indigenous to Latin America), which wrought havoc in the village grain stores. Organophosphorus compounds failed to control it, but as it is closely related to *R. dominica*, against which pyrethroids show high efficacy, trials with pyrethroids were carried out. Makundi (1986) used both topical application and laboratory-formulated powders containing deltamethrin or cyfluthrin with and without PBO. The topical application results listed in Table 16.8 show the very high efficacy levels.

The powder results showed that synergized deltamethrin at 0.25 : 1.25 ppm PBO was sufficient to provide control at above 90% mortality level at 28 days, whereas 1 ppm was required unsynergized. For cyfluthrin the figures were 0.25 : 1.25 ppm and 2 ppm alone.

In Europe, work was carried out in Italy (Molinari, 1990) using deltamethrin at 0.25, 0.5 and 1.0 mg kg^{-1} synergized by PBO (1 : 10) applied to wheat and corn in local storage silos and warehouses. Analysis for residues over 12 months showed very little degradation of the deltamethrin, and residues did not exceed the MRL.

10. MORE RECENT COMPOUNDS

10.1. Cyfluthrin

In addition to the specific testing of cyfluthrin above, work has been carried out to assess its potential for the Australian grain industry. Bengston *et al.* (1987) trialled cyfluthrin and cypermethrin, using laboratory bioassays of malathion-resistant insects on bulk samples of wheat from silos over 8–9

Table 16.8. Toxicity of two pyrethroids to *P. truncatus* by topical application

Pyrethroid	LD$_{50}$ (mg per insect)	95% Fiducial limits	Factor of synergism
Deltamethrin	0.00063	0.0005–0.0007	
Deltamethrin + PBO (1:5)	0.00034	0.0003–0.0004	1.85
Cyfluthrin	0.00217	0.0019–0.0025	
Cyfluthrin + PBO (1:5)	0.00116	0.0010–0.0013	1.87

Reproduced with permission from Makundi (1986).

months storage. Both the following combinations were perceived to have potential commercial use:

- Cyfluthrin (2 mg kg^{-1}) plus PBO (10 mg kg^{-1}).
- Cypermethrin (4 mg kg^{-1}) plus PBO (10 mg kg^{-1}), but failed to control one strain of *S. oryzae* after 1.5 months.

Cyfluthrin was tested in the USA by Arthur (1994) on wheat against *S. oryzae* and *R. dominica*, at rates of 0.5, 1.0, 1.5 and 2 ppm, and also with 8.0 ppm PBO or with 6.0 ppm PBO plus 6.0 ppm chlorpyrifos-methyl. With cyfluthrin alone there was survival of *R. dominica* (0–37%) and *S. oryzae* (94.5% at 0.5 ppm and 43.5% at 1.0 ppm) after 10 months; there was no survival with any formulations containing PBO. These results indicated that the addition of PBO to cyfluthrin was essential for long-term effectiveness, and that the further addition chlorpyrifos-methyl provided no extra advantage.

Synergized cyfluthrin is the main candidate protectant to compete with deltamethrin, in terms of high solo activity, and has been reviewed by Pospis-chil and Smith (1994). They concluded that a dosage rate of 2 mg kg^{-1} synergized by 10 mg kg^{-1} of PBO was sufficient to give at least 9 months protection even against resistant strains of insects.

11. PYRETHROID RESISTANCE

As resistance to malathion led to the introduction of the synthetic pyrethroids and combination treatments, examination of potential resistance to pyrethroids is underway, particularly in Australia.

In 1984 resistance to DDT, lindane, organophosphorus compounds, carbaryl, bioresmethrin and pyrethrins was detected in *O. surinamensis* taken from the field (Attia and Frecker, 1984). The resistance factor was low for pyrethrins (2.3) and bioresmethrin (3.8) but very high for fenitrothion (>160). The inclusion of PBO in the assays reduced the latter to 10 or less.

Experiments described previously (Samson *et al.*, 1990) referred to the selection of deltamethrin resistance and the possible role of PBO in overcoming this. Collins (1990) examined field strains of *T. castaneum*, one of which (QTC279) was taken from a population showing tolerance to a field application of cyfluthrin. This was further selected by exposing it to cyfluthrin until the resistance no longer increased (10 generations). This strain was also resistant to carbaryl and to some extent to organophosphorus compounds. PBO suppressed the resistance.

Further field surveys (Collins *et al.*, 1993) covering the most common storage pests of Queensland found high levels of resistance in *O. surinamensis* to fenitrothion and chlorpyrifos-methyl in both farm and central storage. In three central storages there was a high level of resistance in *R. dominica* to synergized bioresmethrin.

12. SYNERGISM OF OTHER INSECTICIDE GROUPS

12.1. Carbamates

Synergism of carbamate insecticides by PBO has been described for public health pests (Moorefield, 1958; Metcalf *et al.*, 1966), and for a range of insects using carbaryl (1 : 5 ratio), including *Tenebrio molitor* L., synergistic ratio 3.6 (Brattsten and Metcalf, 1970), and *Dermestes ater* De Geer, synergistic ratio 31.8 (Brattsten and Metcalf, 1973).

12.2. Organophosphorus Compounds

Rowlands (1966) demonstrated that wheat grains soaked in PBO prevented the oxidation of malathion and bromophos to their thiolate and phosphate analogues, respectively, when treated at 38°C for 6 hours. Some oxidation occurred at 16–24 hours exposure. The use of PBO with these compounds could increase their persistence (Rowlands, 1967).

   Larvae of multiple organophosphorus-resistant strains of *P. interpunctella* were pretreated with PBO prior to treatment with malathion, dichlorvos, pirimiphos-methyl, diazinon or fenitrothion (Attia *et al.*, 1980). All were equally synergized except malathion suggesting that a mixed function oxidase mechanism was present. Against susceptible strains there was mild antagonism, except for fenitrothion.

   Some authors (Duguet *et al.*, 1990) have suggested that PBO, in combination with deltamethrin plus an organophosphate, synergizes the former but antagonizes the latter. Daglish (1994) found that at 8 mg kg^{-1} it did not affect the efficacy of chlorpyrifos-methyl against *S. oryzae* on paddy rice, and it certainly synergized deltamethrin.

   More research work is desirable on the role of PBO and organophosphorus compounds, both with and without the addition of pyrethroids.

12.3. Juvenile Hormone Analogues and Growth Regulators

There has been some work suggesting that some synergists, including PBO, may have morphogenic effects similar to juvenile hormone analogues (Bowers, 1968). This has not been substantiated (Redfern *et al.*, 1969) and it is most likely that the effect is due to mixed function oxidase inhibition interfering with the terminal steps in juvenile hormone synthesis (Staal, 1986).

   The activity of methoprene against *T. molitor* has been shown to be increased by the addition of PBO (Solomon *et al.*, 1973). Resistant *T. castaneum* adults have been found to metabolize PBO and juvenile hormone mimics more quickly than susceptible strains by attacking the methylene group (Rowlands and Dyte, 1979). The larvae were more resistant to both the hormone mimic and high levels of PBO.

13. DISCUSSION AND FUTURE TRENDS

In 1994 Desmarchelier examined trends and developments for grain protectants which stated that:

- Estimates of daily intake of pesticides based on total dietary studies are soundly based.
- Although there are a number of chemicals that would appear to be useful grain protectants, both synthetic and natural materials, most will not be registered because of the very high cost of obtaining the data required.
- The benefits and limitations of grain protectants on tropical cereal crops and grain legumes have clearly been established.
- Insect growth regulators require work leading to registration, if they are to be used in practice.

The overall picture is one of increasing costs of registration, emphasis on residue reduction, and control of resistant pest strains. The use of the synergistic action of PBO with synthetic pyrethroids, and with other classes of compounds where appropriate, meets those requirements with known toxicity and efficacy profiles built up over years of use. Resistance requires monitoring, as we have seen from the effects of widespread malathion usage. Combinations of treatments, including aeration and cooling as well as synergism by PBO, may be required to extend the life of deltamethrin and other synthetic pyrethroids at low dosage rates in warmer climates.

The immediate future depends on the success of improvements being made to the formulation characteristics of synergized deltamethrin products, which are certain to become a mainstay treatment because of their cost-efficacy benefits, compared with current multi-compound mixtures. Cyfluthrin, also synergized by PBO, offers a potential alternative.

The major turning point for the long-term future role of PBO in grain protection arrived with the 1996 revision by JMPR (Joint Meeting of the FAO Panel of Experts on Pesticide Residues in Food and the Environment and a WHO Expert Group on Pesticide Residues) of the ADI from 0.03 to 0.2 mg $kg^{-1} day^{-1}$. As grain prices spiral and demand increases, the requirement for safe and efficient grain protectants will undoubtedly continue. Synergism by PBO meets these requirements by maximizing the efficiency of the safest protectants currently available, and those under consideration for the foreseeable future.

REFERENCES

Adams, J.M. (1977). A review of the literature concerning losses in stored cereals and pulses, published since 1964. *Tropical Sci.* **19**, 1–28.

Anon. (1995). World grain and feed review. *World Grain* **13**, 26–28.

Ardley, J.H. and Desmarchelier, J.M. (1975). Investigation into the use of resmethrin and bioresmethrin as potential grain protectants. In: *Proceedings of the 1st International*

Working Conference on Stored-Product Entomology, Savannah, Georgia, 7–11 October 1974, pp. 511–516.

Ardley, J.H. and Desmarchelier, J.M. (1978). Field trials of bioresmethrin and fenitrothion combinations as potential grain protectants. *J. Stored Prod. Res.* **14**, 65–67.

Ardley, J.H. and Sticka, R. (1977). The effectiveness of fenitrothion and malathion as grain protectants under bulk storage conditions in New South Wales, Australia. *J. Stored Prod. Res.* **13**, 159–168.

Arthur, F.H. (1994). Cyfluthrin applied with and without piperonyl butoxide and piperonyl butoxide plus chlorpyrifos-methyl for protection of stored wheat. *J. Econ. Entomol.* **87**, 1707–1713.

Attia, F.I. and Frecker, T. (1984). Cross-resistance spectrum and synergism studies in organophosphorus-resistant strains of *Oryzaephilus surinamensis* (L.) (Coleoptera: Cucujidae) in Australia. *J. Econ. Entomol.* **77**, 1367–1370.

Attia, F.I., Shanahan, G.J. and Shipp, E. (1980). Synergism studies with organophosphorus resistant strains of the Indian meal moth. *J. Econ. Entomol.* **73**, 184–185.

Baker, G.J. (1963). The 'dual synergist system' of piperonyl butoxide and MGK 264. *Pyreth. Post* **7**, 16–18.

Banks, H.J. and Desmarchelier, J.M. (1978). New chemical approaches to pest control in stored grain. *Chem. Aust.* **45**, 276–281.

Bengston, M. (1979). Potential of pyrethroids as grain protectants. In: *Australian Contributions to the Symposium on the Protection of Grain Against Insect Damage During Storage*, Moscow, 1978 (Evans, D.E., ed.), pp. 88–98. CSIRO Division of Entomology, Canberra.

Bengston, M., Cooper, L.M. and Grant-Taylor, F.J. (1975). A comparison of bioresmethrin, chlorpyrifos-methyl and pirimiphos-methyl as grain protectants against malathion resistant insects in wheat. *Queensland. J. Agric. Anim.* Sci. **32**, 51–78.

Bengston, M., Cooper, L.M., Davies, R.A.H., Desmarchelier, J.M., Hart, R.J. and Phillips, M.P. (1983a). Grain protectants for the control of malathion-resistant insects in stored sorghum. *Pestic. Sci.* **14**, 385–398.

Bengston, M., Davies, R.A.H., Desmarchelier, J.M., Henning, R., Murray, W., Simpson, B.W., Snelson, J.T., Sticka, R. and Wallbank, B.E. (1983b). Organophosphorothioates and synergised synthetic pyrethroids as grain protectants on bulk wheat. *Pestic. Sci.* **14**, 373–384.

Bengston, M., Davies, R.A.H., Desmarchelier, J.M., Phillips, M.P. and Simpson, B.W. (1984). Organophosphorus and synergised synthetic pyrethroids as grain protectants for stored sorghum. *Pestic. Sci.* **15**, 500–508.

Bengston, M., Desmarchelier, J.M., Hayward, B., Henning, R., Moulden, J.H., Noble, R.M., Smith, G., Snelson, J.T., Sticka, R., Thomas, D., Wallbank, B.E. and Webley, D.J. (1987). Synergised cyfluthrin and cypermethrin as grain protectants on bulk wheat. *Pestic. Sci.* **21**, 23–37.

Bengston, M., Koch, K. and Strange, A. (1990). Chemical control methods. In: *Proceedings of the 5th International Working Conference on Stored Product Protection*, Bordeaux, France, 1990, vol. 1, pp. 471–481.

Bitran, E.A. and Campos, T.B. (1975). Specific action of synergised pyrethroids in the control of *Sitophilus zeamais* Motschulsky, and its use possibilities on stored grain protection. *Biologico* **41**, 287–293. (Port.)

Bitran, E.A., Campos, T.B., Oliviera, D.A. and Chiba, Soyako (1983a). Evaluation of the residual action of some pyrethroid and organophosphorus insecticides in the control of *Sitophilus zeamais* Motschulsky, 1855 infestations on stored maize. *Biologico* **49**, 265–273. (Port.).

Bitran, E.A., Campos, T.B., Oliviera, D.A. and Chiba, Soyako (1983b). Evaluation of residual effectiveness of the pyrethroid deltamethrin in stored grains. *Biologico* **49**, 237–246. (Port.).

Bowers, W.S. (1968). Juvenile hormone: activity of natural and synthetic synergists. *Science* **161**, 895–897.

Boxall, R.A. (1986). A Critical Review of the Methodology for Assessing Farm Level Grain Losses after Harvest. National Resources Institute (TDRI), Chatham, UK. Report G191, p. 139.

Brattsten, L.B. and Metcalf, R.L. (1970). The synergistic ratio of carbaryl and piperonyl butoxide as an indicator of the distribution of multifunctional oxidases in the Insecta. *J. Econ. Entomol.* **63**, 101–104.

Brattsten, L.B. and Metcalf, R.L. (1973). Synergism of carbaryl toxicity in natural insect populations. *J. Econ. Entomol.* **66**, 1347–1348.

Carter, S.W., Chadwick, P.R. and Wickham, J.C. (1975). Comparative observations on the activity of pyrethroids against some susceptible and resistant stored products beetles. *J. Stored Prod. Res.* **11**, 135–142.

Chadwick, P.R. (1963a). A comparison of MGK 264 and piperonyl butoxide as pyrethrum synergists. *Pyreth. Post* **7**, 11–15, 48.

Chadwick, P.R. (1963b). The use of pyrethrum synergists: a discussion for formulators. *Pyreth. Post.* **7**, 25–32.

Champ, B.R. and Ryland, G.J. (1986). The role of residual pesticides in stored-product insect control. In: *Proceedings of the 4th International Working Conference on Stored Product Protection*, Tel Aviv, Israel, 1986, pp. 281–287.

Collins, P.J. (1990). A new resistance to pyrethroids in *Tribolium castaneum* (Herbst). *Pestic. Sci.* **28**, 101–115.

Collins, P.J., Lambkin, T.M., Bridgeman, B.W. and Pulvirenti, C. (1993). Resistance to grain-protectant insecticides in coleopterous pests of stored cereals in Queensland, Australia. *J. Econ. Entomol.* **86**, 239–245.

Daglish, G.J. (1994). Efficacy of several mixtures of grain protectants on paddy and maize. In: *Proceedings of the 6th International Working Conference on Stored Product Protection*, Canberra, Australia, 1994, vol. 2, pp. 762–764.

Desmarchelier, J.M. (1977). Selective treatments, including combinations of pyrethroid and organophosphorus insecticides, for control of stored product Coleoptera at two temperatures. *J. Stored Prod. Res.* **13**, 129–137.

Desmarchelier, J.M. (1994). Grain protectants: trends and developments. In: *Proceedings of the 6th International Working Conference on Stored Product Protection*, Canberra, Australia, 1994, vol. 2, pp. 722–728.

Desmarchelier, J.M., Bengston, M. and Sticka, R. (1979). Stability and efficacy of pyrethrins on grain in storage. *Pyreth. Post* **15**, 3–8.

Dove, W.E. (1947a). Piperonyl butoxide, a new and safe insecticide for the household and field. *Pests* **15**, 30.

Dove, W.E. (1947b). Piperonyl butoxide, a new and safe insecticide for the household and field. *Am. J. Trop. Med.* **27**, 339–345.

Dove, W.E. (1949). Progress of non-hazardous methods of insect control. (Paper read at UNESCO International Technical Conference on Protection of Nature, USA, Aug/Sept 1949, UNESCO.) *Pyreth. Post* (1951), **2**, 9–11.

Dove, W.E. (1951). Piperonyl butoxide and pyrethrins for the protection of grain and similar products from insect damage. (Paper read at 9th International Congress on Entomology, Amsterdam, 1951.) *Rev. Appl. Entomol.* (1953) **41**, 234.

Dove, W.E. (1952a). An evaluation of residual insect sprays for mill insect control. *Milling Prod.* **17**, 20.

Dove, W.E. (1952b). Piperonyl butoxide and pyrethrins for the protection of grain and similar products from insect damage. In: *Transactions of the 9th International Congress on Entomology*, Amsterdam, 1951, vol. 1, pp. 875–879.

Dove, W.E. (1953). Piperonyl butoxide. In: *International Technical Proceedings 39th Midyear Meetings Chemical Speciality Manufacturers Association*, May 1953, pp. 104–105.

Dove, W.E. (1958). Protection of stored grains with pyrethrins and piperonyl butoxide. In: *Proceedings of the 10th National Congress on Entomology*, Montreal, 1956, vol. 4, pp. 65–71.

Dove, W.E. and MacAllister, L.C. (1947). Piperonyl butoxide for control of stored grain insects. Paper read at 59th Annual Meeting of the American Association of Economic Entomologists, 1947.

Dove, W.E. and Schroeder, H.O. (1955). Protection of stored grain with sprays of pyrethrins–piperonyl butoxide emulsion. *J. Agric. Food Chem.* **3**, 932–936.

Duguet, J.S., Fleurat-Lessard, F. and Peruzzi, D. (1990). The advantages of mixing deltamethrin and organo-phosphorous insecticides for the protection of stored cereals. A review of recent trials. In: *Proceedings of the 5th International Working Conference on Stored Product Protection*, Bordeaux, France, 1990, vol. 1, pp. 517–525.

Elliott, M., Janes, N.F. and Potter, C. (1978). Pyrethroids in insect control. *Ann. Rev. Entomol.* **23**, 443–469.

Floyd, E.H. (1961). Effectiveness of malathion dust as a protectant for farm-stored corn in Louisiana. *J. Econ. Entomol.* **54**, 900–904.

Giles, P.H. and Ashman, F. (1971). A study of pre-harvest infestation of maize by *Sitophilus zeamais* Motsch. (Coleoptera, Curculionidae) in the Kenya highlands. *J. Stored Prod. Res.* **7**, 69–83.

Gillenwater, H.B. and Burden, G.S. (1973). Pyrethrum for control of household and stored-product insects. In: *Pyrethrum, The Natural Insecticide* (Casida, J.E., ed.), pp. 243–257. Academic Press, New York.

Glynne Jones, G.D. and Chadwick, P.R. (1960). A comparison of four pyrethrum synergists. *Pyreth. Post* **5**, 22–30.

Glynne Jones, G.D. and Green, E.H. (1959). A comparison of the toxicities of pyrethrins and synergised pyrethrins to *Calandra oryzae* L. and *Calandra granaria* L. *Pyreth. Post* **5**, 3–7.

Goodwin-Bailey, K.F. (1959). Field trials of wheat and shelled corn protection. *Pyreth. Post* **9**, 18–19.

Goodwin-Bailey, K.F. (1960). Synergism of piperonyl butoxide with pyrethrins against several species of insect. *Chem. & Ind.* 18 June, 700–702.

Goodwin-Bailey, K.F. and Holborn, J.M. (1952). Laboratory and field experiments with pyrethrins/piperonyl butoxide powders for the protection of grain. *Pyreth. Post* **2**, 7–17.

Greening, H.G. (1980). Chemical control of insects infesting farm-stored grain and harvest machinery. *Pyreth. Post* **15**, 48–53.

Greening, H.G., Wallbank, B.E. and Attia, F.I. (1974). Resistance to malathion and dichlorvos in stored-product insects in New South Wales. In: *Proceedings of the 1st International Working Conference on Stored-Product Entomology*. Savannah, GA, 1974, pp. 608–617.

Hewlett, P.S. (1951). PBO as a constituent of heavy-oil sprays for the control of stored product insects I: PBO as a synergist for pyrethrum and its effect on the persistence of pyrethroid films. *Bull. Entomol. Res.* **42**, 293–310.

Hewlett, P.S. (1969). The toxicity to *Tribolium castaneum* (Herbst) (Coleoptera, Tenebrionidae) of mixtures of pyrethrins and piperonyl butoxide: fitting a mathematical model. *J. Stored Prod. Res.* **5**, 1–9

Howe, R W. (1965). A summary of estimates of optimal and minimal conditions for population increase of some stored products insects. *J. Stored Prod. Res.* **1**, 177–184.

Joubert, P.C. (1965). The toxicity of contact insecticides to seed-infesting insects. *Pyreth. Post* **8**, 6–15.

Joubert, P.C. and DeBeer, P.R. (1968). The Toxicity of Contact Insecticides to Seed-infesting Insects. Tests with Pyrethrum and Malathion on Infested Maize. Technical Communication no. 73, Department of Agriculture, Pretoria, South Africa. South African Government Printer, Pretoria.

Joubert, P.C. and du Toit, D. M. (1968). The Toxicity of Contact Insecticides to Seed-infesting Insects. Tests with Various Pyrethrum Formulations. Technical Communication no. 83, Department of Agriculture, Pretoria, South Africa. South African Government Printer, Pretoria.

Kantack, B.H. and Laudani, H. (1957). Comparative laboratory tests with emulsion and wettable-powder residues against the Indian-meal moth. *J. Econ. Entomol.* **50**, 513–514.

La Hue, D.W. (1965). Evaluation of Malathion, Synergised Pyrethrum, and Diatomaceous Earth as Wheat Protectants in Small Bins. USDA, ARS Marketing Research Report no. 726. 13 pp.

La Hue, D.W. (1966). Evaluation of Malathion, Synergised Pyrethrum, and Diatomaceous Earth on Shelled Corn as Protectants Against Insects in Small Bins. USDA, ARS Marketing Research Report no. 768. 10 pp.

La Hue, D.W. (1967). Evaluation of Malathion, Synergised Pyrethrum, and Diatomaceous Earth as Protectants Against Insects in Sorghum Grain in Small Bins. USDA, ARS Marketing Research Report no. 781. 11 pp.

La Hue, D.W. (1969). Evaluation of Several Formulations of Malathion as a Protectant of Grain Sorghum Against Insects in Small Bins. USDA, ARS Marketing Research Report no. 828.

La Hue, D.W. (1970). Evaluation of Malathion, Diazinon, a Silica Aerogel, and a Diatomaceous Earth as Protectants on Wheat Against Lesser Grain Borer Attack in Small Bins. USDA, ARS Marketing Research Report no. 860.

Le Pelley, R. and Kockum, S. (1954). Insecticides for the protection of grain in storage. *Bull. Entomol. Res.* **45**, 295–311.

Lloyd, C.J. (1973). The toxicity of pyrethrins and five synthetic pyrethroids to *Tribolium castaneum* (Herbst), and susceptible and pyrethrin-resistant *Sitophilus granarius* (L.). *J. Stored Prod. Res.* **9**, 77–92.

Lloyd, C.J. and Hewlett, P.S. (1958). Relative susceptibility to pyrethrum in oil of stored-products insects. *Bull. Entomol. Res.* **49**, 177–185.

McFarlane, J.A. (1969). The effects of synergised pyrethrins and lindane on pre-emergence mortality of *Acanthoscelides obtectus* (Say.) (Coleoptera: Bruchidae). *J. Stored Pros. Res.* **5**, 177–180.

Makundi, R.H. (1986). The toxicity of deltamethrin and cyfluthrin to the Larger Grain Borer, *Prostephanus truncatus* (Horn) (Coleoptera: Bostrichidae). *Int. Pest Control* May/June, 79–81.

Metcalf, R.L., Fukuto, T.R., Wilkinson, C., Fahmy, M.H., Abd El-Aziz, S. and Metcalf, E.R. (1966). Mode of action of carbamate synergists. *J. Agric. Food Chem.* **14**, 555–562.

Molinari, G.P. (1990). Persistence of deltamethrin residues in stored cereals. In: *Proceedings of the 5th International Working Conference on Stored Product Protection*, Bordeaux, France, 1990, vol. 1, pp. 571–577.

Moorefield, H.H. (1958). Synergism of the carbamate insecticides. *Contrib. Boyce Thompson Inst.* **19**, 501–507.

Noble, R.M., Hamilton, D.J. and Osborne, W.J. (1982). Stability of pyrethroids on wheat in storage. *Pestic. Sci.* **13**, 246–252.

Owen, R.W. and Waloff, N. (1946). A note on the efficiency of a pyrethrum spray in controlling *Ephestia elutella* Hb. moths in a granary. *Ann. Appl. Biol.* **33**, 387–389.

Page, A.B.P. and Blackith, R.E. (1950). Stabilisation of pyrethrins. *Pyreth. Post* **2**, 18–20.

Phillips, G.L. (1959). Control of insects with pyrethrum sprays in wheat stored in ships' holds. *J. Econ. Entomol.* **52**, 557–559.

Phipers, R.F. and Wood, M.C. (1957). An investigation into the reported stabilisation of pyrethrins by PBO. *Pyreth. Post* **4**, 11–12.

Pospischil, R. and Smith, G. (1994). Cyfluthrin plus piperonyl butoxide – a promising new stored product protectant. In: *Proceedings of the 6th International Working Conference on Stored Product Protection*, Canberra, Australia, 1994, vol. 2, pp. 830–832.

Redfern, R.E., McGovern, T.P. and Beroza, M. (1969). Juvenile hormone activity of Sesamex and related compounds in tests on the yellow mealworm. *J. Econ. Entomol.* **63**, 540–545.

Rowlands, D.G. (1966). The activation and detoxification of three organic phosphorothionate insecticides applied to stored wheat grains. *J. Stored Prod. Res.* **2**, 105–116.

Rowlands, D.G. (1967). The metabolism of contact insecticides in stored grains. *Residue Rev.* **17**, 105.

Rowlands, D.G. and Dyte, C.E. (1979). The metabolism of two methylenedioxyphenyl compounds in susceptible and resistant strains of *Tribolium castaneum*. In: *Proceedings of the 10th British Insecticide and Fungicide Conference*, vol. 1, pp. 257–264.

Samson, P.R. and Parker, R.J. (1989). Laboratory studies on protectants for control of Coleoptera in maize. *J. Stored Prod. Res.* **25**, 49–55.

Samson, P.R., Parker, R.J. and Jones, A.L. (1989). Laboratory studies on protectants for control of *Sitophilus oryzae* (Coleoptera: Curculionidae) and *Rhyzopertha dominica* (Coleoptera: Bostrichidae) in paddy rice. *J. Stored Prod. Res.* **25**, 39–48.

Samson, P.R., Parker, R.J. and Hall, E.A. (1990). Synergised deltamethrin as a protectant against *Sitophilus zeamais* Motsch. and *S. oryzae* (L.) (Coleoptera: Curculionidae) on stored maize. *J. Stored Prod. Res.* **26**, 155–161.

Saxena, S.C. (1967). Synergised and unsynergised pyrethrum. III. Comparative evaluation of piperonyl butoxide and MGK 264 as pyrethrum synergists. *Appl. Entomol. Zool.* **2**, 158–162.

Solomon, K.R., Bowlus, S.B., Metcalf, R.L. and Katzenellenbogen, J.A. (1973). The effect of PBO and triorthocresyl phosphate on the activity of juvenile hormone mimics and their sulphur isosteres in *Tenebrio molitor* L. and *Oncopeltus fasciatus* (Dallas). *Life Sci.* **13**, 733–742.

Standing Committee on Agriculture (1981). Report of the SCA working party on infestation in grain. In: *Australian Grains Industry Conference*, Canberra, 26–30 October, 1981.

Staal, G.B. (1986). Anti juvenile hormone agents. *Ann. Rev. Entomol.* **31**, 391–429.

Udeaan, A.S. and Kalra, R.L. (1983). Joint action of piperonyl butoxide and lindane in susceptible and resistant strains of *Tribolium castaneum* Herbst. *J. Entomol. Res.* **7**, 21–24.

Walkden, H.H. and Nelson, H.D. (1959). Evaluation of Synergised Pyrethrum for the Protection of Stored Wheat and Shelled Corn from Insect Attack. USDA Market Research Report no. 322. 48 pp.

Watt, M. (1962). Grain protection with malathion. *Agric. Gaz. N.S.W.* **73**, 529.

Watters, F.L. (1961). Effectiveness of lindane, malathion, methoxychlor and pyrethrins-piperonyl butoxide against the hairy spider beetle, *Ptinus villiger. J. Econ. Entomol.* **54**, 397.

Weaving, A.J.S. (1970). Susceptibility of some bruchid beetles of stored pulses to powders containing pyrethrins and piperonyl butoxide. *J. Stored Prod. Res.* **6**, 71–77.

Webley, D.J. (1985). Use of pesticides in systems for central storage of grain. Proceedings of the International Seminar 'Pesticides and Humid Tropical Grain Storage Systems', Manila, Philippines, 1985. *ACIAR Proc.* **14**, 327–334.

Webley, D.J. (1994). Grain protectants and pesticide residues. In: *Proceedings of the 6th International Working Conference on Stored Product Protection*, Canberra, Australia, 1994, vol. 2, pp. 857–862.

Wilbur, D.A. (1952). Effectiveness of dusts containing piperonyl butoxide and pyrethrins in protecting wheat against insects. *J. Econ. Entomol.* **45**, 913–920.

17

The Use of Piperonyl Butoxide in Household Formulations

LINDSAY R. SHOWYIN

1. INTRODUCTION

Piperonyl butoxide (PBO) is a very important material for the formulator of household insecticides. It has stood the test of time and is widely used in a variety of formulations in most parts of the world.

2. EXPECTATION BY CONSUMERS

Insecticides, when used by the consumer in the home, are usually expected to act very quickly to be regarded as efficacious. Efficacy is therefore judged more on how quickly an annoying fly, mosquito or cockroach is knocked down rather than how quickly it dies, although there is always the expectation that death automatically follows knockdown.

Our experience, particularly with aerosol household insecticides, also shows most users to be 'hunters' in that chasing and spraying an insect is quite common. The expectation is to shoot down or stop the offending insect and the more quickly this occurs the 'better' the product is judged. Even today, in the era of more awareness of the environment, this mode of action and expectation still seems to persist.

The task of the formulator has therefore been to attempt to satisfy these needs. This chapter aims to explore the background and history of household insecticides using PBO. It draws predominantly on the Australian experience and background, partly because of the household insect problems that exist in Australia, and partly because Australian household insecticides use some of the highest active levels in the world. This is not to say that similar demands do not exist elsewhere in the world. The Aussie salute, the bushfly, flies and Mortein®, are very much part of the Australian culture and heritage and per capita Australians use aerosol insecticides at the highest rate, or as high as anywhere in the world.

PIPERONYL BUTOXIDE
ISBN 0-12-286975-3

3. BACKGROUND

PBO was first made commercially available by US Industrial Chemicals (USI) in 1948. H. Wachs was the chemist most responsible for its development (Glynne Jones, 1989). PBO is generally associated with pyrethrum, although pyrethrum products were available long before PBO. The reason for this association is that they are generally used together in a formulation.

4. WHY PBO?

Pyrethrins, extracted from the pyrethrum flower (*Chrysanthemum cinerariae-folium*), are good knockdown and kill agents against a variety of insects. Supply has been traditionally rather erratic because of climatic conditions and competition crops. Pyrethrum extract is also relatively expensive, and so searches for alternatives or diluents have been necessary.

DDT was seen as a cheap alternative toxicant to pyrethrins but the advent of synergists saw a major change in outlook. A synergist is a chemical which increases the biological activity of another chemical. The latter is usually active in its own right, but the synergist itself may have no or little insecticidal activity on its own. Most cases of synergism result from the ability of the synergist to interfere with the metabolic detoxification of the insecticide (Wilkinson, 1973). One of the first synergists for pyrethrins was sesame oil. The ability of sesame oil to synergize pyrethrins is due to the fact it contains two methylene-dioxyphenyl (MDP) compounds, sesamin and sesamolin. PBO also contains the MDP grouping, and when used with pyrethrins in a formulation, similar or better performance is obtained at a significantly reduced cost than with pyrethrins alone. In particular, faster knockdown than with straight pyrethrum is evident. A side benefit is the good solvency properties of PBO, particularly for many of the other plant extractives in pyrethrum extract, thus reducing the possibility of aerosol valve blockage. PBO is also claimed to have a protective action on pyrethrins exposed to sunlight.

Synergists do not have to contain the MDP grouping. Other compounds showing similar activity include aryl propynyl ethers, propynyl oxime ethers, propynyl phosphonate esters and benzyl thiocyanates.

In the late 1950s four of the synergists in popular use contained the MDP group:

- PBO
- propyl isome – dipropyl 5,6,7,8-tetrahydro-7-methylnaphtho[2,3-*d*]-1,3-dioxole-5,6 dicarboxylate
- sulfoxide – 2-(1,3-benzodioxol-5-yl) = ethyl octyl sulfoxide
- 'Tropital' – 5[bis[2-(2-butoxyethoxy) = ethoxy]methyl]-1,3 benzodioxole (Wilkinson, 1973).

Of these only PBO is in use today. This is because it is an excellent cost-efficient material. In addition it has a low toxicity profile, which in a world of ever-increasing regulation makes an attractive material in terms of the provision of the required toxicological data.

MGK® 264 (*N*-octyl-bicycloheptene-dicarboximide) is another widely used synergist. It does not have the same structure and activity as the above compounds and when used alone with pyrethrum it is not as effective against house flies. A three-way combination of pyrethrum, MGK® 264 and PBO has therefore often been used (Baker, 1963).

Pyrethrins are not the only materials that can be synergized by PBO – synthetic pyrethroids, carbamates and others can also be synergized.

The ratio of PBO to pyrethrins (100%) or other pyrethroids varies considerably. *The Pesticide Manual* indicates a range of 5 : 1 to 20 : 1 PBO : pyrethrins (Worthing and Hance, 1991). The following examples of early commercial household products show that the range is even wider.

The actual ratios used by formulators are usually determined with an eye on the economic ratio of the costs of the actives and PBO as well as the biological performance and the target insects.

5. EARLY HOUSEHOLD PRODUCTS

Exact dates of first usage of PBO in household formulations are unclear. However the following products are known from the 1950s and 1960s. Some examples are shown in Table 17.1.

6. AEROSOLS

Lyle Goodhue and Bill Sullivan of USA were the inventors of the first aerosol insecticide in USA in 1941. This became the 'bug bomb' for use in the Pacific Islands by American troops during World War II. The first commercial aerosol (an insecticide) followed after the war.

Goodhue and Sullivan's initial patent (US 2.321,023) granted in 1943 on aerosol insecticides shows a mix of pyrethrins and sesame oil with CFC (chlorofluorocarbon) propellant (Sullivan, 1974). The CFC, being liquefied under pressure, vaporized immediately when delivered to the atmosphere to ensure extreme break-up of the insecticide/solvent mix. This gave fine particles which lingered much longer in the air than those dispensed by a flit gun, as well as dispensing considerably less product.

In order to illustrate the variety of active ingredients used with PBO over a lengthy period, the Australian market leader, Reckitt & Colman's Mortein®, is used as an example. Mortein® was first established around 1880 as a powdered insecticide. It was extended to a liquid in 1928 and to an aerosol in 1953 (Edwards, 1982). The Mortein® red can (Pressure Pak®/Fast Knockdown) has always been the flagship insecticide for this brand. It has contained PBO in a variety of combinations for over 30 years. Examples are shown in Table 17.2.

The personal insect repellent Aerogard® was launched in Australia in 1964 and

Table 17.1.   Early household products containing PBO

Name	Formulation	Ratio
Liquids		
Taylor's Anti Fly Non-Poisonous Insect Spray	500 mg pyrethrin per 100 cm^3 300 mg PBO per 100 cm^3	0.6 : 1 PBO : pyr
Mortein® Dairy Spray	0.061% w/w pyrethrins 0.532%w/wPBO 0.306% w/w MGK® 264	8.7 : 1 PBO : pyr 5.0 : 1 MGK®264 : pyr
Pea Beu® Liquid	1.06 g kg^{-1} pyrethrins 3.2 g kg^{-1} PBO	3 : 1 PBO : pyr
Powders		
Mortein® Insect Powder	45 g kg^{-1} pyrethrins 15 g kg^{-1} PBO 40 g kg^{-1} terpene alcohol thiocyanyl acetate (Thanite) min. 82% isobornyl thiocyanoacetate and max. 18% related compounds	1 : 3 PBO : pyr
Pea Beu® Ant & Roach Dusting Powder	1.0 g kg^{-1} pyrethrins 3.0 g kg^{-1} PBO 500 g kg^{-1} boric acid	1 : 3 pyr : PBO

pyr, pyrethrins.

Table 17.2.   PBO in a variety of combinations over the years in Mortein®, Pressure Pak®/Fast Knockdown

Year	Product
1960s	0.2% w/w pyrethrins, 1.5% w/w PBO, 1.5% methoxychlor
Late-1960s	0.2% w/w pyrethrins, 1.5% w/w PBO, 0.75% MGK® 264
1974	0.4% w/w pyrethrins, 0.8% w/w PBO, 0.025% bioresmethrin
1978	Pyrethrum partially substituted with tetramethrin
1980s	2.70 g kg^{-1} allethrin 20:80, 2.41 g kg^{-1} tetramethrin, 10.83 g kg^{-1} PBO
1990s	2.70 g kg^{-1} allethrin 20:80, 2.41 g kg^{-1} tetramethrin, 4.09 g kg^{-1} PBO, 6.59 g kg^{-1} MGK® 264

contained PBO in combination with pyrethrins, MGK® 326 (dipropyl isocin-chomeronate), MGK® 264 and EHD (ethyl hexane diol). The product contained PBO for around 25 years and demonstrated safe usage of the material in direct topical applications to human skin.

Not surprisingly, water-based aerosol insecticides were developed. Probably the motivating factor was cost reduction, but it must be acknowledged that solvent-based pyrethrum/PBO formulae were sometimes regarded as irritant to the nose and throat. Debate ensued on the chief offender. However, over the years, development in refining techniques has led to lower odour and less irritant pyrethrum whilst Endura has developed their Ultra PBO which has less odour and a higher active level.

Water-based aerosols were developed by MGK® and first commercialized by S.C. Johnson with Raid Home & Garden® in USA in 1963 (Glynne Jones, 1989). In Australia, use of PBO in water-based products was made by Richardson Vick in Pea Beu® in 1976 (0.45% w/w pyrethrins/1.20% w/w PBO). In addition there was a Pea Beu® Pine Fresh Low Irritant (0.27% w/w tetramethrin, 0.095% w/w d-phenothrin, 1.08% w/w PBO) and S.C. Johnson had Protector® Fast Knock-down (0.44% w/w tetramethrin, 0.4% w/w PBO, 0.045% w/w d-phenothrin) in a water base.

Again, it can be seen that PBO can be utilized with a wide selection of active ingredients.

7. PBO AND THE OFFICIAL TEST AEROSOL

The first OTA (Official Test Aerosol) was adopted by the Chemical Specialities Manufacturers Association (CSMA) in the USA in 1953. This contained pyrethrins and DDT, but replacement became necessary because of the development of house fly resistance to many insecticides. A product containing 0.2% pyrethrins, 1.6% PBO was proposed as the Tentative Official Test Aerosol in 1970. It was formally accepted as the Official Test Aerosol in 1973 and was designated OTA-11(6) (Anon., 1979).

The OTA has remained in existence and variations exist in other parts of the world. Results using the OTA as a marker are quite often required for registration purposes.

8. SUMMARY

PBO is widely used in a variety of formulations throughout the world. From a consumer standpoint it has helped to provide an economical and efficient product. From a formulator viewpoint it has provided flexibility and an excellent cost-performance profile, as well as good solvency properties. Its continued predominance is largely unchallenged, as it has no equal nor any direct challenger in sight.

REFERENCES

Anon. (1979). Testing standards for insecticides, aerosols and pressurized space sprays, 1976 revision. *Chemical Times & Trends,* 50–53.
Baker, G.J. (1963). The dual synergist system of piperonyl butoxide and MGK 264. *Pyreth. Post* **7**, 16–18.
Edwards, J. (1982). Out of the Blue. *The Dominion Press* **75**.
Glynne Jones, G.D. (1989). Important Dates & Famous People Associated with Pyrethrum, the Natural Insecticide with a History Extending over 150 years. 7–9.
Sullivan, W.N. (1974). History of the insecticidal aerosol. *Maintenance Supplies,* 68–69.
Wilkinson, C.F. (1973). Insecticide synergism. *Chemtech,* 493–495.
Worthing, C.R. and Hance, R.J. (1991). *The Pesticide Manual,* 9th edn, p. 686. The British Crop Protection Council, 686–687, Surrey.

18

The Use of Piperonyl Butoxide in Formulations for the Control of Pests of Humans, Domestic Pets and Food Animals

PAUL KEANE

1. INTRODUCTION

The synergistic action of piperonyl butoxide (PBO) with the major classes of insecticides – pyrethrins, pyrethroids, carbamates, organophosphates and many insect growth regulators – has ensured its continued use in the control of insects affecting humans and animals.

The safety and immediate efficacy of pyrethrins synergized with PBO has been recognized for many decades, and they remain the most common combinations of insecticide/acaricide products for direct application to both humans and animals. Lice and crabs have to be controlled with safe, effective products since they involve direct application to humans. Similarly flea and tick control involves direct application to pets, many of which are sensitive to pesticides.

Products based on synergized pyrethrins offer the physician, farmer, pet owner and veterinarian safe and effective control of insect pests without the inclusion of toxic additives, and without restrictions on repeat treatments. Since PBO is not a cholinesterase inhibitor, it may be used in conjunction with flea or tick dips, collars and oral medications.

In addition to having a low mammalian toxicity and a long record of safety, PBO rapidly degrades in the environment.

1.1. The Benefits of PBO

These can be summarized as follows.

- It permits insect control at economic levels with labile compounds.
- It stabilizes pyrethrum formulations.

PIPERONYL BUTOXIDE
ISBN 0-12-286975-3

- It lowers both the initial dose and the level of residues.
- PBO interferes with the detoxification of the insecticide by inhibiting the functions of the insects' 'mixed function oxidase' (MFO) enzymes (Casida, 1970) and plays an important role in preventing or delaying the onset of resistance.
- Approved food additive tolerances for PBO and pyrethrins mean that these products can be used in food processing areas for meat, poultry, milk and eggs.

2. PESTS OF HUMANS

The importance of insect-borne diseases fully recognized in Africa and India has often been underestimated in developed countries, and mosquitoes, flies and cockroaches may be regarded more as nuisance pests rather than carriers of diseases.

The potential for serious outbreaks of disease always exists and happens with surprising frequency, as shown recently by numerous outbreaks of the tick-borne Lyme disease, and mosquito-transmitted arboviruses. Malaria, sleeping sickness, leishmaniasis, elephantiasis, and many other major human diseases are caused by parasites that spend part of their life cycles in people, and partly in vectors such as insects (see Table 18.1).

A report on 'Emerging Infections' prepared for the American Institute of Medicine points out that although insect-borne infections were more or less controlled in North America and Europe by the mid-1970s, they could return. Malaria is one example since the vectors remain and the health authorities are now less vigilant in controlling the disease.

Every year over 100 million people get ill from malaria and more than 2 million die. Although most live in developing countries, more than 10 000 cases were diagnosed in Europe last year, and over 1000 cases in the USA.

Table 18.1. A global summary of the major insect-borne diseases

Type of pest/vector	Disease transmitted	Number of countries endemic	Number of people at risk	Number of cases annually	Mortality per year
Cockroaches and flies	Diarrhoea	200	5 200 000 000	5 200 000 000	10 000 000
Mosquitoes	Arbovirus	130	3 000 000 000	1 000 000	10 000
	Malaria	107	1 800 000 000	10 000 000 000	2 000 000
	Filariasis	90	905 000 000		
Sandflies	Leishmaniasis	86	5 000 000 000	1 200 000	50 000
Blackflies	Onchocerciasis	36	200 000 000	20 000 000	–
Bugs	Chagas	19	65 000 000	24 500 000	100 000
Tsetse	Sleeping sickness	38	10 000 000	10 000	1 000
Fleas/rodents	Plague	41	100 000 000	1 000	100
Lice, fleas, mites, ticks	Typhus	100	100 000 000	100 000	1 000
Snails	Schistosomiasis	69	1 000 000 000	230 000 000	100 000

The impact of tick-borne diseases in the USA is enormous, not only in direct medical costs, which have been reported as high as $50 000 a year for treatment, but also in peripheral issues such as lost work time and even reduced real estate values in infested areas. Some estimates indicate that tick-borne diseases cost society more than US $1 billion a year (D.L. Weld, personal communication).

2.1. Mosquitoes and PBO

PBO has been used successfully in mosquito control programmes for over 50 years. The availability of PBO led to the development of synergized pyrethrin cold foggers, which produced aerosol droplets and changed the control emphasis from larvicidal to adulticidal programmes.

The addition of PBO conferred greater mortality and efficacy at a reduced cost and made it possible for pyrethrins to compete economically for the first time with the established organophosphate adulticiding materials, such as malathion and Dibrom (registered trademark of Valent USA Corporation). Rates as low as 0.002 lb acre^{-1} (2.24 g hectare), synergized 1 : 5 with PBO, were found to be effective (Baker, 1973).

In the early 1970s, field trials in malarial areas of El Salvador with ground-applied ULV (ultra-low volume) sprays of synergized pyrethrins demonstrated a dramatic decrease in the natural population of the malarial vector as well as high mortality in adult mosquitoes. Surveillance studies showed that the ULV spraying had a marked effect on reducing the transmission of this disease (Hobbs, 1976). ULV synergized pyrethrin sprays (5.00% pyrethrins, 25.00% PBO applied to give pyrethrins dosages of 0.0011 and 0.0018 lb active ingredient (a.i) per acre (1.23 g/hectare and 2.02 g/hectare) also provided effective control of *Aedes* mosquitoes involved in the transmission of dengue haemorrhagic fever in the tropics (Yap *et al.*, 1978).

Synergized pyrethrins sprays are relatively expensive for the treatment of large areas, and their use has largely been superseded by synergized pyrethroids, which offer greater efficacy at a lower cost. The addition of PBO enhances the insecticidal activity of pyrethroids such as resmethrin, reducing the amount of toxicant used to about half the usual rate without the loss of knockdown or mortality. Permethrin and PBO and deltamethrin and PBO also show considerable promise in these applications. Resmethrin synergized at the rate of one part with three parts of PBO (Scourge: registered trademarek of Roussel Bio Corporation) is particularly useful because of its high efficacy, low mammalian toxicity, and desirable environmental characteristics. These include low toxicity to wildlife and nontarget species, rapid degradation, and a lack of accumulation in the environment.

Since the development of insect resistance to organophosphate and carbamate insecticides, there has been a major resurgence of interest in both natural pyrethrum and pyrethroids synergized with PBO. Resmethrin/PBO formulations are now an accepted part of integrated programmes of resistance management and have replaced many of the adulticiding functions of currently available organophosphate and carbamate insecticides.

Pyrethrin or pyrethroid sprays synergized with PBO are used to kill mosquitoes and other disease vectors on international flights to those countries requiring complete aircraft disinsectization.

Lifestyle changes with outdoor living has stimulated the need for products to control biting midges, gnats and mosquitoes, particularly at dusk, and PBO is a significant component of most consumer indoor/outdoor spray and fogger formulations. PBO is frequently added to vaporizing mat formulations. In these applications, PBO improves the physical release of the toxicant such as bioallethrin.

2.2. Tsetse Fly

Tsetse flies are vectors of human trypanosomiasis and nagana and are a major factor in limiting the agricultural, economic and tourist potential of many countries in Africa. Estimates show that nearly 25% of tourist destinations in Kenya and Tanzania may be infected. Game animals form the reservoir and continuously provide the vector with trypanosomes, making the disease endemic and persistent and the control measures very difficult and inefficient. Topically applied pyrethrum synergized with PBO has a high toxicity to the tsetse fly, and aerially applied ULV sprays in Tanzania have reduced fly populations by 95% in high-density areas (Lee *et al.*, 1968). Small-scale aircraft applications of synergized pyrethrum against high-density populations of tsetse flies in dense thickets have also indicated that synergized pyrethrins might be used cost-effectively and economically from the air (Kuria and Bwogo, 1986).

2.3. Bugs, Ticks, Lice, Mites, and Wasps

The debilitating Chagas disease of humans, transmitted by a triatomid bug *Rhodnius prolixus* endemic throughout much of South America, can be controlled by premise sprays of pyrethrins, permethrin, and PBO. New formulations with synergized deltamethrin show considerable promise in the battle to eradicate this disease.

Tick-borne Lyme disease is a serious threat to humans in many parts of the USA. In the Northeastern states, other tick-borne diseases including babesiosis and human granulocytic ehrlichiosis (HGE) have been reported regularly, with over 100 cases in 1995. In addition, ticks are the vectors of Rocky Mountain Fever.

Pet sprays containing PBO, pyrethrins and permethrin in a water-based formulation are now commonly applied to pets to reduce infestation levels of the tick vectors. The benefits of PBO in enhancing the performance of pyrethrins as an acaricide are shown in a unique cost-effective use for the control of bee-infesting mites. Trials in Germany using 0.08% pyrethrins synergized at a 1 : 4 ratio with PBO was an effective remedy for the control of bee mites (*Varroa jacobsoni*). Additional advantages noted were the absence of bee mortality and the inhibition of resistance built up in the mites (Nijhuis *et al.*, 1987).

Water-based pressurized sprays containing PBO and pyrethrins have been used against bedbugs for several decades in Africa, and in tests in Kenya, beds and frames sprayed with pyrethrins and PBO provided good protection for one month. Dust formulations containing synergized pyrethrins and amorphous silica gel are also effective against bed bugs.

Head louse infestations periodically cause disruptions at schools all over the world. The treatment with pyrethrins and PBO is relatively cheap, and effective. The most common formulations contain either 0.15% or 0.3% pyrethrins synergized with 3.0% PBO formulated into nonfoaming shampoos (water-based μm-sized emulsions) that are applied directly to the affected area and repeated after 1 week to kill any newly hatched lice. The pediculicidal and ovicidal efficacy of synergized pyrethrins are enhanced in shampoo formulations as a result of the surface-tension-lowering action of the formulation and are preferable to direct applications of lotions.

In recent years there has been a steady increase in the incidence of infestation by *Phthirus pubis*, a sexually transmitted louse. Nonprescription liquid pediculicides containing 0.3% pyrethrins synergized with 3.0% PBO completely eradicate these adult lice and nymphs in one 10-minute application with no treatment-induced side effects (Newsome *et al.*, 1979).

Human infection from exposure to pets infected with scabies mite can occur and are controlled by ointments containing synergized pyrethrins.

Stinging insects such as wasps and hornets are an increasing nuisance and a worldwide hazard to utility workers and others who work outdoors. A recent formulation, ProControl by AgrEvo UK, contains tetramethrin, permethrin and PBO specially formulated to give very rapid knockdown and kill using a directional spray jet that can be safely directed at the wasp nest from a distance by the professional pest controller.

2.4. Pests of Livestock and Food Animals

On farms while cattle are the most affected by insects, poultry and sheep have many specific pest problems. However, all livestock share common pests. The most serious of these are flies, grubs, ticks and lice.

Biting flies cause about US $1340 million in damage annually from blood loss, irritation and annoyance resulting in reduced weight and milk yields and such insect-transmitting diseases as encephalitis, infectious anaemia and blue tongue.

Hypoderma bovis (cattle grubs), which enter the hide and cause annual losses of about $330 million in the USA, and considerably more in other major cattle-producing countries, can be controlled in nonlactating cattle by insecticide applications in sprays, pour-ons and spot-ons.

The face fly and house fly are currently controlled by insecticides and sanitation. Numerous species of ticks cause extensive losses to cattle producers and are controlled by repeated insecticide treatments, as are cattle lice, mange, scab and itch mites.

Significant advances have been achieved in both ULV and controlled droplet application (CDA) systems and further development of the charged particle sprayer (electrostatic sprayers) will certainly influence the application of routine animal sprays, particularly in poultry houses. PBO works well in all of these developmental techniques.

Two-thirds of the economic losses caused by insects are suffered by the cattle industry and over 84% of beef farmers and virtually all dairy farmers use insecticides to control insect pests. Fly control methods in confinement beef rely heavily on sanitation and insecticide sprays. Because residual sprays are adversely affected by rain, high temperatures and ultraviolet light, area sprays of short residual, quick knockdown insecticides are used as low concentrate fine mists.

Cattle sheds, barns, stables, and dairies are typically sprayed or misted with pyrethrin sprays mixed with PBO, and the animals may also be treated with the spray. The treatments can both kill and repel flies.

Dairy and livestock sprays were usually oil-based at 0.10% pyrethrins and 1.00% PBO, and are typically applied as a fog or fine mist, using about 2 oz of spray per 1000 ft^3 (209 ml/100 m^3) of space. Microemulsion aqueous sprays are now preferred.

Dairy aerosols use high concentrations of pyrethrins, typically 0.50%, synergized with PBO, and in addition to the high usage level, the unit itself is larger – most of the cylinders are 2 lb ($\sim$ 1 kg) or 8 lb ($\sim$ 4 kg) cylinders.

Dairy and farm space sprays containing 0.5% pyrethrins and 4.00% PBO together with the repellent Di-n-propyl isocinchomeronate (MGK 326) are used to kill and repel fly pests in dairy, beef and hog operations.

Ear tags, made of plastic and impregnated with pyrethroids, soon induced resistance as the dose available to the insect progressively declined.

Field tests in Hawaii with a membrane-based ear tag incorporating permethrin and PBO have demonstrated that up to 95% of the total load can be released at a constant rate for each component. This invention has resulted in reduced insecticide resistance and increased the duration of control (Herbig and Smith, 1988).

Experimental acaricidal ear tags containing 5% permethrin and 5.5% PBO were found to be effective in reducing tick infestation of *Rhipicephalus appendiculatus* in tests on cattle in Kenya in 1992 (Young, 1992).

In South Africa, a commercial cattle pour-on and dip formulation containing PBO in combination with cypermethrin and alphamethrin and an organophosphate has been introduced to control resistant blue ticks, and the same formulation is also effective for the control of face flies and biting and sucking lice.

Biting flies are a serious pest of hogs because both the face fly and house fly transmit hog cholera and these pests breed in tremendous numbers in hog lots. Insecticides are also used to control the hog louse and hog mites. Fogging with 1.00% pyrethrins synergized with 2.00% PBO is a standard fly treatment in hog operations, and the same formulation can be applied directly to the animals. Other fogging formulations containing 0.50% pyrethrins and 1.00% PBO in a dual-synergist system are also frequently used.

Approximately 68% of sheep farmers use insecticides to control the major pest species – blow flies, lice, keds and scabies. Face flies transmit eyeworms and biting gnats transmit blue tongue. All of these pests are readily controlled with synergized pyrethrin sprays.

A recent study in Australia demonstrated that PBO effectively synergizes the toxicity of the pyrethroids cypermethrin, deltamethrin, cyhalothrin and alpha-cypermethrin against the sheep body louse, *Bovicola ovis* (Levot, 1994). Cypermethrin synergized with PBO is now an established means of controlling resistant sheep body lice in Australia.

About 71% of poultry (including turkeys and ducks) farmers use insecticides to control such pests as northern fowl mite, chicken mites, lice, fleas and chiggers. Flies in poultry houses can be controlled by ULV fogging, usually a formulation derived from concentrates containing 5.0% pyrethrins and 25.00% PBO. Typically the space spray treatments are designed for use in mechanical sprayers, which produce particles of aerosol size, but they can also be applied in fogging equipment. General purpose aqueous insecticides with 0.1% pyrethrins and 1.00% PBO control poultry lice, bed bugs and mites when sprayed on perches, walls and nests or applied directly to the birds in a fine mist.

Dust formulations containing 0.10% pyrethrins and 1.00% PBO in an amorphous silica gel and petroleum oil base are used to control ectoparasites on poultry and caged pet birds. Synergized pyrethrins diluted to contain 0.01% pyrethrins and 0.04% PBO have been shown to be very effective in controlling infestations of the red poultry mite (*Dermanyssus avium*) (Westermarck, 1967). Although toxic to fish, pyrethrins have a long history of usage in protecting dried fish from insect infestation and are used because of their low mammalian toxicity. Techniques developed in Norway in the 1960s for the protection of drying codfish from blowfly damage have been extended to sun-dried Tilapia lake fish in Malawi and the method is used in rural areas of East Africa. A 97% reduction in blow fly damage was achieved using a formulation of 0.125% pyrethrins and 0.125% PBO as a dip prior to drying (McLellan, 1963). Protection of dried fish from beetle infestation using a pyrethrum dipping treatment was shown by McLellan (1964).

A novel use for synergized pyrethrins as a delousing agent in salmon aquaculture has been developed (De Pauw and Joyce, 1991). Pyrethrins and PBO at a 1 : 1 ratio in an oil-based formulation selectively penetrated the lice *Lepoeptheirus salmonis* and *Caligus elongatus*, providing up to 96% efficacy (Boxaspen and Holm, 1991).

2.5. Horses

Insect and tick-transmitted diseases of horses cause losses of more than $150 million per year in the US alone. Face flies, horse flies, stable flies, house flies, mosquitoes and biting gnats cause constant irritation and repeated use of insecticides is currently the recommended solution. As a general rule, horses receive very gentle treatment in the control of insect pests. The most frequently used formulations are pyrethrins synergized with PBO to provide both a control and a repellent effect. Typical formulations are available as sprays, wipe-on ointments, lotions or mists, and usually contain 0.15% pyrethrins and 1.50% PBO.

Repellent sprays are an important part of horse care products. These can be relatively simple formulations containing PBO and pyrethrins at a ratio of 0.1% pyrethrins to 1.0% PBO together with citronella oil at 1.0%, or much more complex products. This type of spray provides a protective hair coating against biting flies and is usually formulated to impart a high sheen to the coat when brushed out.

3. CONTROLLING PESTS OF PETS

Pet products are among the fastest growing segment of consumer insecticides, and this growth is attributed to the huge increase in the number of pets, and heightened awareness of insect-related problems in pets.

Within the pet care market over 74% of liquids and aerosol products now contain PBO, usually as a synergist for pyrethrins, but also with other active ingredients. Flea control is a major problem.

The brown dog tick, *Rhipicephalus sanguineus*, is a serious pest of dogs in the southern and eastern USA. This tick is capable of transmitting canine piroplasmosis and canine rickettsias as well as being the vector of several diseases of humans and animals. The brown dog tick was the first major tick species to have developed well-defined resistance to insecticides. Pyrethrins synergized with PBO at a 1 : 10 ratio provided the most effective control of 26 insecticides tested against resistant ticks in screening tests (Gladney, 1972).

3.1. Flea and Tick Products

Many species of fleas are confined to a particular host species, but some can live on a variety of hosts. Dog and cat fleas, for example, are a cosmopolitan pest of humans and other mammals. Under favourable conditions the entire life cycle may be completed in 3–6 weeks, and the adult may live dormant within the cocoon for months before emerging.

The existence of four different life stages makes effective control difficult without continuous applications of pesticides, and the mobility and small size of the adult flea increases the difficulty of control. Fogging the entire premises is the most effective treatment for killing adult and pre-adult fleas.

Multi-room foggers with a PBO synergized fenvalerate formulation, used twice at 10- to 14-day intervals, provides extensive control of fleas and other flying and crawling insects in both home and industry uses. The actuator cap of this type of product allows the can to spray continuously until empty.

For comprehensive control of fleas, ticks and biting flies it is essential to control both the eggs and the larvae that exist both indoors and outdoors. To achieve this result either premise or surface sprays are used in conjunction with direct application sprays. Pyrethrins at 0.05%, synergized with PBO at 1.00%, in addition to 1.00% propoxur will kill adult fleas and larvae when applied to carpets and furnishings.

Mists for flea and tick control normally have a high ratio of synergism –

usually 1 : 10 with pyrethrins and PBO. These sprays kill fleas, lice and ticks and also provide a temporary repellent effect against gnats, flies and mosquitoes. These same formulations can be used in pet sleeping quarters and bedding.

In addition to the targeted market where PBO synergized products have already established a leadership role, multipurpose products are under development that can be applied directly to pets and vegetable gardens, and which are also effective flying insect killers for outdoor use.

Tick-killing products usually have more than one active ingredient in conjunction with PBO. One very successful veterinary product combines two insecticides (permethrin and pyrethrins) in a dual-synergist system (PBO and MGK 264) and two repellents to provide effective control of not only ticks and fleas, but also of lice, chiggers, gnats, and mosquitoes.

The most popular consumer formulations for direct application to the animal are normally pyrethrum-based and incorporate either PBO alone, or a dual-synergist system with both PBO and MGK 264. Sprays, shampoos, powders and dusts are available for the control of ear mites in pets, and the majority of these contain PBO.

A desiccant dust containing amorphous silica gel in combination with 1.00% pyrethrins and 10.00% PBO can be applied as a crack and crevice treatment against fleas and lice and may also be applied directly to the animal. A companion formulation containing 0.50% pyrethrins and 5.00% PBO in a silica gel base is designed for use on zoo and laboratory animals including monkeys, chimpanzees and rodents. Both products kill lice, fleas, and ticks on contact by a combination of physical dehydration and chemical action.

Lotions and mists for the control and repellency of ear mites of dogs/puppies, cats/kittens, and rabbits are frequently formulated with PBO. A typical formulation will contain 0.15% pyrethrins synergized at a 1 : 10 ratio with PBO.

Since synergized pyrethrins are not cholinesterase inhibitors, they may be used in conjunction with flea and tick dips, collars and oral medications. Products based on pyrethrum and PBO offer veterinarians safe, effective residual action without the inclusion of toxic additives and without restrictions on repeat applications or concomitant treatments with other insecticides or anthelmintics (MacDonald and Miller, 1986).

4. NEWER METHODS

4.1. Microencapsulated Technology

Microencapsulation techniques have been developed to enhance the residual activity of the formulations by placing a thin chemical shell around the synergized pyrethrins. This allows for a sustained release of the insecticide by diffusion through the shell wall for 30 to 60 days.

Commercial formulations using these techniques have extended the residual activity of synergized pyrethrins to German cockroaches under field conditions to 1 month (Bennet and Rea, 1978). This feature, coupled with low mammalian toxicity, gives these formulations advantages over many other chemicals in a variety of field conditions.

Tests in 1984 with synergized pyrethrins in a microencapsulated polymer system in an isopropanol-based spray demonstrated the benefits of the system in prolonging the residual effects of synergized pyrethrins and resulted in the development of a successful commercial product (MacDonald and Miller, 1986).

Timed-release formulations have made a major impact on the traditional market for pet care products by offering both immediate kill and extended protection. Typically, tick and flea reinfestations of cats and dogs depend on the residual activity of the product, but in most cases an improved residual activity is correlated with an increase in mammalian toxicity.

Residual action against fleas for up to 8 days can be obtained with a formulation containing 0.15% pyrethrins and 0.70% PBO in a dual-synergist, microencapsulated system.

A proprietary microencapsulated technology developed by 3M and sold under the Sectrol brand contains the highest level of synergized pyrethrins currently available in a pet spray (Sectrol is a registered trademark of the 3M Company). The active ingredients are contained within the microcapsules, which are inactive while suspended within the liquid, but which become active after spraying on a surface, providing residual activity over a period of weeks and months. Small amounts of pyrethrins are constantly present on the surface of the capsule, and as the surface insecticide gradually breaks down on exposure to light and air, more insecticide diffuses to the surface to replace it, giving the capsules their time-released quality.

4.2. Foams

Cats and kittens may be more sensitive to many insecticides than other pets, and are often nervous about treatments. For this reason, foams that can be applied to the palm of the hand before application and then massaged into the hair are the preferred treatments, and pyrethrins and PBO are the most commonly used active ingredients. Foam products, containing 0.15% pyrethrins and 0.70% PBO in a dual-synergist system, provide rapid flea kill and at least 8 days of residual control of fleas on cats.

A further microencapsulated pet and household formulation containing 0.11% pyrethrins, 0.22% PBO and 0.37% MGK 264 can be used for the control of fleas and ticks indoors as well as applied directly to the pet.

5. PBO AND INSECT GROWTH REGULATORS (IGRS)

Because of its compatibility, synergistic action and solvent action, PBO is frequently used in formulations containing IGRs. PBO is known to synergize the activity of many of the most commonly used IGRs, including methoprene, hydroprene, and fenoxycarb.

Several indoor flea treatment sprays now combine PBO, pyrethrins and chlorpyrifos with an IGR such as methoprene. This type of treatment is designed to kill visible adult fleas and the residual action coupled with the IGR will prevent eggs from developing into adult fleas for up to 30 weeks. Extensions of this formulation are used as carpet sprays, shampoos and powders as well as premise sprays in the pet area.

Total-release aerosols containing synergized pyrethrins provide adequate control of adult fleas, but the addition of methoprene and fenoxycarb considerably extends control by preventing larval development and emergence of adults. An aerosol-containing methoprene, tetramethrin and PBO in a dual-synergist system is currently one of the leading consumer flea control products. The formulation is designed as a nonstaining fogger that kills fleas on contact, and prevents eggs from hatching for up to 120 days. Total-release formulations of pyrethrins, PBO and fenoxycarb have been shown to control fleas for at least 60 days (Osbrink *et al.*, 1986).

Pyrethrins and PBO in combination with methoprene showed 100% control of feline fleas 35 days after treatment, with a satisfactory residual effect up to 2 months (Donahue, 1992). In addition there is evidence that PBO is an IGR in its own right. In a laboratory study PBO added to three major cotton plant allelochemicals in the diet of tobacco budworm larvae significantly decreased the growth rate of the 5-day-old larvae, and was found to be toxic to the 1- and 3-day-old larvae (Heydin *et al.*, 1988).

6. PBO WITH LINALOOL AND D-LIMONENE

PBO synergizes the activity of linalool, an oil extracted from citrus peel that is effective in controlling the cat flea. Formulations with linalool and PBO can be used on both the pet to control adult fleas or applied to the carpet or bedding to control eggs, larvae and pupae (Hink *et al.*, 1988).

D-Limonene, another citrus extract with insecticidal properties, is also extensively synergized with PBO. Formulations with PBO are toxic to all life stages of the cat flea and produce a more rapid mortality of the adult fleas when applied directly to the pet (Hink and Fee, 1986).

REFERENCES

Baker, G.J. (1973). Mosquito adulticiding with pyrethrum. *Pyreth. Post* **12**, 12–14.
Bennet, G.W. and Rea, D.L. (1978). Evaluation of encapsulated pyrethrins (Sectrol) for German cockroach and cat flea control. *Pyreth. Post* **14**, 68–71.
Boxaspen, K. and Holm, J.C. (1991). *Aquaculture and the Environment*. Abstracts of contributions presented at the International Conference, Aquaculture Europe '91, Dublin, Ireland, 10–12 June, 1991.
Casida, J.E. (1970). Mixed function oxidase involvement in the biochemistry of insecticide synergists. *J. Agric. Food Chem.* **18**, 753–772.
De Pauw, N. and Joyce, J. (1991). A new treatment against sea lice. In: *Aquaculture and the Environment*. Abstracts of contributions presented at the International Conference, Aquaculture Europe '91, Dublin, Ireland, 10–12 June, 1991.
Donahue, W.A. (1992). Evaluation of a synergised pyrethrins/methoprene spray against feline flea infestation. *Vet. Medicine* **87**, 999–1007.

Gladney, W.J. (1972). Insecticides tested for control of nymphal brown dog ticks. *Pyreth. Post* **11**, 132–134.

Herbig, S.M. and Smith, K.L. (1988). A membrane-based cattle insecticide eartag. *J. Cont. Rel.* **8**, 63–72.

Heydin, P.A., Parrott, W.L., Jenkins, J.N., Mulroon, J.E. and Menn, J.J. (1988). Elucidating mechanisms of tobacco budworm resistance to allelochemicals by dietary tests with insecticide synergists. *Pestic. Biochem. Physiol.* **32**, 55–61.

Hink, W.F. and Fee, B.J. (1986). Toxicity of D-limonene, the major component of citrus peel oil, to all stages of the cat flea, *Ctenocephalides felis* (Siphonaptera: Pulicidea). *J. Med. Entomol.* **23**, 400–404.

Hink, W.F., Liberati, T.A. and Collart, M.G. (1988). Toxicity of linalool to life stages of the cat flea, *Ctenocephalides felis* (Siphonaptera: Pulicidea) and its efficacy in carpet and on animals. *J. Med. Entomol.* **25**, 1–4.

Hobbs Jesse, H. (1976). A trial of ultra-low volume pyrethrin spraying as a malarial control measure in El Salvador. *Pyreth. Post.* **13**, 143–147.

Kuria, J.N. and Bwogo, R.K. (1986). Aerial applications of pyrethrum ULV formulations for the control of *Glossina pallidipes*. *Pyreth. Post* **16**, 52–60.

Lee, C.W. (1968). Aerial applications of pyrethrum aerosol to control tsetse fly. *Pyreth. Post* **9**, 37–40.

Levot, G. (1994). Pyrethroid synergism by Piperonyl Butoxide in *Bovicola ovis*. *J. Aust. Entomol. Soc.* **33**; 123–126.

MacDonald, J. and Miller, T.A. (1986). Dynamics of natural flea infestation and evaluation of a control program. *Pyreth. Post.* **16**, 84–88.

McLellan, R.H. (1963). The use of a pyrethrum dip as protection for drying fish in Uganda. *Pyreth. Post* **7**, 8.

McLellan, R.H. (1964). A pyrethrum dipping treatment to protect dried fish from beetle infestation. *Pyreth. Post* **7**, 30.

Miller, T.A. (1984). Maximizing the potency of nature's own flea and tick insecticide pyrethrin. *Canine Prac.* **11**.

Newsome, J.H., Fiore, J.L., Jr, and Hackett, E. (1979). Treatment of infestation with *Phthirus pubis*: comparative efficacies of synergised pyrethrins and gamma-benzene hexachloride. *Sex Transm. Dis.* **6**, 203–205.

Nijhuis, H., Enss, K., VoB, S., Tietgen, I. and Tietgen, W. (1987). The use of pyrethrum extract for the control of bee mites, *Varroa jacobsoni,* in apiculture. *Pyreth. Post* **16**, 118–119.

Osbrink, W.L.A., Rust, M.K. and Reerson, D.A. (1986). Distribution and control of cat fleas in homes in Southern California (Siphonaptera: Pulicidae). *J. Econ. Etomol.* **79**, 135–140.

Westermarck, H. (1967). The use of pyrethrins against red mite in poultry in Finland. *Pyreth. Post* **9**, 10–11.

Yap, H.H., Thiruvengadam, V. and Yap, K.-H. (1978). A preliminary field evaluation of ULV synergised pyrethrins for the control of *Aedes aegypti* (Linnaeus) and *A. albopictus* (Skuse) in urban areas of West Malaysia. *Pyreth. Post* **14**, 98–102.

Young, A.S. (1992). Immunisation of cattle against theileriosis in Nakuru District of Kenya by infection and treatment and the introduction of unconventional tick control. *Vet. Parasitol.* **42**, 225–240.

19

Some Lesser Known Properties and Potential New Uses of Piperonyl Butoxide

ANTONIA GLYNNE JONES

1. INTRODUCTION

There are numerous references to PBO in the scientific literature, most of which refer to its action as an insecticide synergist or its use to control resistance. Amongst the remainder, there are a few which describe other properties of PBO and these are presented, in abstract form, in this chapter.

2. EFFECTS ON THE SNAIL *LYMNAEA ACUMINATA*

Singh and Agarwal (1989) demonstrated that treatments of the snail *Lymnaea acuminata* with only 0.23% of 40% and 80% of the 48-hour LC_{50} dose of carbaryl mixed with five times the concentration of PBO induced the same lethal effects as 40% and 80% of the LC_{50} dose alone. Neither carbaryl nor PBO by themselves caused any major changes in the tissues of the snails when applied in the concentrations used in the carbaryl–PBO mixtures.

The actual molar concentration of carbaryl + synergist mixture was 1/70th of the carbaryl concentration needed for the same lethality. Since PBO is rapidly biodegradable and thus short-lived in the environment, a reduction of 70% in the amount of carbaryl used in snail baits would provide considerable benefits to birds and wildlife.

3. PBO REDUCES TERMITE FEEDING

Heist and Henderson (1996) showed that at a concentration of 0.8% PBO reduced feeding in Formosian termites to 27% compared with 58% for 0.1% PBO and 55% with acetone alone. In addition, at the higher concentration level termites exposed to PBO sometimes turned brown and shiny. They concluded

PIPERONYL BUTOXIDE
ISBN 0-12-286975-3

302   *Piperonyl Butoxide*

that PBO does deter feeding. PBO possibly affects the termites metabolically over time and may prove to be lethal as well as a deterrent.

PBO is now added to commercial termite formulations based on permethrin in Japan. The PBO performs two functions: firstly, it acts as a solvent so that unacceptable volatile solvents are excluded from treatments applied on or near dwelling houses; and secondly, it acts as an effective but also very persistent synergist, since anaerobic conditions apply where the permethrin is impregnated in wood (see Chapter 6, this volume).

4. THE EFFECT OF PBO ON MITES INFESTING ANIMALS

For many years the British Pharmaceutical Codex (BP Vet.) has listed PBO as an acaracide for veterinary use and a PBO formulation used to control mites in horses' ears.

Recent trials done by Breathnach (1996) at University College Dublin revealed that concentrations varying from 0.05% to 0.20% w/v of PBO alone effectively controlled the mite *Psoroptes cuniculi* infesting rabbits' ears. The experimental data are summarized graphically in Fig. 19.1.

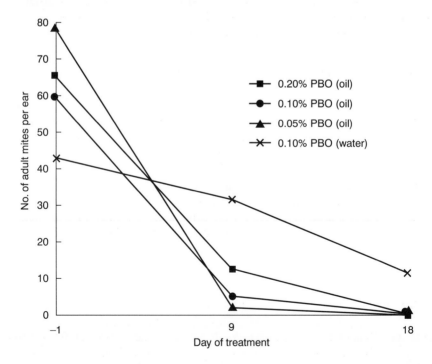

Figure 19.1 Graph of mean number of adult mites per ear with varying treatments of PBO at days −1, 9 and 18 of treatment (with permission from Breathnach, 1996)

5. PBO AND THE CONTROL OF *VARROA JACOBSONI*, THE MITE ATTACKING HONEY BEES

Varroa jacobsoni has become a serious threat to the maintenance of honey bee colonies so essential for the pollination of a variety of crops. Taufluvalinate was used to control this pest, but resistance first appeared in Sicily in 1991 and has spread to other parts of Europe. Nijhuis *et al.* (1987) demonstrated that pyrethrum extract synergized with PBO is very effective as a remedy against *Varroa jacobsoni*.

Hillesheim *et al.* (1996) reported on experiments where PBO was added to taufluvalinate, resulting in a significant decrease in the LC_{50} values of the susceptible and resistant mites. Of special interest were their measurements of the toxicity of PBO alone to both the susceptible and resistant mite; these are represented graphically in Fig. 19.2.

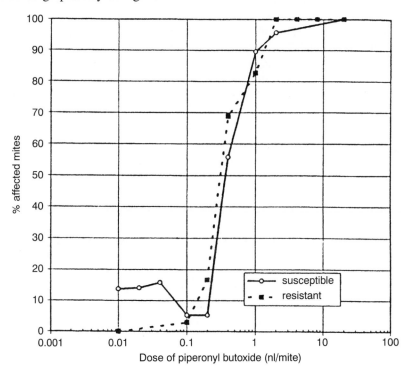

Figure 19.2 Percentage of affected susceptible and resistant *Varroa* mites (dead and paralysed mites) after 48-hour treatment with synergist PBO. Susceptible mites, $n = 418$; resistant mites, $n = 270$ (with permission from Hillesheim *et al.*, 1996).

6. THE EFFECT OF PBO ON PLANT-FEEDING MITES

Guirguis *et al.* (1977), working in Egypt, published data on the toxicity of PBO to a number of strains of the female mite *Tetranychus arabicus* and suggest an LD_{50} in the region of 100 ppm.

7. THE EFFECT OF PBO ON HOUSE DUST MITES

Whilst studying the effects of permethrin on two species of house mites, *Dermatophagoides pteronyssinus* and *D. farinae*, Eremina and Stepanova (1995) measured the action of PBO alone and with permethrin. They found PBO alone to be of the same order of toxicity as permethrin to *D. farinae*, the American house dust mite, but less toxic to *D. pteronyssinus*, the European house dust mite.

8. TOXICITY OF PBO TO THE CATTLE TICK *BOOPHILUS MICROPLUS*

Schuttner *et al.* (1974) tested PBO alone and with carbaryl against this important pest of cattle, whilst working at the CSIRO laboratory in Queensland, Australia. In addition to demonstrating an increased effect from adding PBO to carbaryl, they made a series of important observations on the acaricidal properties of PBO used alone. They suggested that this tick was vitally dependent on a mixed function oxidase system which was inhibited by PBO.

Mention was also made of PBO in certain aqueous colloids being particularly lethal to tick larvae of both resistant and susceptible strains. Unfortunately only a summary of these important results was published.

In addition, Connat and Nepa (1990) examined various anti-juvenile hormone agents on the fecundity of the female tick *Ornithodoros moubata* and demonstrated that 500 µg of PBO applied topically induced rapid death.

9. SYNERGISTIC EFFECTS ON NEEM EXTRACTS

Lang (1983) demonstrated very significant synergism by PBO when added to various extracts of neem seeds and applied to the 4th instars of the diamondback moth, *Plutella xylostella* L., and the Colorado beetle, *Leptinotarsa decemlineata*. In addition to an increase in kill, there was an acceleration of the onset of mortality, an added advantage with a slow-acting toxicant.

Songkittisuntorn (1989) increased mortality in the rice leafhopper, *Nephotettix virescens*, by 42.5% by adding 0.1% PBO to 5% neem oil.

10. USE OF PBO TO CONTROL RESISTANCE IN HERBICIDES

PBO is known to enhance the activity of several herbicides, but when this action was first described detailed studies on the metabolic fate of PBO in plants had not been undertaken, nor had a full evaluation been made of its environmental fate, all essential for commercial use. Consideration has also been given to using PBO to overcome resistance in herbicides.

Varsano and Rubin (1991) and Raja Rao *et al.* (1995) are representative of the several references in the published literature on this topic.

11. PBO AS A PROTECTANT AGAINST POTATO SPINDLE TUBER VIROID INFECTION

Singh (1977) tested various oils and PBO for their effect on local lesions of *Scopolia sinensis* plants inoculated with the potato spindle tuber viroid (PSTV). Local lesions were inhibited 100% by PBO. At a concentration of 1% PBO prevented infection of potato with PSTV in sprayed leaves for up to 4 days. Inhibition by PBO was observed in several potato cultivars, but not in the tomato plants.

In one greenhouse trial, equal numbers of PSTV-infected and healthy potato plants were intermingled and 'cultivated' regularly. In one group sprayed each week with PBO, three of the healthy plants became infected with PSTV, while seven plants did so in the control group.

12. INHIBITION OF INDUCED RESPIRATORY TRACT CARCINOGENESIS BY PBO IN HAMSTERS

Schuller and McMahon (1985) looked at the influence of pretreating Syrian golden hamsters with PBO on the biological effects *in vivo* of *N*-nitrosodiethylamine (DEN). PBO pretreatment significantly reduced covalent binding of *N*-(ethyl-1-14C)DEN to tissue macromolecules in trachea, lung and liver, but it did not change the tissue distribution of the parent compound.

In a chronic experiment, hamsters treated with PBO before each DEN injection did not develop any tumours or precancerous changes in the lungs, while 60% of the animals given DEN alone developed lung tumours with the morphology of Clara cells and endocrine cells. Tumour incidence in the trachea was also significantly reduced by PBO, but to a lesser extent than in the lungs.

13. ANTICONVULSANT ACTIVITY AND NEUROTOXICITY OF PBO IN MICE

Ater *et al.* (1984) documented the anticonvulsant properties of PBO and compared them with those of clinically established antiepileptic drugs. PBO administered intraperitoneally to mice exerted peak antimaximal electroshock activity and peak neurotoxicity at 5 and 7 hours, respectively. The median neurotoxic dose was 1.690 mg kg^{-1}. In the maximal electroshock seizure test the median effective dose (ED_{50}) was 457 mg kg^{-1} and the protective index (PI) was 3.69. In the subcutaneous pentylenetetrazol (PTZ) test the ED_{50} was 443 mg kg^{-1} and PI was 3.81. PBO prevented seizure spread and elevated seizure threshold, and its PI compared favourably with those of clinically useful anticonvulsants.

14. THE SYNERGISM OF CARBAMATE INSECTICIDES BY PBO

The ability of PBO to synergize cholinesterase-inhibiting insecticides has often been ignored when the mode of action of PBO is discussed. Moorefield (1958) presented detailed data on this subject and showed that both aromatic and heterocyclic carbamates can be potentiated with PBO. A graphical presentation of data is reproduced in Fig. 19.3 on the effect of adding PBO to two carbamate insecticides to control house flies. This is of particular interest in that application was only made to female flies by individual measured drops – much more precise and relevant to effective control than spray applications to flies of both sexes. This was a classic demonstration of the powerful synergistic action of PBO.

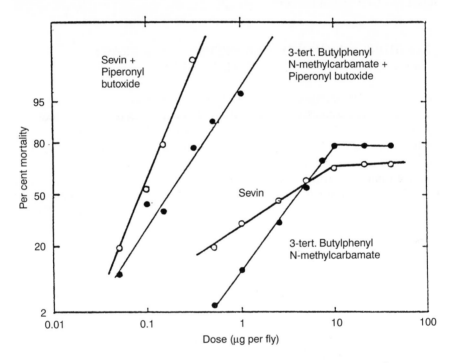

Figure 19.3 Effect of synergizing carbamate insecticides on dosage-mortality response of the house fly. Ratio of carbamate to PBO, 1:5. PBO applied alone effected no mortality at the doses used. Reproduced with permission from Moorefield (1958).

15. INCREASED FACTORS OF SYNERGISM WITH THE DEVELOPMENT OF RESISTANCE TO INSECTICIDES

Glynne Jones (1983) drew attention to an additional feature of the synergistic action of PBO, namely an increased effect on insects which have developed resistance to certain insecticides. There had been some evidence in the literature but this important effect was fully clarified by Dr Sun, of the National University, Taiwan, and published by Sun *et al.* (1985). These data showed increased factors of synergism with resistant as compared to susceptible strains of the diamondback moth. The data are represented graphically in Fig. 19.4.

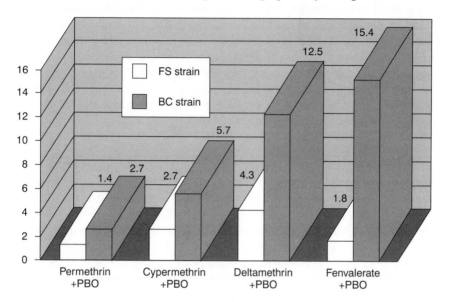

Figure 19.4. Synergism of four pyrethroids by PBO in susceptible (FS) and resistant (BC) strains of DBN. Reproduced with permission from Sun *et al.* (1985).

16. EFFECT OF ADDING PBO TO THE SPEED OF ONSET OF RESISTANCE DEVELOPMENT IN MOSQUITOES

Thomas *et al.* (1991) induced 1449-fold resistance to deltamethrin through continuous larval selections of *Qulex quinquefasciatus* for 40 generations. When the larvae were subjected to selection pressure using deltamethrin and PBO (1:5) the speed of selection for deltamethrin resistance in the larvae slowed down considerably by 17 to 63%.

In a parallel selection study, a 137-fold deltamethrin-resistant strain when subject to continuous selection pressure with synergized deltamethrin, showed 76% reversion in resistance in the first generation and significantly retarded the development of resistance in subsequent generations. Thomas *et al.* concluded that even though high larval resistance develops under intense laboratory selection, the addition of PBO can drastically reduce the speed of development of deltamethrin resistance in this species of mosquito.

17. EFFECT OF ADDING PBO TO ADULT *HELICOVERPA ARMIGERA* RESISTANT TO FENVALERATE

Daly and Fisk (1993) looked at pyrethroid resistance in adult *Helicoverpa armigera* in both laboratory and field-selected strains. PBO synergized the effects of the synthetic pyrethroid fenvalerate 1000-fold and eliminated resistance entirely.

The contribution of this adult life stage to the evolution of resistance had not been fully evaluated and adults are exposed to insecticide during aerial application and through contact with treated crops. From field trials of commercial application rates Daly and Fisk concluded that selective mortality in the adult life stage contributes to the evolution of pyrethroid resistance.

18. INCREASED EFFICACY OF PYRETHROIDS AGAINST THE BOOKLOUSE WITH PBO

The booklouse, *Liposcelis bostrychophila,* has become an increasing pest in the food industry, as well as an important pest in libraries and archives. Unlike other stored product insect pests, they are not easily killed by pyrethroids.

Turner *et al.* (1991) undertook trials with pyrethroids, organophosphates and a carbamate against this pest. The pyrethroids were generally less effective than the organophosphates (the carbamate was in-between), but the situation was altered when the insects were pretreated with PBO – the efficiency of permethrin was dramatically increased and mixtures of permethrin and PBO produced high mortality rates. It was presumed that this insect's tolerance of pyrethroids is due to a pronounced ability to detoxify the permethrin using mixed function oxidases.

19. INCREASED MORTALITY OF PHOTOTOXINS ON MOSQUITO LARVAE WITH PBO

Fields *et al.* (1991) noted that the thiophenes alpha-terthienyl and methyl-alpha-terthienyl are found in many species of the family Asteracene and are highly phototoxic to mosquito larvae. These compounds (including a synthetic analogue) controlled *Aedes intrudens* in field trials at application rates between 10 and 40 g ha^{-1}. PBO greatly increased the mortality of mosquito larvae at low application rates of the most potent phototoxin, cyano-alpha-terthienyl.

20. PBO AS A TOOL IN AQUATIC TOXICOLOGICAL RESEARCH WITH ORGANOPHOSPHATE INSECTICIDES

Ankley *et al.* (1991) discovered that the co-administration of PBO effectively reduced the acute toxicity of four metabolically activated organophosphate insecticides (parathion, methyl parathion, diazinon and malathion) to three cladoceran test species *Ceridaphnia dubia, Daphnia magna* and *D. pulex*. The toxicity of diclorvos, chlorfenvinphos and mevinphos not requiring metabolic activation was not affected. The results suggested that PBO could be an effective tool in toxicological research focused upon identifying specific compounds responsible for toxicity in complex aqueous mixtures.

21. INTERACTION BETWEEN THE ANTIHELMINTIC FENBENDAZOLE AND PBO IN SHEEP AND GOATS

Benchaoui and Mckellar (1996) assessed the efficacy of the combination of PBO and fenbendazole in sheep against two species of benzimidazole-resistant abomasal nematodes, *Ostertagia circumcincta* and *Haemonchus contortus*. PBO alone had no effect on the nematodes. For *O. circumcincta* fenbendazole had less than 8% efficacy, but with PBO the combination gave over 97% control; for *H. contortus* the combination increased the efficacy from 84% to 99%.

Pretreatment of goats with PBO caused a greater than three-fold increase in the relative bioavailability of fenbendazole and fenbendazole sulfoxide.

The *in vitro S*-oxidation of fenbendazole and fenbendazole sulfoxide was also studied using microsomal preparations from rat liver. PBO significantly inhibited the sulfoxidation and sulfonation of fenbendazole.

Benchaoui and Mckellar (1996) concluded that PBO inhibited the oxidative conversion of fenbendazole into inactive metabolites and this resulted in a potentiated anthelmintic action.

REFERENCES

Ankley, G.T., Dierkes, J.R., Jensen, D.A. and Peterson, G.S. (1991). Piperonyl butoxide as a tool in aquatic toxocological research with organophosphate insecticides. *Ecotoxicol Environ. Safety* **21**, 266–274.

Ater, A.B., Swinyard, E.A., Tolman, K.G. and Franklin, M.R. (1984). Anticonvulsant activity and neurotoxicity of piperonyl butoxide in mice. *Epilepsia* **25**, 551.

Benchaoui, H.A. and Mckellar, Q.A. (1996). Interaction between fenbendazole and piperonyl butoxide: pharmacokinetic and pharmacodynamic implications. *J. Pharm. Pharmacol.* **48**, 753–759.

Breathnach, R. (1996). Clinical trial to investigate the efficacy of piperonyl butoxide in the treatment of *Psoroptes cuniculi* infestation in rabbits. Private study.

Connat, J.L. and Nepa, M. (1990). Effects of different anti-juvenile hormone agents on the fecundity of the female tick *Ornithodoros moubata*. *Pestic. Biochem. Physiol.* **37**, 266–374.

Daly, J.C. and Fisk, J.H. (1993). Expression of pyrethroid resistance in adult *Helicoverpa armigera* (Lepidoptera: Noctuidae) and selective mortality in field populations. *Bull. Entomol. Res.* **83**, 23–28.

Eremina, O.Yu. and Stepanova, G.N. (1995). Study of the mechanism of permethrin action on house dust mites. *Biology Bull.* **22**, 60–66.

Fields, P.G., Arnason, J.T., Philogène, B.J.R, Aucoin, R.R., Morand, P. and Soucy-Breau, C. (1991). Phototoxins as insecticides and natural plant defences. *Mem. Entomol. Soc. Canad.* **159**, 29–38.

Glynne Jones, G.D. (1983) The use of piperonyl butoxide to increase the susceptibility of insects which have become resistant to pyrethroids and other insecticides. *Int. Pest Control* Jan/Feb 1983, **25**, 14–15, 21.

Guirguis, M.W., Mohamed, I.I. and El-Raham, A. (1977) The effect of piperonyl butoxide on the toxicity of different Acaricides to the adult female mite of *Tetranuchus arabicus. Agric. Res. Rev.* **55**, 41–48.

Heist, D. and Henderson, G. (1996). Piperonyl butoxide reduces feeding in Formosan Termites. *LPCA News*, p. 30. LSU Agricultural Centre, Louisiana State University, USA.

Hillesheim, E., Ritter, W. and Bassand, D. (1996). First data on resistance mechanisms of *Varroa jacobsoni* (OUD) against tau-fluvalinate. *Exp. Appl. Acarol.* **20**, 283–296.

Lang, W. (1983). Piperonyl butoxide: synergistic effects on different neem seed extracts. In: *Proceedings of the 2nd International Neem Conference*, Rauischoizhausen, pp. 129–140.

Moorefield, H.H. (1958). Synergism of the carbamate insecticides. *Contrib. Boyce Thompson Inst.* **19**, 501–508.

Nijhuis, H., Enss, K., VoB, S., Tietgen, I. and Tietgen, W. (1987). The use of pyrethrum-extract for control of bee mites, *Varroa jacobsoni* in apiculture. *Pyreth. Post* **16**, 118–119.

Raja Rao, S., Feng, P.C.C. and Schafer, D.E. (1995). Enhancement of thiazopyr bioefficacy by inhibitors of monooxygenases. *Pestic. Sci.* **45**, 209–213.

Schuller, M. and McMahon, J.B. (1985). Inhibition of *N*-nitrosodiethylamine-induced respiratory tract carcinogenesis by piperonyl butoxide in hamsters. *Cancer Res.* **45**, 2807–2812.

Schuttner, C.A., Roulston, W.J. and Wharton, R.H. (1974). *Nature* **249**, 386.

Singh, R.P. (1977). Piperonyl butoxide as a protectant against potato spindle tuber viroid infection. *Phytopathology* **67**, 933–935.

Singh, D.K. and Agarwal, R.A. (1989). Toxicity of piperonyl butoxide–carbaryl synergism on the snail *Lymnaea acuminata. Int. Revue. Ges. Hydrobiol.* **74**, 689–699.

Songkittisuntorn, U. (1989). The efficacy of neem oil and neem-extracted substances on the rice leafhopper *Nephotettix virescens. J. Nat. Res. Council Thailand* **21**, 37–58.

Sun, C.N., Wu, T.K., Chen, J.S. and Lee, W.T. (1985). Insecticide resistance in diamondback moth. In: *Proceedngs of the 1st International Workshop on Diamondback Moth Management* (Talekar, N.S. and Griggs, T.D., (eds), pp. 359–371. Asian Vegetable Research and Development Centre, Shanhua, Taiwan.

Thomas, A., Kumar, S. and Pillai, M.K.K. (1991). Piperonyl butoxide as a countermeasure for deltamethrin resistance in *Culex quinquefasciatus* Say. *Advancement of Entomology* **16**, 1–10.

Turner, B., Maude-Roxby, H. and Pike, V. (1991). Control of the domestic insect pest *Liposcelis bostrychophila* (Badonnel) (Psocoptera): an experimental evaluation of the efficiency of some insecticides. *Int. Pest Cont.* Nov/Dec 1991, **33**, 153–157.

Varsano, R. and Rubin. B. (1991). Increased herbicidal activity of triazine herbicides by piperonyl butoxide. *Phytoparasitica* **19**, 225–236.

20

The Response of Industry to the Data Requirements of the United States Environmental Protection Agency – The PBO Task Force

RICHARD A. MILLER

In 1983, the United States Environmental Protection Agency (USEPA) issued a Data Call-In (DCI) to all registrants of piperonyl butoxide (PBO) requiring them to produce a substantial amount of product chemistry, acute and chronic toxicity data to support the re-registration of PBO. This was done under the authority of Section 3(c) 2(B) of the Federal Insecticide, Fungicide and Rodenticide Act (FIFRA – the law governing the sale of pesticides in the USA).

In response to the DCI, the manufacturers and sellers of PBO technical advised the USEPA that they would develop the necessary data. These companies were Endura Spa, Takasago USA, Inc., Hardwicke Chemical Company Division of Ethyl Corporation, McLaughlin GormLey King Company, Fairfield American Company and its parent, Wellcome Foundation, Prentiss Drug & Chemical Company, Inc. and Penick Corporation. The companies formed the PBO Task Force (PBTF), a joint venture under US Law, to produce the data. The PBTF is the owner of the data and all rights in the data, worldwide.

Later, in response to an amendment to FIFRA that was passed in 1988, the USEPA issued a further DCI, which required a very large amount of data concerning effects on wildlife, fish and aquatic organisms, environmental fate, and crop residue data. Since some of the members of the original PBO Task Force did not wish to continue developing data, while additional companies wished to join in the development of the new data, a new joint venture, the PBO Task Force II (PBTFII) was formed. The members of PBTFII are now Endura Spa, Takasago International Corp. (USA), AgrEvo Environmental Health, McLaughlin Gormley King Company, Prentiss Incorporated and S.C. Johnson & Son, Inc. As with the PBTF, this joint venture is also the owner of the data and all rights in the data, worldwide.

PIPERONYL BUTOXIDE
ISBN 0-12-286975-3

These groups have produced more than US$12.5 million worth of data in support of PBO. Much of this data has been summarized in this book. For further information, please contact Mr John D. Conner, Jr, Esq., McKenna & Cuneo, 1900 K Street NW, Washington, DC 20006, USA.

Glossary of Terms

ADI	Acceptable daily intake
AOAC	Association of Official Analytical Chemists, USA
AOAC Methods	Official Methods of Analysis of the Association of Official Analytical Chemists
BCPC	British Crop Protection Council
BS	British Standard
BUN	Blood Urea Nitrogen
CIPAC	Collaborative International Pesticides Analytical Council Limited
DT$_{50}$	Time for 50% loss; half-life
EC$_{50}$	Median effective concentration
EPA	Environmental Protection Agency (a US Federal body)
FDA	Food and Drug Administration (USA)
FAO	Food and Agricultural Organization (of the United Nations)
FOS	Factor of synergism
GC-MS	Combined gas chromatography–mass spectrometry
GLC	Gas–liquid chromatography
HEPES	4-(2-hydroxyethyl)-1-piperazine ethan acid
HPLC	High-performance liquid chromatography
IARC	International Agency for Cancer Research
IPM	Integrated pest management
ISO	International Organization for Standardization
IUPAC	International Union of Pure and Applied Chemistry
JMPR	Joint Meeting of the FAO Panel of Experts on Pesticide Residues and the Environment and the WHO Expert Group on Pesticide Residues
LC$_{50}$	Concentration which is lethal to or affects 50% of the tested group in a set time period (i.e. 48 or 96 h)
LD	Lethal dose
LD$_{50}$	Median lethal dose – dose required to kill 50% of the tested group
Limit dose	Limited dose or exposure level – in a toxicology study this is a maximum level of treatment or exposure beyond which testing is not required

LOEC	Lowest observed effect concentration
LT_{50}	Time to 50% knockdown for a fixed dose
MAFF	Ministry of Agriculture Fisheries and Food (England and Wales)
MATC	Maximum acceptable toxicant concentration. MATC is the geometric mean of the LOEC and NOEC and is a commonly used expression of chronic aquatic toxicity
MEL	Maximum exposure limit
MLD	Median lethal dose – dose required to kill 50% of the test population
MOS	Margin of safety (MOS, effects/exposure) is defined here as the appropriate effects value (e.g. LC_{50} or chronic NOEC for the most sensitive species tested), divided by, the appropriate exposure value (EEC). This is also sometimes called the margin of exposure (MOE)
MRL	Maximum residue limit
NADPH	Nicotinamide-adenine dinucleotide phosphate (reduced)
NOAEL	1. No observable *adverse* effect level – the highest dose at which there are no statistically significant or biologically significant increases in the frequency or severity of *adverse* effects in the exposed population relative to controls. Effects such as enzyme induction would not be considered adverse effects. Hence it would be appropriate to speak of a NOEL with regard to enzyme induction.
	2. No effect level/no toxic effect level/no adverse effect level – the maximum dose used in a test which produces no adverse effects.
NOEC	no observable effect concentration
NOEL	1. No observable effect level – the highest dose at which there are no statistically significant increases in the frequency or severity of biological or physiological effects in the exposed population relative to controls. Often this is lower than the NOAEL
	2. the dose level (quantity of a substance administered to a group of experimental animals) which demonstrates the absence of adverse effects observed or measured at higher dose levels. This NOEL should produce no biologically significant differences between the group of chemically exposed animals and an unexpected control group maintained under identical conditions
PBO	Piperonyl butoxide
PBTF	Piperonyl Butoxide Task Force
RBC	Red blood cell
RP-TLC	Reversed phase thin-layer chromatography
TID	Thermionic detector

TLC	Thin-layer chromatography
TL$_m$	Median tolerance limit
ULV	Ultra-low volume
WBC	White blood cell
WHO	World Health Organization

Index

Note: *Italic* page numbers refer to figures and tables; main discussions are indicated by **bold** page numbers.